Categorical Homotopy Theory

This book develops abstract homotopy theory from the categorical perspective, with a particular focus on examples. Part I discusses two competing perspectives by which one typically first encounters homotopy (co)limits: either as derived functors definable when the appropriate diagram categories admit compatible model structures or through particular formulae that give the right notion in certain examples. Riehl unifies these seemingly rival perspectives and demonstrates that model structures on diagram categories are unnecessary. Homotopy (co)limits are explained to be a special case of weighted (co)limits, a foundational topic in enriched category theory. In Part II, Riehl further examines this topic, separating categorical arguments from homotopical ones. Part III treats the most ubiquitous axiomatic framework for homotopy theory – Quillen's model categories. Here Riehl simplifies familiar model categorical lemmas and definitions by focusing on weak factorization systems. Part IV introduces quasi-categories and homotopy coherence.

EMILY RIEHL is a Benjamin Peirce Fellow in the Department of Mathematics at Harvard University and a National Science Foundation Mathematical Sciences Postdoctoral Research Fellow.

NEW MATHEMATICAL MONOGRAPHS

All the titles listed below can be obtained from good booksellers or from Cambridge University Press. For a complete series listing visit www.cambridge.org/mathematics.

Categorical Homotopy Theory

EMILY RIEHL
Harvard University

CAMBRIDGE
UNIVERSITY PRESS

32 Avenue of the Americas, New York NY 10013-2473, USA

Cambridge University Press is part of the University of Cambridge.

It furthers the University's mission by disseminating knowledge in the pursuit of education, learning and research at the highest international levels of excellence.

www.cambridge.org
Information on this title: www.cambridge.org/9781107048454

First published 2014

A catalogue record for this publication is available from the British Library

Library of Congress Cataloguing in Publication data
Riehl, Emily.
Categorical homotopy theory / Emily Riehl, Harvard University.
pages   cm. – (New mathematical monographs)
Includes bibliographical references and index.
ISBN 978-1-107-04845-4 (hardback)
1. Homotopy theory.   2. Algebra, Homological.   I. Title.
QA612.7.R45   2015
514'.24–dc23   2013049898

ISBN 978-1-107-04845-4 Hardback

To my students, colleagues, friends who inspired this work.

What we are doing is finding ways for *people* to understand and
think about mathematics.

– William P. Thurston, "On proof
and progress in mathematics"

Contents

Preface

The viewpoint taken by William Thurston's essay – that mathematical progress is made by advancing human understanding of mathematics and not only through the proof of new theorems – succinctly describes the character and focus of the course that produced this book. Although certain results appearing in this volume may surprise working homotopy theorists, the mathematical content of this text is not substantially new. Instead, the central value of this account derives from the more qualitative insights provided by its perspective. The theorems and topics discussed here illustrate how categorical formalisms can be used to organize and clarify a wealth of homotopical ideas.

The central project of homotopy theory, broadly defined, is to study the objects of a category up to a specified notion of "weak equivalence." These weak equivalences are morphisms that satisfy a certain closure property vis-à-vis composition and cancellation that is also satisfied by the isomorphisms in any category – but weak equivalences are not generally invertible. In experience, it is inconvenient to work directly in the **homotopy category**, constructed by formally inverting these maps. Instead, over the years, homotopy theorists have produced various axiomatizations that guarantee that certain "point-set level" constructions respect weak equivalences and have developed models in which weak constructions behave like strict ones. By design, this patchwork of mathematical structures can be used to solve a wide variety of problems, but they can be rather complicated for the novice to navigate. The goal of this book is to use category theory to illuminate abstract homotopy theory and, in particular, to distinguish the formal aspects of the theory, principally having to do with enrichments, from techniques specific to the homotopical context.

The ordering of topics demands a few words of explanation. Rather than force the reader to persevere on good faith through pages of prerequisites, we wanted to tell one of the most compelling stories right away. Following [22], we introduce a framework for constructing derived functors between categories equipped with a reasonable notion of weak equivalence that captures all the essential features of, but is much more general than, their construction in model category theory.

Why bother with this generalization? First, it exhibits the truth to the slogan that the weak equivalences are all that matter in abstract homotopy theory, showing that particular notions of cofibrations/fibrations and cofibrant/fibrant objects are irrelevant to the construction of derived functors – any notion will do. Second, and perhaps most important, this method for producing derived functors extends to settings, such as categories of diagrams of a generic shape, where appropriate model structures do not necessarily exist. In the culmination of the first part of this book, we apply this theory to present a uniform general construction of homotopy limits and colimits that satisfies both a local universal property (representing homotopy coherent cones) and a global one (forming a derived functor).

A further advantage of this approach, which employs the familiar two-sided (co)bar construction, is that it generalizes seamlessly to the enriched context. Any discussion of homotopy colimits necessarily encounters enriched category theory; some sort of topology on the ambient hom-sets is needed to encode the local universal property. These notes devote a fair amount of isolated attention to enriched category theory because this preparation greatly simplifies a number of later proofs. In general, we find it clarifying to separate the categorical aspects of homotopy theory from the homotopical ones. For instance, certain comparisons between models of homotopy colimits actually assert an isomorphism between the representing objects, not just the homotopy types. It is equally interesting to know when this is not the case.

Classical definitions of homotopy colimits, as in [10], are as **weighted colimits**. An ordinary colimit is an object that represents cones under a fixed diagram, whereas a homotopy colimit is an object representing "homotopy coherent" cones. The functor that takes an object in the diagram to the appropriately shaped homotopy coherent cone above it is called the **weight**. We believe that weighted limits and colimits provide a useful conceptual simplification for many areas of mathematics, and thus we begin the second part of this book with a thorough introduction, starting with the **Set**-enriched case, which already contains a number of important ideas. As we expect this topic to be unfamiliar, our approach is quite leisurely.

Our facility with enriched category theory allows us to be quite explicit about the role enrichment plays in homotopy theory. For instance, it is well known that the homotopy category of a simplicial model category is enriched over the homotopy category of spaces. Following [79], we present a general framework that detects when derived functors and more exotic structures, such as weighted homotopy colimits, admit compatible enrichments. Enrichment over the homotopy category of spaces provides a good indication that these definitions are "homotopically correct." Our formalism also allows us to prove that in an appropriate general context, total derived functors of left adjoints, themselves enriched over the homotopy category of spaces, preserve homotopy colimits.

We conclude this part with an interesting observation due to Michael Shulman: in the setting for these derived enrichment results, the weak equivalences can be productively compared with another notion of "homotopy equivalence" arising directly from the enrichment. Here we are using "homotopy" very abstractly; for instance, we do not require an interval object. Nonetheless, in close analogy with classical homotopy theory, the localization at the weak equivalences factors through the localization at the homotopy equivalences. Furthermore, the former homotopy category is equivalent to a restriction of the latter to the "fibrant–cofibrant" objects, between which these two notions of weak equivalence coincide.

After telling this story, we turn in the third part, perhaps rather belatedly, to the model categories of Daniel Quillen. Our purpose here is not to give a full account – this theory is well documented elsewhere – but rather to emphasize the clarifying perspective provided by weak factorization systems, the constituent parts in a model structure that are in some sense orthogonal to the underlying homotopical structure visible to the axiomatization of [22]. Many arguments in simplicial homotopy theory and in the development of the theory of quasi-categories take place on the level of weak factorization systems and are better understood in this context.

The highlight of this section is the presentation of a new variant of Quillen's small object argument due to Richard Garner [31] that, at essentially no cost, produces functorial factorizations in cofibrantly generated model categories with significantly better categorical properties. In particular, we show that a cofibrantly generated simplicial model category admits a fibrant replacement monad and a cofibrant replacement monad that are simplicially enriched. Related observations have been made elsewhere, but we do not suspect that this precise statement appears in the literature.

The proofs of these results introduce ideas with broader applicability. A main theme is that the functorial factorizations produced by Garner's construction have a much closer relationship to the lifting properties that characterize the cofibrations and fibrations in a model structure. Indeed, observations related to this "algebraic" perspective on the cofibrations and fibrations can be used to produce functorial factorizations for non–cofibrantly generated model structures [3].

Our construction of enriched functorial factorizations is complemented by a discussion of enriched lifting properties. There are notions of enriched weak factorization systems and enriched cofibrant generation, and these behave similarly to the familiar unenriched case. In the model structure context, this leads to a notion of an enriched model category that is reminiscent of but neither implies nor is implied by the usual axioms. This theory, which we believe is not found in the literature (the *n*Lab aside), illuminates the distinction between the (classical) Quillen-type and Hurewicz-type model structures on the category of chain complexes over a commutative ring: the latter is an enrichment of the former. Indeed, the *same* sets of generating cofibrations and trivial cofibrations produce both model structures! We find it particularly interesting to note that the Hurewicz-type model structure, which is not cofibrantly generated in the traditional sense, is cofibrantly generated when this notion is enriched in the category of modules over the commutative ring (see [2]).

The section on model categories concludes with a brief exposition of Reedy category theory, which makes use of weighted limits and colimits to simplify foundational definitions. This chapter contains some immediate applications, proving that familiar procedures for computing homotopy limits and colimits in certain special cases have the same homotopy type as the general formulae introduced in Part I. Further applications of Reedy category theory follow later in our explorations of various "geometric" underpinnings of quasi-category theory.

In the final part of this book, we give an elementary introduction to quasi-categories, seeking, wherever possible, to avoid repeating things that are clearly explained in [49]. After some preliminaries, we use a discussion of homotopy coherent diagrams to motivate a translation between quasi-categories and simplicial categories, which are by now more familiar. Returning our attention to simplicial sets, we study isomorphisms within and equivalences between quasi-categories, with a particular focus on inverting edges in diagrams. The last chapter describes geometrical and 2-categorical motivations for definitions encoding the category theory of quasi-categories, presenting a number of not-yet-published insights of Dominic Verity. This perspective will be developed much more fully in [74, 75]. A reader interested principally in quasi-categories would do well to read Chapters 7, 11, and 14 first. Without this preparation, many of our proofs become considerably more difficult.

Finally, the very first topic is the author's personal favorite: Kan extensions. Part of this choice has to do with Harvard's unique course structure. The first week of each term is "shopping period," during which students pop in and out of a number of courses prior to making their official selections. Anticipating a number of students who might not return, it seemed sensible to discuss a topic that is reasonably self-contained and of the broadest interest – indeed, significant applications appear throughout this text.

Prerequisites

An ideal student might have passing acquaintance with some of the literature on this subject: simplicial homotopy theory via [32, 55]; homotopy (co)limits via [10]; model categories via one of [24, 36, 38, 58, 65]; quasi-categories via [40, 49]. Rather than present material that one could easily read elsewhere, we chart a less-familiar course that should complement the insights of the experienced and provide context for the naïve student who might later read the classical accounts of this theory. The one prerequisite on which we insist is an acquaintance with and affinity for the basic concepts of category theory: functors and natural transformations; representability and the Yoneda lemma; limits and colimits; adjunctions; and (co)monads. Indeed, we hope that a careful reader with sufficient categorical background will emerge from this book confident that he or she fully understands each of the topics discussed here.

While the categorical prerequisites are essential, acquaintance with specific topics in homotopy theory is merely desired and not strictly necessary. Starting from Chapter 2, we occasionally use the language of model category theory to suggest the right context and intuition to those readers who have some familiarity with it, but these remarks are inessential. For particular examples appearing in the following, some acquaintance with simplicial sets in homotopy theory would also be helpful. Because these combinatorial details are essential for quasi-category theory, we give a brief overview in Chapter 15, which could be positioned earlier, were it not for our preference to delay boring those for whom this is second nature.

Dual results are rarely mentioned explicitly, except in cases where there are some subtleties involved in converting to the dual statement. In Chapters 1 and 2, we make casual mention of 2-categories before their formal definition – categories enriched in **Cat** – is given in Chapter 3. Note all 2-categories that appear are strict. Interestingly for a monograph devoted to the study of a weakened notion of equivalence between objects, we have no need for the weaker variants of 2-category theory.

Notational Conventions

We use boldface for technical terms that are currently being or will soon be defined and quotation marks for nontechnical usages meant to suggest particular intuition. Italics are for emphasis.

We write $\emptyset$ and $*$ for initial and terminal objects in a category. In a symmetric monoidal category, we also use $*$ to denote the unit object, whether or not the unit is terminal. We use $1, 2, \ldots$ for ordinal categories; for example, 2 is the category $\bullet \to \bullet$ of the "walking arrow." Familiar categories of sets, pointed sets, abelian groups, k-vector spaces, categories, and so on are denoted by **Set**, **Set**$_*$, **Ab**, **Vect**$_k$, **Cat**, and so on; **Top** should be a convenient category of spaces, as treated in Section 6.1. Generally, the objects of the category so denoted are suggested by a boldface abbreviation, and the morphisms are left unmentioned, assuming the intention is the obvious one.

We generally label the composite of named morphisms through elision but may use a $\cdot$ when the result would be either ambiguous or excessively ugly. The hom-set between objects x and y in a category C is most commonly denoted by $C(x, y)$, although $\mathrm{hom}(x, y)$ is also used on occasion. An underline, for example $\underline{C}(x, y)$ or $\underline{\mathrm{hom}}(x, y)$, signals that extra structure is present; the form this structure takes depends on what sort of enrichment is being discussed. In the case where the enrichment is over the ambient category itself, we frequently use exponential notion y^x for the internal hom-object. For instance, $\mathcal{D}^C$ denotes the category of functors $C \to \mathcal{D}$.

Natural transformations are most commonly denoted with a double arrow $\Rightarrow$ rather than a single arrow. This usage continues in a special case: a natural transformation $f \Rightarrow g$ between diagrams of shape 2, that is, between morphisms f and g, is simply a commutative square with f and g as opposing faces. The symbol $\rightrightarrows$ is used to suggest a parallel pair of morphisms, with common domain and codomain. Given a pair of functors $F: C \rightleftarrows \mathcal{D}: G$, use of a reversed turnstile $F \dashv G$ indicates that F is left adjoint to G.

Displayed diagrams should be assumed to commute, unless explicitly stated otherwise. The use of dotted arrows signals an assertion or hypothesis that a particular map exists. Commutative squares decorated with a $\ulcorner$ or a $\lrcorner$ are pushouts or pullbacks, respectively. We sometimes use $\sim$ to decorate weak equivalences. The symbol $\cong$ is reserved for isomorphisms, sometimes simply denoted with an equality. The symbol $\simeq$ signals that the abutting objects are equivalent in whatever sense is appropriate, for example, homotopy equivalent or equivalent as quasi-categories.

Certain simplicial sets are given the following names: Δ^n is the standard (represented) n-simplex; $\partial \Delta^n$ is its boundary, the subset generated by non-degenerate simplices in degree less than n; Λ_k^n is the subset with the kth codimension-one face also omitted. We follow the conventions of [32] and write d^i and s^j for the elementary simplicial operators (maps in Δ between $[n]$ and $[n-1]$). The contrasting variance of the corresponding maps in a simplicial set is indicated by the use of lower subscripts – d_i and s_j – though whenever practical, we prefer instead to describe these morphisms as right actions by the simplicial operators d^i and s^j. This convention is in harmony with the Yoneda lemma: the map d^i acts on an n-simplex x of X, represented by a morphism $x: \Delta^n \to X$, by precomposing with $d^i: \Delta^{n-1} \to \Delta^n$.

Acknowledgments

I consulted many sources while preparing these notes and wish to apologize for others deserving of mention that were inadvertently left out. More specific references appear throughout this text.

I would like to thank several people at Cambridge University Press for their professionalism and expertise: Diana Gillooly, my editor; Louis Gulino and Dana Bricken, her editorial assistants; and Josh Penney, the production editor. I am also deeply appreciative of the eagle-eyed copyediting services provided by Holly T. Monteith and Adrian Pereira at Aptara, Inc.

The title and familiar content from Chapter 1 were of course borrowed from [50]. The presentation of the material in Chapters 2, 5, and 9–10 was strongly influenced by a preprint of Michael Shulman [79] and subsequent conversations with its author. His paper gives a much more thorough account of this story than is presented here; I highly recommend the original. The title of Chapter 3 was chosen to acknowledge its debt to the expository paper [46]. The content of Chapter 4 was surely absorbed by osmosis from my advisor, Peter May, whose unacknowledged influence can also be felt elsewhere. Several examples, intuitions, and observations appearing throughout this text can be found in the notes [18]; in the books [32, 36, 58]; or on the *n*Lab, whose collaborative authors deserve accolades. The material on weighted limits and colimits is heavily influenced by current and past members of the Centre of Australian Category Theory. I was first introduced to the perspective on model categories taken in Chapter 11 by Martin Hyland. Richard Garner shared many of the observations attributed to him in Chapters 12 and 13 in private conversation, and I wish to thank him for enduring endless discussions on this topic. It is not possible to overstate the influence Dominic Verity has had on this presentation of the material on quasi-categories. Any interesting unattributed results on that topic should be assumed to be due to him.

Finally, and principally, I wish to thank those who attended and inspired the course. This work is dedicated to them. Comments, questions, and observations from Michael Andrews, Omar Antolín Camarena, David Ayala, Tobias Barthel, Kestutis Cesnavicius, Jeremy Hahn, Markus Hausmann, Gijs Heuts, Akhil Mathew, Luis Pereira, Chris Schommer-Pries, Kirsten Wickelgren, Eric Wofsey, and Inna Zakharevich led to direct improvements in this text. Tobias Barthel and Moritz Groth made several helpful comments and caught a number of typos. Philip Hirschhorn, Barry Mazur, and Sophia Roosth were consulted on matters of style. I am grateful for the moral support and stimulating mathematical environment provided by the Boston homotopy theory community, particularly Mike Hopkins and Jacob Lurie at Harvard and Clark Barwick, Mark Behrens, and Haynes Miller at MIT. I would also like to thank Harvard

University for giving me the opportunity to create and teach this course and the National Science Foundation for support through their Mathematical Sciences Postdoctoral Research Fellowship, award number DMS-1103790. Last, but not least, I am grateful for years of love and encouragement from friends and family, who made all things possible.

Part I

Derived functors and homotopy (co)limits

1

All concepts are Kan extensions

Given a pair of functors $K: \mathcal{C} \to \mathcal{D}$, $F: \mathcal{C} \to \mathcal{E}$, it may or may not be possible to extend F along K. Obstructions can take several forms: two arrows in $\mathcal{C}$ with distinct images in $\mathcal{E}$ might be identified in $\mathcal{D}$, or two objects might have empty hom-sets in $\mathcal{C}$ and $\mathcal{E}$ but not in $\mathcal{D}$. In general, it is more reasonable to ask for a best approximation to an extension taking the form of a universal natural transformation pointing either from or to F. The resulting categorical notion, quite simple to define, is surprisingly ubiquitous throughout mathematics, as we shall soon discover.

1.1 Kan extensions

Definition 1.1.1   Given functors $F: \mathcal{C} \to \mathcal{E}$, $K: \mathcal{C} \to \mathcal{D}$, a **left Kan extension** of F along K is a functor $\mathrm{Lan}_K F: \mathcal{D} \to \mathcal{E}$ together with a natural transformation $\eta: F \Rightarrow \mathrm{Lan}_K F \cdot K$ such that for any other such pair $(G: \mathcal{D} \to \mathcal{E}, \gamma: F \Rightarrow GK)$, γ factors uniquely through η, as illustrated:[1]

Dually, a **right Kan extension** of F along K is a functor $\mathrm{Ran}_K F: \mathcal{D} \to \mathcal{E}$ together with a natural transformation $\epsilon: \mathrm{Ran}_K F \cdot K \Rightarrow F$ such that for any

---

[1]   Writing α for the natural transformation $\mathrm{Lan}_K F \Rightarrow G$, the right-hand **pasting diagrams** express the equality $\gamma = \alpha_K \cdot \eta$, i.e., that γ factors as $F \overset{\eta}{\Longrightarrow} \mathrm{Lan}_K F \cdot K \overset{\alpha_K}{\Longrightarrow} GK$.

$(G: \mathcal{D} \to \mathcal{E}, \delta: GK \Rightarrow F)$, δ factors uniquely through ϵ, as illustrated:

Remark 1.1.2   This definition makes sense in any 2-category, but for simplicity, this discussion is relegated to the 2-category **Cat** of categories, functors, and natural transformations.

The intuition is clearest when the functor K of Definition 1.1.1 is an inclusion; assuming certain (co)limits exist, when K is fully faithful, the left and right Kan extensions do in fact extend the functor F along K; see 1.4.5. However in general, this need not be the case:

Exercise 1.1.3   Construct a toy example to illustrate that if F factors through K along some functor H, it is not necessarily the case that $(H, 1_F)$ is the left Kan extension of F along K.

Remark 1.1.4   In unenriched category theory, a **universal property** is encoded as a representation for an appropriate **Set**-valued functor. A left Kan extension of $F: \mathcal{C} \to \mathcal{E}$ along $K: \mathcal{C} \to \mathcal{D}$ is a representation for the functor

$$\mathcal{E}^{\mathcal{C}}(F, - \circ K): \mathcal{E}^{\mathcal{D}} \to \mathbf{Set}$$

that sends a functor $\mathcal{D} \to \mathcal{E}$ to the set of natural transformations from F to its restriction along K. By the Yoneda lemma, any pair (G, γ) as in Definition 1.1.1 defines a natural transformation

$$\mathcal{E}^{\mathcal{D}}(G, -) \stackrel{\gamma}{\Longrightarrow} \mathcal{E}^{\mathcal{C}}(F, - \circ K).$$

The universal property of the pair $(\mathrm{Lan}_K F, \eta)$ is equivalent to the assertion that the corresponding map

$$\mathcal{E}^{\mathcal{D}}(\mathrm{Lan}_K F, -) \stackrel{\eta}{\Longrightarrow} \mathcal{E}^{\mathcal{C}}(F, - \circ K)$$

is a natural isomorphism, that is, that $(\mathrm{Lan}_K F, \eta)$ represents this functor.

Extending this discussion, it follows that if, for fixed K, the left and right Kan extensions of any functor $\mathcal{C} \to \mathcal{E}$ exist, then these define left and right adjoints to the precomposition functor $K^* \colon \mathcal{E}^{\mathcal{D}} \to \mathcal{E}^{\mathcal{C}}$:

$$
\mathcal{E}^{\mathcal{D}}(\mathrm{Lan}_K F, G) \cong \mathcal{E}^{\mathcal{C}}(F, GK) \quad \mathcal{E}^{\mathcal{C}} \xleftarrow{K^*} \mathcal{E}^{\mathcal{D}} \quad \mathcal{E}^{\mathcal{C}}(GK, F) \cong \mathcal{E}^{\mathcal{D}}(G, \mathrm{Ran}_K F)
$$

with Lan_K above and Ran_K below.

$$(1.1.5)$$

The 2-cells η are the components of the unit for $\mathrm{Lan}_K \dashv K^*$, and the 2-cells ϵ are the components of the counit for $K^* \dashv \mathrm{Ran}_K$. The universal properties of Definition 1.1.1 are precisely those required to define the value at a particular object $F \in \mathcal{E}^{\mathcal{C}}$ of a left and right adjoint to a specified functor, in this case K^*.

Conversely, by uniqueness of adjoints, the objects in the image of any left or right adjoint to a precomposition functor are Kan extensions. This observation leads to several immediate examples.

Example 1.1.6 A small category with a single object and only invertible arrows is precisely a (discrete) group. The objects of the functor category $\mathbf{Vect}_k^G$ are G-representations over a fixed field k; arrows are G-equivariant linear maps. If H is a subgroup of G, restriction $\mathbf{Vect}_k^G \to \mathbf{Vect}_k^H$ of a G-representation to an H-representation is simply precomposition by the inclusion functor $i \colon H \hookrightarrow G$. This functor has a left adjoint, induction, which is left Kan extension along i. The right adjoint, coinduction, is right Kan extension along i:

$$
\mathbf{Vect}_k^G \xrightarrow{\;\mathrm{res}\;} \mathbf{Vect}_k^H
$$

with ind_H^G above and coind_H^G below.

$$(1.1.7)$$

The reader unfamiliar with the construction of induced representations need not remain in suspense for very long; see Theorem 1.2.1 and Example 1.2.9. Similar remarks apply for G-sets, G-spaces, based G-spaces, or indeed G-objects in any category – although in the general case, these adjoints might not exist.

Remark 1.1.8 This example can be enriched (cf. 7.6.9): extension of scalars, taking an R-module M to the S-module $M \otimes_R S$, is the $\mathbf{Ab}$-enriched left Kan extension along an $\mathbf{Ab}$-functor $R \to S$ between one-object $\mathbf{Ab}$-categories, more commonly called a ring homomorphism.

Example 1.1.9 Let $\triangle$ be the category of finite non-empty ordinals $[0], [1], \ldots$ and order-preserving maps. **Set**-valued presheaves on $\triangle$ are called **simplicial sets**. Write $\triangle_{\leq n}$ for the full subcategory on the objects $[0], \ldots, [n]$. Restriction along the inclusion functor $i_n \colon \mathbb{\Delta}_{\leq n} \hookrightarrow \mathbb{\Delta}$ is called n-truncation. This functor has both left and right Kan extensions:

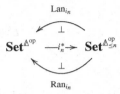

The composite comonad on $\mathbf{Set}^{\mathbb{\Delta}^{\mathrm{op}}}$ is sk_n, the functor that maps a simplicial set to its n-**skeleton**. The composite monad on $\mathbf{Set}^{\mathbb{\Delta}^{\mathrm{op}}}$ is cosk_n, the functor that maps a simplicial set to its n-**coskeleton**. Furthermore, sk_n is left adjoint to cosk_n, as is the case for any comonad and monad arising in this way.

Example 1.1.10 The category $\mathbb{\Delta}$ is a full subcategory containing all but the initial object $[-1]$ of the category $\mathbb{\Delta}_+$ of finite ordinals and order-preserving maps. Presheaves on $\mathbb{\Delta}_+$ are called **augmented simplicial sets**. Left Kan extension defines a left adjoint to restriction,

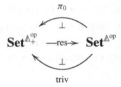

that augments a simplicial set X with its set $\pi_0 X$ of path components. Right Kan extension assigns a simplicial set the trivial augmentation built from the one-point set.

A final broad class of examples has a rather different flavor.

Example 1.1.11 In good situations, the composite of a functor $F \colon \mathcal{C} \to \mathcal{D}$ between categories equipped with subcategories of "weak equivalences" and the localization functor $\mathcal{D} \to \mathrm{Ho}\mathcal{D}$ admits a right or left Kan extension along the localization functor $\mathcal{C} \to \mathrm{Ho}\mathcal{C}$, called the **total left derived functor** or **total right derived functor**, respectively. This is the subject of Chapter 2.

1.2 A formula

Importantly, if the target category $\mathcal{E}$ has certain limits and colimits, then right and left Kan extensions for any pair of functors exist and furthermore can be computed by a particular (co)limit formula. Recall that a category is **small** if it has a mere set of morphisms and **locally small** if it has a mere set of morphisms between any fixed pair of objects.

Theorem 1.2.1 ([50, X.4.1–2]) *When C is small, D is locally small, and $\mathcal{E}$ is cocomplete, the left Kan extension of any functor $F : C \to \mathcal{E}$ along any functor $K : C \to D$ is computed at $d \in D$ by the colimit*

$$\mathrm{Lan}_K F(d) = \int^{c \in C} D(Kc, d) \cdot Fc \qquad (1.2.2)$$

and in particular necessarily exists.

Some explanation is in order. The "·" is called a **copower** or a **tensor**: if S is a set and $e \in \mathcal{E}$, then $S \cdot e$ is the S-indexed coproduct of copies of e. Assuming these coproducts exist in $\mathcal{E}$, the copower defines a bifunctor $\mathbf{Set} \times \mathcal{E} \to \mathcal{E}$.

The integral $\int^C$, called a **coend**, is the colimit of a particular diagram constructed from a functor that is both covariant and contravariant in C. Given $H : C^{\mathrm{op}} \times C \to \mathcal{E}$, the coend $\int^C H$ is an object of $\mathcal{E}$ equipped with arrows $H(c, c) \to \int^C H$ for each $c \in C$ that are collectively universal with the property that the diagram

$$\begin{array}{ccc}
H(c', c) & \xrightarrow{\ f_* \ } & H(c', c') \\
{\scriptstyle f^*}\Big\downarrow & & \Big\downarrow \\
H(c, c) & \longrightarrow & \int^C H
\end{array} \qquad (1.2.3)$$

commutes for each $f : c \to c'$ in C. Equivalently, $\int^C H$ is the coequalizer of the diagram

$$\coprod_{f \in \mathrm{arr}\, C} H(\mathrm{cod}\, f, \mathrm{dom}\, f) \underset{f_*}{\overset{f^*}{\rightrightarrows}} \coprod_{c \in \mathrm{ob}\, C} H(c, c) \dashrightarrow \int^C H \qquad (1.2.4)$$

Remark 1.2.5 If $H : C^{\mathrm{op}} \times C \to \mathcal{E}$ is constant in the first variable, that is, if H is a functor $C \to \mathcal{E}$, then the coequalizer (1.2.4) defines the usual colimit of H.

Remark 1.2.6 Assuming these colimits exist, the coend (1.2.2) is isomorphic to the colimit of the composite $K/d \xrightarrow{U} C \xrightarrow{F} \mathcal{E}$ of F with a certain forgetful functor. The domain of U is the **slice category**, a special kind of **comma category** whose objects are pairs $(c \in C, Kc \to d \in D)$ and whose morphisms are arrows in C that make the obvious triangle in D commute. Both formulas encode a particular **weighted colimit** of F in a sense that is made precise in Chapter 7. In particular, we prove that these formulas agree in 7.1.11.

Exercise 1.2.7   Let C be a small category, and write $C^{\triangleright}$ for the category obtained by adjoining a terminal object to C. Give three proofs that a left Kan extension of a functor $F: C \to \mathcal{E}$ along the natural inclusion $C \to C^{\triangleright}$ defines a colimit cone under F: one using the defining universal property, one using Theorem 1.2.1, and one using the formula of 1.2.6.

Dually, the **power** or **cotensor** e^S of $e \in \mathcal{E}$ by a set S is the S-indexed product of copies of e, defining a bifunctor $\mathbf{Set}^{\mathrm{op}} \times \mathcal{E} \to \mathcal{E}$ that is contravariant in the indexing set. For $H: C^{\mathrm{op}} \times C \to \mathcal{E}$, an **end** $\int_C H$ is an object in $\mathcal{E}$ together with morphisms satisfying diagrams dual to (1.2.3) and universal with this property.

Exercise 1.2.8   Let $F, G: C \rightrightarrows \mathcal{E}$, with C small and $\mathcal{E}$ locally small. Show that the end over C of the bifunctor $\mathcal{E}(F-, G-): C^{\mathrm{op}} \times C \to \mathbf{Set}$ is the set of natural transformations from F to G.

Example 1.2.9   Let us return to Example 1.1.6. In the category $\mathbf{Vect}_k$, finite products and finite coproducts coincide: these are just direct sums of vector spaces. If V is an H-representation and H is a finite index subgroup of G, then the end and coend formulas of Theorem 1.2.1 and its dual both produce the direct sum of copies of V indexed by left cosets of H in G. Thus, for finite index subgroups, the left and right adjoints of (1.1.7) are the same; that is, induction from a finite index subgroup is both left and right adjoint to restriction.

Example 1.2.10   We can use Theorem 1.2.1 to understand the functors sk_n and cosk_n of Example 1.1.9. If $m > n$ and $k \leq n$, each map in $\Delta^{\mathrm{op}}([k], [m]) = \Delta([m], [k])$ factors uniquely as a non-identity epimorphism followed by a monomorphism.[2] It follows that every simplex in $\mathrm{sk}_n X$ above dimension m is degenerate; indeed, $\mathrm{sk}_n X$ is obtained from the n-truncation of X by freely adding back the necessary degenerate simplices.

Now we use the adjunction $\mathrm{sk}_n \dashv \mathrm{cosk}_n$ to build some intuition for the n-coskeleton. Suppose $X \cong \mathrm{cosk}_n X$. By adjunction, an $(n+1)$-simplex corresponds to a map $\mathrm{sk}_n \Delta^{n+1} = \partial \Delta^{n+1} \to X$. In words, each $(n+1)$-sphere in an n-coskeletal simplicial set has a unique filler. Indeed, any m-sphere in an n-coskeletal simplicial set, with $m > n$, has a unique filler. More precisely, an m-simplex is uniquely determined by the data of its faces of dimension n and below.

Exercise 1.2.11   Directed graphs are functors from the category with two objects E, V and a pair of maps $s, t: E \rightrightarrows V$ to $\mathbf{Set}$. A natural transformation between two such functors is a graph morphism. The forgetful functor

---

[2] This is the content of the Eilenberg–Zilber lemma [28, II.3.1, pp. 26–27]; cf. Lemma 14.3.7.

DirGph → **Set** that maps a graph to its set of vertices is given by restricting along the functor from the terminal category $\mathbb{1}$ that picks out the object V. Use Theorem 1.2.1 to compute left and right adjoints to this forgetful functor.

1.3 Pointwise Kan extensions

A functor $L \colon \mathcal{E} \to \mathcal{F}$ **preserves** $(\mathrm{Lan}_K F, \eta)$ if the whiskered composite $(L\mathrm{Lan}_K F, L\eta)$ is the left Kan extension of LF along K:

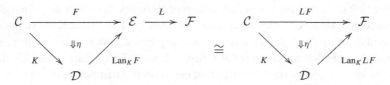

Example 1.3.1 The forgetful functor $U \colon$ **Top** → **Set** has both left and right adjoints and hence preserves both limits and colimits. It follows from Theorem 1.2.1 that U preserves the left and right Kan extensions of Example 1.1.6.

Example 1.3.2 The forgetful functor $U \colon \mathbf{Vect}_k \to$ **Set** preserves limits but not colimits because the underlying set of a direct sum is not simply the coproduct of the underlying sets of vectors. Hence it follows from 1.2.1 and 1.1.6 that the underlying set of a G-representation induced from an H-representation is not equal to the G-set induced from the underlying H-set.

Even when we cannot appeal to the formula presented in 1.2.1 adjoint functors preserve compatibly handed Kan extensions:

Lemma 1.3.3 *Left adjoints preserve left Kan extensions.*

Proof Suppose given a left Kan extension $(\mathrm{Lan}_K F, \eta)$ with codomain $\mathcal{E}$ and suppose further that $L \colon \mathcal{E} \to \mathcal{F}$ has a right adjoint R with unit ι and counit ν. Then, given $H \colon \mathcal{D} \to \mathcal{F}$, there are natural isomorphisms

$$\mathcal{F}^{\mathcal{D}}(L\mathrm{Lan}_K F, H) \cong \mathcal{E}^{\mathcal{D}}(\mathrm{Lan}_K F, RH) \cong \mathcal{E}^{\mathcal{C}}(F, RHK) \cong \mathcal{F}^{\mathcal{C}}(LF, HK).$$

Taking $H = L\mathrm{Lan}_K F$, these isomorphisms act on the identity natural transformation, as follows:

$$1_{L\mathrm{Lan}_K F} \mapsto \iota_{\mathrm{Lan}_K F} \mapsto \iota_{\mathrm{Lan}_K F \cdot K} \cdot \eta \mapsto \nu_{L\mathrm{Lan}_K F \cdot K} \cdot L\iota_{\mathrm{Lan}_K F \cdot K} \cdot L\eta = L\eta.$$

Hence $(L\mathrm{Lan}_K F, L\eta)$ is a left Kan extension of LF along K. $\square$

Unusually for a mathematical object defined by a universal property, generic Kan extensions are rather poorly behaved. We see specific examples of this insufficiency in Chapter 2, but for now we have to rely on expert opinion. For instance, Max Kelly reserves the name "Kan extension" for pairs satisfying the condition we presently introduce, calling those of our Definition 1.1.1 "weak" and writing that "our present choice of nomenclature is based on our failure to find a single instance where a weak Kan extension plays any mathematical role whatsoever" [46, §4]. By the categorical community's consensus, the important Kan extensions are **pointwise** Kan extensions.

Definition 1.3.4   When $\mathcal{E}$ is locally small, a right Kan extension is a **pointwise** right Kan extension[3] if it is preserved by all representable functors $\mathcal{E}(e, -)$.

Because covariant representables preserve all limits, it is clear that if a right Kan extension is given by the formula of Theorem 1.2.1, then that Kan extension is pointwise; dually, left Kan extensions computed in this way are pointwise. The surprise is that the converse also holds. This characterization justifies the terminology: a pointwise Kan extension can be computed pointwise as a limit in $\mathcal{E}$.

Theorem 1.3.5 ([50, X.5.3])   *A right Kan extension of F along K is pointwise if and only if it can be computed by*

$$\mathrm{Ran}_K F(d) = \lim \left(d/K \xrightarrow{U} \mathcal{C} \xrightarrow{F} \mathcal{E} \right)$$

in which case, in particular, this limit exists.

Proof   If $\mathrm{Ran}_K F$ is pointwise, then by the Yoneda lemma and the defining universal property of right Kan extensions,

$$\mathcal{E}(e, \mathrm{Ran}_K F(d)) \cong \mathbf{Set}^{\mathcal{D}}(\mathcal{D}(d, -), \mathcal{E}(e, \mathrm{Ran}_K F))$$
$$\cong \mathbf{Set}^{\mathcal{C}}(\mathcal{D}(d, K-), \mathcal{E}(e, F-)).$$

The right-hand set is naturally isomorphic to the set of cones under e over the functor FU; hence this bijection exhibits $\mathrm{Ran}_K F(d)$ as the limit of FU. $\square$

Remark 1.3.6   Most commonly, pointwise Kan extensions are found whenever the codomain category is cocomplete (for left Kan extensions) or complete (for right), but this is not the only case. In Chapter 2, we see that the most common construction of the total derived functors defined in 1.1.11 produces pointwise Kan extensions, even though homotopy categories have notoriously few limits and colimits (see Proposition 2.2.13).

---

[3] A functor $K : \mathcal{C} \to \mathcal{D}$ is equally a functor $K : \mathcal{C}^{\mathrm{op}} \to \mathcal{D}^{\mathrm{op}}$, but the process of replacing each category by its opposite reverses the direction of any natural transformations; succinctly, "op" is a 2-functor $(-)^{\mathrm{op}} : \underline{\mathbf{Cat}}^{\mathrm{co}} \to \underline{\mathbf{Cat}}$. A left Kan extension is **pointwise**, as we are in the process of defining, if the corresponding right Kan extension in the image of this 2-functor is pointwise.

1.4 All concepts

The following examples justify Saunders Mac Lane's famous assertion that "the notion of Kan extensions subsumes all the other fundamental concepts of category theory" [50, §X.7].

Example 1.4.1   Consider Kan extensions along the unique functor $K : C \to$ $\mathbb{1}$ to the terminal category. A functor $G : \mathbb{1} \to \mathcal{E}$ picks out an object of $\mathcal{E}$; precomposing with K yields the constant functor $C \to \mathcal{E}$ at this object. Hence the universal property (1.1.5) specifies that $\mathrm{Lan}_K F$ represents the set of natural transformations from $F : C \to \mathcal{E}$ to a constant functor, that is, that $\mathrm{Lan}_K F$ represents cones under F, that is, that $\mathrm{Lan}_K F$ is the colimit of F. Dually, $\mathrm{Ran}_K F$ is the limit. In the pointwise case, this can be deduced directly from 1.2.1 and 1.2.5.

Example 1.4.2   If $F : C \rightleftarrows \mathcal{D} : G$ is an adjunction with unit $\eta : 1 \Rightarrow GF$ and counit $\epsilon : FG \Rightarrow 1$, then (G, η) is a left Kan extension of the identity functor at C along F and (F, ϵ) is a right Kan extension of the identity functor at $\mathcal{D}$ along G. Conversely, if (G, η) is a left Kan extension of the identity along F and if F preserves this Kan extension, then $F \dashv G$ with unit η.

Exercise 1.4.3   Prove these assertions by writing down the appropriate diagram chase. As a hint, note that an adjunction $F \dashv G$ induces an adjunction

$$\mathcal{E}^C \underset{F^*}{\overset{G^*}{\underset{\perp}{\rightleftarrows}}} \mathcal{E}^{\mathcal{D}}$$

that is, for any $H : C \to \mathcal{E}, K : \mathcal{D} \to \mathcal{E}, \mathcal{E}^{\mathcal{D}}(HG, K) \cong \mathcal{E}^C(H, KF)$.

Example 1.4.4   From the defining universal property, the right Kan extension of a functor F along the identity is (isomorphic to) F. In the case $F : C \to \mathbf{Set}$, we can apply Theorem 1.2.1 and Example 1.2.8 to deduce that

$$Fc \cong \mathrm{Ran}_{1_C} F(c) \cong \int_{x \in C} Fx^{C(c,x)} \cong \int_{x \in C} \mathbf{Set}(C(c, x), Fx) \cong \mathbf{Set}^C(C(c, -), F)$$

the right-hand side being the set of natural transformations from the functor represented by c to F. This is the Yoneda lemma.

Corollary 1.4.5   *If $\mathcal{E}$ is complete and K is fully faithful, then* $\mathrm{Ran}_K F \cdot K \cong F$.

Proof   Mac Lane [50, X.3.3] proves a more precise version: that the counit of the right Kan extension gives the asserted isomorphism. When K is fully faithful, $\mathcal{C}(c, x) \cong \mathcal{D}(Kc, Kx)$ for each pair $c, x \in \mathcal{C}$. By Theorem 1.2.1 and the Yoneda lemma,

$$\mathrm{Ran}_K F(Kc) \cong \int_{x \in \mathcal{C}} Fx^{\mathcal{D}(Kc, Kx)} \cong \int_{x \in \mathcal{C}} Fx^{\mathcal{C}(c, x)} \cong Fc. \qquad \square$$

Example 1.4.6   Equally, $\mathrm{Lan}_{1_c} F \cong F$, from which we deduce the **co-Yoneda lemma**:

$$Fc \cong \int^{x \in \mathcal{C}} \mathcal{C}(x, c) \cdot Fx. \qquad (1.4.7)$$

More precise analysis shows that the canonical cone under the coend diagram is a colimit cone. In the case $F : \mathcal{C} \to \mathbf{Set}$, the copower "$\cdot$" is symmetric, and so the coend (1.4.7) is isomorphic to a coend in which the sets Fx and $\mathcal{C}(x, c)$ are swapped. Letting c vary, we conclude that F is canonically a colimit of representable functors, a fact that is frequently called the **density theorem**. We describe this colimit more precisely in 7.2.7.

Exercise 1.4.8   Use Theorem 1.2.1, the Yoneda lemma, and the co-Yoneda lemma to deduce another form of the density theorem: that the left Kan extension of the Yoneda embedding $\mathcal{C} \to \mathbf{Set}^{\mathcal{C}^{op}}$ along itself is the identity functor. This says that the representable functors form a **dense subcategory** of the presheaf category $\mathbf{Set}^{\mathcal{C}^{op}}$.

1.5 Adjunctions involving simplicial sets

We close this chapter with a homotopical application of the theory of Kan extensions. We show that all adjoint pairs of functors whose left adjoints have the category of simplicial sets as their domain arise in the same way. Succinctly, this is the case because the Yoneda embedding establishes the functor category $\mathbf{Set}^{\Delta^{op}}$, henceforth denoted by $\mathbf{sSet}$, as the free colimit completion of Δ.[4]

Construction 1.5.1   Let $\mathcal{E}$ be any cocomplete, locally small category, and let $\Delta^{\bullet} : \Delta \to \mathcal{E}$ be any covariant functor. We write Δ^n for the image of $[n]$; similarly, for $X \in \mathbf{sSet}$, we write X_n for the set of n-**simplices** and $X_{\bullet} : \Delta^{op} \to \mathbf{Set}$ when we wish to emphasize its role as a functor. Define $L : \mathbf{sSet} \to \mathcal{E}$ to

---

[4] None of this analysis depends on the nature of Δ; any small category would suffice.

be the left Kan extension of $\Delta^\bullet$ along the Yoneda embedding:

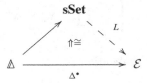

Because $\mathcal{E}$ is assumed to be cocomplete, the functor L is defined on objects by the coend

$$LX := \int^{n \in \Delta} \mathbf{sSet}(\Delta^n, X) \cdot \Delta^n \cong \int^{n \in \Delta} X_n \cdot \Delta^n$$

$$\cong \mathrm{coeq} \left(\coprod_{[n] \to [m] \in \Delta} X_m \cdot \Delta^n \rightrightarrows \coprod_{[n] \in \Delta} X_n \cdot \Delta^n \right). \qquad (1.5.2)$$

The leftmost Δ^n in (1.5.2) is the representable functor $\Delta(-, [n]) \colon \Delta^{\mathrm{op}} \to \mathbf{Set}$; the first congruence is by the Yoneda lemma, which establishes a bijection between maps $\Delta^n \to X$ of simplicial sets and n-simplices of X. The functor L is defined on arrows by the universal properties of these colimits. Uniqueness of the universal property will imply that L is functorial, as is always the case when one uses a colimit construction to define a functor. By Corollary 1.4.5, $L\Delta^n \cong \Delta^n$, somewhat justifying this abuse of notation.

Because L is defined by a colimit and colimits commute with each other, L preserves colimits. Hence, your favorite adjoint functor theorem implies that L has a right adjoint $R \colon \mathcal{E} \to \mathbf{sSet}$. From the desired adjoint correspondence and the Yoneda lemma, for any $e \in \mathcal{E}$,

$$(Re)_n \cong \mathbf{sSet}(\Delta^n, Re) \cong \mathcal{E}(L\Delta^n, e) \cong \mathcal{E}(\Delta^n, e).$$

Thus we define the n-simplices of Re to be the maps in $\mathcal{E}$ from Δ^n to e. The face and degeneracy maps for this simplicial set are given by precomposition by the appropriate maps in the cosimplicial object $\Delta^\bullet$. Levelwise postcomposition defines a map of simplicial sets for each $e \to e' \in \mathcal{E}$ and makes R a functor.

Example 1.5.3 There is a natural functor $\Delta \to \mathbf{Top}$ such that the nth space Δ^n is the standard topological n-simplex $\Delta^n = \{(x_0, \ldots, x_n) \in \mathbb{R}^{n+1} \mid x_i \geq 0, \sum_i x_i = 1\}$. The associated right adjoint $S \colon \mathbf{Top} \to \mathbf{sSet}$ is called the **total singular complex functor**. Following the preceding prescription, its left adjoint

$|-|$: **sSet** $\to$ **Top** is defined at a simplicial set X by[5]

$$|X| = \int^n X_n \times \Delta^n = \text{colim} \left(\coprod_{f:\, [n]\to[m]} X_m \times \Delta^n \overset{f_*}{\underset{f^*}{\rightrightarrows}} \coprod_{[n]} X_n \times \Delta^n \right)$$

and is called its **geometric realization**.

Here, again, Corollary 1.4.5, also applicable in each of the examples to follow, implies that $|\Delta^n| = \Delta^n$, somewhat justifying this notational conflation.

Exercise 1.5.4   Prove that geometric realization is left adjoint to the total singular complex functor by demonstrating this fact for any adjunction arising from the construction of 1.5.1.

Example 1.5.5   Let $\Delta^\bullet : \Delta \to$ **Cat** be the functor that sends $[n]$ to the ordinal category

$$0 \to 1 \to \cdots \to n$$

with $n + 1$ objects and n generating non-identity arrows together with their composites and the requisite identities. Each order-preserving map in Δ defines the object function of a unique functor between categories of this type, and all functors arise this way. Thus $\Delta^\bullet$ is a full embedding $\Delta \hookrightarrow$ **Cat**. The right adjoint is the **nerve** functor $N :$ **Cat** $\to$ **sSet**. By 1.5.1, an n-simplex in NC is a functor $[n] \to C$, that is, a string of n composable arrows in C. The ith degeneracy map inserts the appropriate identity arrow at the ith place, and the ith face map leaves off an outside arrow if i is 0 or n and composes the ith and $(i + 1)$th arrows otherwise.

The left adjoint $h :$ **sSet** $\to$ **Cat** maps a simplicial set to its homotopy category. Interpreting the formula (1.5.2) in **Cat** yields the following explicit description of the category hX. Its objects are the 0-simplices, also called **vertices**, of X. Its arrows are freely generated from the 1-simplices – any "composable path" of 1-simplices oriented in the correct direction represents a morphism in hX – subject to relations witnessed by 2-simplices: if there is a 2-simplex in X with zeroth face k, first face ℓ, and second face j, then $\ell = kj$ in hX. We see in Chapter 15 that there is a simpler description of this functor on the subcategory of quasi-categories.

Example 1.5.6   Construction 1.5.1 shows that **sSet** is **cartesian closed**; that is, for every simplicial set Y, the functor $- \times Y :$ **sSet** $\to$ **sSet** has a right

---

[5] The copower $X_n \cdot \Delta^n$ in topological spaces is isomorphic to the cartesian product $X_n \times \Delta^n$, where the set X_n is given the discrete topology (compare with Remark 4.0.1).

adjoint, which we refer to as the **internal hom**. Indeed, every category of **Set**-valued presheaves is cartesian closed, and the construction of the right adjoint given here is the usual one.

We desire an adjunction

$$\mathbf{sSet}(X \times Y, Z) \cong \mathbf{sSet}(X, Z^Y).$$

Taking X to be representable and applying the Yoneda lemma, this becomes

$$\mathbf{sSet}(\Delta^n \times Y, Z) \cong \mathbf{sSet}(\Delta^n, Z^Y) \cong (Z^Y)_n,$$

so we define the set of n-simplices of the internal hom Z^Y to be the set of maps $\Delta^n \times Y \to Z$. The full adjunction is a consequence of the density theorem, or of 1.5.4, which amounts to the same thing.

Example 1.5.7   The **subdivision** sdΔ^n of the nth represented simplicial set is defined to be the simplicial set formed by taking the nerve of the poset of inclusions of non-empty subsets of $[n]$. To illustrate, sdΔ^2 has the following vertices and non-degenerate 1-simplices:

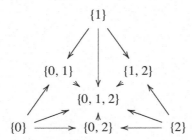

together with six non-degenerate 2-simplices filling the evident triangles. Left Kan extension defines a subdivision functor that has a right adjoint called **extension**:

$$\mathrm{sd} \colon \mathbf{sSet} \; \underset{\longleftarrow}{\overset{\longrightarrow}{\perp}} \; \mathbf{sSet} \colon \mathrm{ex}$$

The "last vertex" map sd$\Delta^n \to \Delta^n$ defines a natural transformation from the cosimplicial simplicial set defining the subdivision to the Yoneda embedding. It follows that there is a natural transformation sd $\Rightarrow$ id between the left Kan extensions. Write $\eta \colon$ id $\Rightarrow$ ex for the adjunct natural transformation. The colimit of the sequence of natural transformations η_{ex^n} defines a functor called ex$^\infty$, which has a number of applications in simplicial homotopy theory; cf., e.g., [32, III.4.8].

Example 1.5.8   Write **sCat** for the category of small categories and functors **enriched over sSet** in a sense we make precise in Chapter 3. A particular cosimplicial object in **sCat**, which can be regarded as a simplicial thickening or cofibrant replacement of the discrete simplicial categories [*n*], is used to define an adjunction

$$\mathfrak{C}: \textbf{sSet} \; \underset{\longleftarrow}{\overset{\longrightarrow}{\perp}} \; \textbf{sCat}: \mathfrak{N}$$

whose right adjoint is the **homotopy coherent nerve**. See Chapter 16 for a considerably more detailed description of this adjunction.

2

Derived functors via deformations

In common parlance, a construction is **homotopical** if it is invariant under weak equivalence. A generic functor frequently does not have this property. In certain cases, the functor can be approximated by a **derived functor**, a notion first introduced in homological algebra, which is a universal homotopical approximation either to or from the original functor.

The definition of a total derived functor is simple enough: it is a Kan extension whose handedness unfortunately contradicts that of the derived functor, along the appropriate localization (see Example 1.1.11). But, unusually for constructions characterized by a universal property, generic total derived functors are poorly behaved: for instance, the composite of the total left derived functors of a pair of composable functors is not necessarily a total left derived functor for the composite. The problem with the standard definition is that total derived functors are not typically required to be pointwise Kan extensions. In light of Theorem 1.3.5, this seems reasonable, because homotopy categories are seldom complete or cocomplete.

One of the selling points of Daniel Quillen's theory of model categories is that they highlight classes of functors – the left or right Quillen functors – whose left or right derived functors can be constructed in a uniform way, making the passage to total derived functors pseudofunctorial. However, it turns out a full model structure is not necessary for this construction, as suggested by the slogan that "all that matters are the weak equivalences."

In this chapter, following [22], we consider functors between **homotopical categories** that are equipped with some reasonable collection of arrows called "weak equivalences." We define left or right deformations associated to a particular functor and describe an analogous construction of point-set level left or right derived functors whose homotopical universal property is independent of the deformation used. The total derived functors produced in this manner are pointwise and indeed **absolute** Kan extensions, explaining why derived functors obtained in this manner behave better than generic ones satisfying a weaker universal property.

A major benefit to producing well-behaved derived functors without a model structure is the application to colimit and limit functors of generic shapes. The corresponding diagram categories frequently lack an appropriate model structure. We end this chapter by previewing a result, proven in Chapter 5, that constructs homotopy colimit and homotopy limit functors of any shape in the general setting of a simplicial model category.

2.1 Homotopical categories and derived functors

The ideas that follow were first introduced in [22], although, for aesthetic reasons, we have departed slightly from their terminology. A good summary can be found in [79, §§2–4].

Definition 2.1.1 A **homotopical category** is a category $\mathcal{M}$ equipped with a wide[1] subcategory $\mathcal{W}$ such that, for any composable triple of arrows

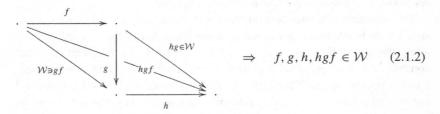

$$\Rightarrow \quad f, g, h, hgf \in \mathcal{W} \quad\quad (2.1.2)$$

if hg and gf are in $\mathcal{W}$, so are f, g, h, and hgf.

The arrows in $\mathcal{W}$ are called **weak equivalences**; the condition (2.1.2) is called the **2-of-6 property**.

Remark 2.1.3 The 2-of-6 property implies, but is stronger than, the usual **2-of-3 property**: if any two of f, g, and gf are in $\mathcal{W}$, so is the third. Nonetheless, by Remark 2.1.9 and Lemma 2.1.10, the weak equivalences of any model category satisfy the 2-of-6 property. Hence any model category has an underlying homotopical category (cf. Definition 11.3.1).

Examples include spaces with homotopy equivalences or weak homotopy equivalences; chain complexes with chain homotopy equivalences or quasi-isomorphisms; simplicial sets with simplicial homotopy equivalences, weak homotopy equivalences, or equivalences of quasi-categories; categories or groupoids with equivalences; and many others.

---

[1] **Wide** means containing all the objects; some prefer the term "lluf." This condition is a cheeky way to say that all identities, and thus, by the 2-of-6 property, all isomorphisms, are in $\mathcal{W}$.

Example 2.1.4   Any category can be regarded as a **minimal** homotopical category taking the weak equivalences to be the isomorphisms. To justify this, we must show that the class of isomorphisms in any category satisfies the 2-of-6 property. Consider a composable triple f, g, h such that gf and hg are isomorphisms. The map g has right inverse $f(gf)^{-1}$. Because hg is an isomorphism, g is monic, so this right inverse is also a left inverse. Hence g and therefore also f, h, and hgf are isomorphisms.

If there is no other obvious notion of weak equivalence, we use this choice as the default.

Digression 2.1.5 (homotopy equivalences are weak homotopy equivalences)   The fact that the isomorphisms in any category satisfy the 2-of-6 property is used to prove that the maps forming a homotopy equivalence $f \colon X \rightleftarrows Y \colon g$ of spaces are weak homotopy equivalences. Because $gf \simeq 1$ and $fg \simeq 1$ and π_n is homotopy invariant, the horizontal group homomorphisms

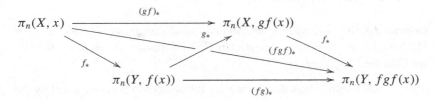

are isomorphisms. By the 2-of-6 property, it follows that the left-hand map is also an isomorphism.

Definition 2.1.6   The **homotopy category** $\mathrm{Ho}\mathcal{M}$ of a homotopical category $(\mathcal{M}, \mathcal{W})$ is the formal localization of $\mathcal{M}$ at the subcategory $\mathcal{W}$.

The following explicit construction is due to Pierre Gabriel and Michel Zisman [28, 1.1]. The category $\mathrm{Ho}\mathcal{M}$ has the same objects as $\mathcal{M}$. Its morphisms are equivalence classes of finite zigzags of morphisms in $\mathcal{M}$, with only those arrows in $\mathcal{W}$ permitted to go backward, subject to the following relations:

- adjacent arrows pointing in the same direction may be composed
- adjacent pairs $\xleftarrow{w} \xrightarrow{w}$ or $\xrightarrow{w} \xleftarrow{w}$ with $w \in \mathcal{W}$ may be removed
- identities pointing either forward or backward may be removed

There is a canonical identity-on-objects localization functor

$$\mathcal{M} \xrightarrow{\ \gamma\ } \mathrm{Ho}\mathcal{M}$$

characterized by the following universal property: precomposition with γ induces a bijective correspondence between functors $\text{Ho}\mathcal{M} \to \mathcal{N}$ and functors $\mathcal{M} \to \mathcal{N}$ that send weak equivalences to isomorphisms.

Example 2.1.7 The homotopy category of a minimal homotopical category is isomorphic to that category.

Example 2.1.8 The homotopy category of the homotopical category of topological spaces and weak homotopy equivalences is equivalent to the **homotopy category of spaces**, that is, to the category of CW complexes and homotopy classes of maps. Quillen generalized this result to characterize the homotopy category of any model category (see 11.3.13). A further generalization will be given in Theorem 10.5.1.

Remark 2.1.9 It is not generally the case that all arrows of $\mathcal{M}$ that become isomorphisms in $\text{Ho}\mathcal{M}$ are weak equivalences. When this is true, the homotopical category $\mathcal{M}$ is called **saturated**. Quillen shows that all model categories are saturated [65, Proposition 5.1].

Lemma 2.1.10 *Let $\mathcal{M}$ be a category equipped with any collection of arrows $\mathcal{W}$. If the localization, constructed as by Gabriel-Zisman, is saturated, then $\mathcal{W}$ satisfies the 2-of-6 property.*

Proof Saturation means that the weak equivalences in $\mathcal{M}$ are created by the isomorphisms in $\text{Ho}\mathcal{M}$, which satisfy the 2-of-6 property by the discussion in Example 2.1.4. □

A major obstacle to understanding the category $\text{Ho}\mathcal{M}$ is that it is not necessarily locally small: one could easily imagine non-equivalent zigzags from x to y in $\mathcal{M}$ snaking through each object of $\mathcal{M}$. Quillen proves that homotopy categories associated to model categories are locally small, but even so, it is often preferable to try to avoid working in the homotopy category at all and to seek instead to understand which **point-set level** constructions are homotopically meaningful, that is, which descend to functors between the appropriate homotopy categories.

To be precise, we say a functor between homotopical categories is **homotopical** if it preserves weak equivalences. By the universal property of localization, a homotopical functor F induces a unique functor

$$\begin{array}{ccc} \mathcal{M} & \xrightarrow{\ F\ } & \mathcal{N} \\ \gamma \downarrow & & \downarrow \delta \\ \text{Ho}\mathcal{M} & \dashrightarrow & \text{Ho}\mathcal{N} \\ & F & \end{array}$$

commuting with the localizations. We now have sufficient terminology to discuss an important subtle point.

Remark 2.1.11   The universal property of the localization functor $\gamma : \mathcal{M} \to$ Ho$\mathcal{M}$ is 2-categorical: natural transformations between homotopical functors $\mathcal{M} \to \mathcal{N}$ taking values in a minimal homotopical category $\mathcal{N}$ correspond bijectively to natural transformations between the associated functors Ho$\mathcal{M} \to \mathcal{N}$. But if $\mathcal{N}$ is not a minimal homotopical category, we must distinguish between functors taking values in $\mathcal{N}$ and in Ho$\mathcal{N}$. A natural transformation $\alpha : F \Rightarrow F'$ between functors $\mathcal{M} \to \mathcal{N}$ descends to a unique transformation $\delta\alpha$ between functors Ho$\mathcal{M} \to$ Ho$\mathcal{N}$. Conversely, a "downstairs" natural transformation corresponds to a unique map between functors $\mathcal{M} \to$ Ho$\mathcal{N}$, but it might not be possible to lift this natural transformation along $\delta : \mathcal{N} \to$ Ho$\mathcal{N}$.

Let us now describe a few examples.

Example 2.1.12   Familiar "homotopy invariants" such as homotopy groups π_n, homology groups H_n, or cohomology rings H^* are homotopical functors from the homotopical category of spaces with homotopy equivalences to the appropriate minimal homotopical category of groups or rings or to the appropriate algebraic gadget. Tautologically, the functor π_* that takes a space to its $\mathbb{N}$-indexed family of higher homotopy groups is a homotopical functor from the homotopical category of spaces and weak homotopy equivalences to the minimal homotopical category of graded groups (or graded sets).

Example 2.1.13   Any functor F equipped with a natural weak equivalence to or from the identity functor is homotopical by the 2-of-3 property. For instance, the functor that maps a space X to the cylinder $X \times I$, where $I = [0, 1]$ is the standard interval, is homotopical because the canonical projection $X \times I \xrightarrow{\sim} X$ is a natural weak equivalence. Another example is the path space functor, mapping a space to the space X^I of continuous functions from I to X with the compact-open topology, which is homotopical on account of the natural weak equivalence $X \xrightarrow{\sim} X^I$ that picks out the subspace of constant paths.

Example 2.1.14   An additive functor between abelian categories gives rise to a functor between the respective categories of chain complexes. If the original functor is exact, the induced functor preserves quasi-isomorphisms and is therefore homotopical in this sense.

Example 2.1.15   Let $F : \mathcal{A} \to \mathcal{B}$ be a not necessarily exact additive functor between abelian categories. Write $\mathbf{Ch}_{\geq 0}(\mathcal{A})$ and $\mathbf{Ch}_{\geq 0}(\mathcal{B})$ for the categories of chain complexes in $\mathcal{A}$ and $\mathcal{B}$ concentrated in non-negative degrees. The induced functor $F_\bullet : \mathbf{Ch}_{\geq 0}(\mathcal{A}) \to \mathbf{Ch}_{\geq 0}(\mathcal{B})$ is homotopical when we take the weak equivalences to be the chain homotopy equivalences but not necessarily homotopical when we instead take the weak equivalences to be the quasi-isomorphisms. For a counterexample, take $\mathcal{A} = \mathcal{B} = \mathbf{Ab}$ and consider the quasi-isomorphism

$$
\begin{array}{ccccccccccc}
A_\bullet & \cdots & \longrightarrow & \mathbb{Z}/2 & \hookrightarrow & \mathbb{Z}/4 & \xrightarrow{0} & \mathbb{Z}/2 & \hookrightarrow & \mathbb{Z}/4 & \longrightarrow & 0 \\
f_\bullet \downarrow & & & \downarrow & & \downarrow & & \downarrow & & \downarrow & & \\
B_\bullet & \cdots & \longrightarrow & 0 & \longrightarrow & \mathbb{Z}/2 & \longrightarrow & 0 & \longrightarrow & \mathbb{Z}/2 & \longrightarrow & 0
\end{array}
$$

Applying the functor $\hom_\mathbb{Z}(\mathbb{Z}/2, -)$ pointwise yields

$$
\begin{array}{ccccccccccc}
\hom_\mathbb{Z}(\mathbb{Z}/2, A)_\bullet & \cdots & \longrightarrow & \mathbb{Z}/2 & \xrightarrow{\cong} & \mathbb{Z}/2 & \xrightarrow{0} & \mathbb{Z}/2 & \xrightarrow{\cong} & \mathbb{Z}/2 & \longrightarrow & 0 \\
\hom_\mathbb{Z}(\mathbb{Z}/2, f)_\bullet \downarrow & & & \downarrow & & \downarrow 0 & & \downarrow & & \downarrow 0 & & \\
\hom_\mathbb{Z}(\mathbb{Z}/2, B)_\bullet & \cdots & \longrightarrow & 0 & \longrightarrow & \mathbb{Z}/2 & \longrightarrow & 0 & \longrightarrow & \mathbb{Z}/2 & \longrightarrow & 0
\end{array}
$$

which is not a quasi-isomorphism; indeed, these chain complexes are not quasi-isomorphic.

Example 2.1.16 Let $\mathcal{D}$ be the category $\bullet \longleftarrow \bullet \longrightarrow \bullet$. Consider **Top** as a homotopical category with weak homotopy equivalences and $\mathbf{Top}^\mathcal{D}$ as a homotopical category with weak equivalences defined **pointwise**;[2] that is, weak equivalences are natural transformations between diagrams whose components are weak equivalences of spaces.

The category **Top** has pushouts, defining a functor $\mathbf{Top}^\mathcal{D} \to \mathbf{Top}$. Consider the natural weak equivalences whose components are displayed vertically:

$$
\begin{array}{ccccc}
D^n & \longleftarrow\!\!\!\supset & S^{n-1} & \hookrightarrow & D^n \\
\wr \downarrow & & \downarrow 1 & & \wr \downarrow \\
* & \longleftarrow & S^{n-1} & \longrightarrow & *
\end{array}
$$

Here the top maps are the boundary inclusions. The pushout along the top is S^n, which has non-trivial nth homology, in contrast with the one-point space $*$, the pushout along the bottom. This shows that the functor $\operatorname{colim}: \mathcal{M}^\mathcal{D} \to \mathcal{M}$ is not homotopical.

Indeed, many interesting functors between homotopical categories are not themselves homotopical. Derived functors are defined to be the closest homotopical approximation in some sense. Because the associated comparison

---

[2] Unfortunately, this terminology has no relation to pointwise Kan extensions. Synonyms in this context include "objectwise," which, while more precise, is a tad clunky, and "levelwise," which we have decided has no meaning outside of special cases such as $\mathcal{D} = \Delta$.

natural transformations point in one direction or another, we obtain dual left- and right-handed notions. Of course, a given functor might have neither left nor right derived functors. In practice, a functor seldom has both.

Recall the definition of a total derived functor from Example 1.1.11:

Definition 2.1.17 A **total left derived functor** $\mathbf{L}F$ of a functor F between homotopical categories $\mathcal{M}$ and $\mathcal{N}$ is a *right* Kan extension $\mathrm{Ran}_\gamma \delta F$:

$$
\begin{array}{ccc}
\mathcal{M} & \xrightarrow{\;\;F\;\;} & \mathcal{N} \\[4pt]
{\scriptstyle \gamma}\downarrow & \Uparrow & \downarrow{\scriptstyle \delta} \\[4pt]
\mathrm{Ho}\mathcal{M} & \underset{\mathbf{L}F}{- \,-\, \rightarrow} & \mathrm{Ho}\mathcal{N}
\end{array}
\qquad (2.1.18)
$$

where γ and δ are the localization functors for $\mathcal{M}$ and $\mathcal{N}$.

Dually, a total *right* derived functor $\mathbf{R}F$ is a *left* Kan extension $\mathrm{Lan}_\gamma \delta F$. By the universal property of γ, $\mathbf{L}F$ is equivalently a homotopical functor $\mathbf{L}F \colon \mathcal{M} \to \mathrm{Ho}\mathcal{N}$, which is often called the "left derived functor" of F. Here we will reserve this terminology for a particular case: sometimes, although by no means always, there exists a lift of a left derived functor along δ; (cf. 2.1.11). Shulman [79] appropriately calls these lifts "point-set derived functors," but we call them simply **derived functors** because they will be the focus of much of what is to follow.

Definition 2.1.19 A **left derived functor** of $F \colon \mathcal{M} \to \mathcal{N}$ is a homotopical functor $\mathbb{L}F \colon \mathcal{M} \to \mathcal{N}$ equipped with a natural transformation $\lambda \colon \mathbb{L}F \Rightarrow F$ such that $\delta\lambda \colon \delta \cdot \mathbb{L}F \Rightarrow \delta \cdot F$ is a total left derived functor of F.

2.2 Derived functors via deformations

There is a common setting in which derived functors exist and admit a simple construction. Such categories have a collection of "good" objects on which the functor of interest becomes homotopical and a functorial reflection into this full subcategory. The details are encoded in the following axiomatization due to [22].

Definition 2.2.1 A **left deformation** on a homotopical category $\mathcal{M}$ consists of an endofunctor Q together with a natural weak equivalence $q \colon Q \xRightarrow{\sim} 1$.

Remark 2.2.2 As in 2.1.13, the functor Q is necessarily homotopical. Let $\mathcal{M}_Q$ be any full subcategory of $\mathcal{M}$ containing the image of Q. The inclusion

$\mathcal{M}_Q \to \mathcal{M}$ and the left deformation $Q: \mathcal{M} \to \mathcal{M}_Q$ induce an equivalence between $\text{Ho}\mathcal{M}$ and $\text{Ho}\mathcal{M}_Q$.

The notation is meant to evoke the following class of examples.

Example 2.2.3   If $\mathcal{M}$ is a model category with a functorial cofibration–trivial fibration factorization, then the action of this factorization on arrows of the form $\emptyset \to x$ produces a functor Q whose image lands in the subcategory of cofibrant objects together with a natural weak equivalence $q_x: Qx \overset{\sim}{\to} x$.

In [79], the subcategory $\mathcal{M}_Q$ is called a "left deformation retract." We prefer to refer to it as the subcategory of **cofibrant objects**, trusting the reader to understand that because we have not specified any model structures, Quillen's technical definition is not what we mean. We like this term for aesthetic reasons and because it conjures the correct intuition in common examples.

Definition 2.2.4   A **left deformation** for a functor $F: \mathcal{M} \to \mathcal{N}$ between homotopical categories consists of a left deformation for $\mathcal{M}$ such that F is homotopical on an associated subcategory of cofibrant objects. When F admits a left deformation, we say that F is **left deformable**.

Example 2.2.5   The geometric realization functor $| - |: \textbf{Top}^{\Delta^{op}} \to \textbf{Top}$ fails to preserve pointwise homotopy equivalences. In [78, Appendix A], Graeme Segal describes three left deformations for geometric realization into the subcategory of "good" simplicial spaces, in which the inclusions corresponding to the degeneracy maps are closed (Hurewicz) cofibrations.

In the context of model categories, Ken Brown's lemma (proven in 11.3.14) shows that any left Quillen functor is left deformable with respect to the subcategory of cofibrant objects (meant here in the technical sense).

Lemma 2.2.6 (Ken Brown's lemma)   *A left Quillen functor (i.e., a functor that preserves cofibrations, trivial cofibrations, and colimits, or at least the initial object) preserves weak equivalences between cofibrant objects.*

Remark 2.2.7   To state certain results later, it is necessary to associate a chosen subcategory of cofibrant objects to a left deformation on a homotopical category. To that end, we observe that any left deformable functor has a maximal subcategory on which it is homotopical [22, 40.4]; the proof of this fact makes explicit use of the 2-of-6 property.

Henceforth, when we say that $\mathcal{M}$ has a specified left deformation, we tacitly choose a subcategory of cofibrant objects as well, but we do not insist on the maximal choice, which is frequently not the most convenient.

Our first main result proves that left deformations can be used to construct left derived functors.

Theorem 2.2.8 ([22, 41.2–5]) *If* $F \colon \mathcal{M} \to \mathcal{N}$ *has a left deformation* $q \colon Q \overset{\sim}{\Rightarrow} 1$, *then* $\mathbb{L}F = FQ$ *is a left derived functor of* F.

Proof Write $\delta \colon \mathcal{N} \to \mathrm{Ho}\mathcal{N}$ for the localization. We must show that the functor $\delta F Q$ and natural transformation $\delta F q \colon \delta F Q \Rightarrow \delta F$ satisfy the appropriate universal property in $\mathrm{Ho}\mathcal{N}^{\mathrm{Ho}\mathcal{M}}$, or equivalently, by 2.1.11 above, in the full subcategory of $(\mathrm{Ho}\mathcal{N})^{\mathcal{M}}$ spanned by the homotopical functors. Suppose $G \colon \mathcal{M} \to \mathrm{Ho}\mathcal{N}$ is homotopical, and consider $\gamma \colon G \Rightarrow \delta F$. Because G is homotopical and $q \colon Q \Rightarrow 1$ is a natural weak equivalence, $Gq \colon GQ \Rightarrow G$ is a natural isomorphism. Using naturality of γ, it follows that γ factors through $\delta F Q$ as

$$ G \overset{(Gq)^{-1}}{\Longrightarrow} GQ \overset{\gamma_Q}{\Longrightarrow} \delta F Q \overset{\delta F q}{\Longrightarrow} \delta F. $$

To prove uniqueness, suppose γ factors as

$$ G \overset{\gamma'}{\Longrightarrow} \delta F Q \overset{\delta F q}{\Longrightarrow} \delta F. $$

Note that γ'_Q is uniquely determined: q_Q is a natural weak equivalence between objects in $\mathcal{M}_Q$. Because F is homotopical on $\mathcal{M}_Q$, this means that $F q_Q$ is a natural weak equivalence, and thus $\delta F q_Q$ is an isomorphism. Uniqueness of γ' now follows by naturality:

$$
\begin{array}{ccc}
GQ & \overset{\gamma'_Q}{\Longrightarrow} & \delta F Q^2 \\
{\scriptstyle Gq} \Big\Downarrow & & \Big\Downarrow {\scriptstyle \delta F Q q} \\
G & \underset{\gamma'}{\Longrightarrow} & \delta F Q
\end{array}
$$

Because q is a natural weak equivalence and the functors G and $\delta F Q$ are homotopical, the vertical arrows are natural isomorphisms. $\qquad\square$

Write **LDef** for the 2-category whose objects are homotopical categories equipped with specified left deformations and subcategories of cofibrant objects, whose morphisms $(\mathcal{M}, Q) \to (\mathcal{M}', Q')$ are functors that restrict to homotopical functors $\mathcal{M}_Q \to \mathcal{M}'_{Q'}$, and whose 2-cells are arbitrary natural transformations between parallel functors. In the terminology introduced previously, a 1-cell in **LDef** is a left deformable functor with respect to the specified left deformations that preserves cofibrant objects.

Theorem 2.2.9   *There is a pseudofunctor*

$$\mathbf{L}\colon \underline{\mathbf{LDef}} \to \underline{\mathbf{Cat}}$$

that sends a homotopical category to its homotopy category, a left deformable functor to its total left derived functor, and a natural transformation to its derived natural transformation.

Proof   The details are exactly as in [38, 1.3.7–9], which proves the analogous statement for model categories and left Quillen functors. The main point is that there are natural weak equivalences

$$\mathbb{L}G \cdot \mathbb{L}F = GQ' \cdot FQ \overset{Gq'}{\Longrightarrow} GFQ = \mathbb{L}(GF) \qquad \mathbb{L}1_{\mathcal{M}} = Q \overset{q}{\Longrightarrow} 1_{\mathcal{M}}$$

that descend to natural isomorphisms $\mathbb{L}G \cdot \mathbb{L}F \cong \mathbb{L}(GF)$ and $\mathbb{L}1_{\mathcal{M}} \cong 1_{\mathrm{Ho}\mathcal{M}}$. Note the first map is a weak equivalence only because we assumed that F maps $\mathcal{M}_Q$ into $\mathcal{M}'_{Q'}$, a subcategory on which G is homotopical.   □

Remark 2.2.10   The proof of the previous theorem, which fixes a defect mentioned in the introduction to this chapter, depends on our particular construction of total left derived functors. Indeed, it is not true in general that the composite of two total left derived functors is a total left derived functor for the composite.

There are obvious dual notions of **right deformation** $r\colon 1 \overset{\sim}{\Rightarrow} R$ and **right derived functor**. Here the notation is meant to suggest fibrant replacement. As before, we refer to a full subcategory $\mathcal{M}_R$ containing the image of R as the subcategory of **fibrant objects** and trust the reader to remember that we are not presuming that these are the fibrant objects in any model structure. By Ken Brown's lemma, if F is right Quillen, then F preserves weak equivalences between fibrant objects, and hence $\mathbb{R}F = FR$ is a right derived functor.

Left and right Quillen functors frequently occur in adjoint pairs, in which case their total left and right derived functors form an adjunction between the appropriate homotopy categories. An analogous result is true for deformable functors. An adjoint pair $F \dashv G$ with F left deformable and G right deformable is called a **deformable adjunction**.

Theorem 2.2.11 ([22, 44.2])   *If $F\colon \mathcal{M} \underset{\longleftarrow}{\overset{\perp}{\longrightarrow}} \mathcal{N} \colon G$ is a deformable adjunction, then the total derived functors form an adjunction*

$$\mathbf{L}F\colon \mathrm{Ho}\mathcal{M} \underset{\longleftarrow}{\overset{\perp}{\longrightarrow}} \mathrm{Ho}\mathcal{N}\colon \mathbf{R}G$$

Furthermore, the total derived adjunction $\mathbf{L}F \dashv \mathbf{R}G$ *is the unique adjunction compatible with the localizations in the sense that the diagram of hom-sets*

$$\mathcal{N}(Fm, n) \cong \mathcal{M}(m, Gn)$$

$$\delta \downarrow \qquad\qquad \downarrow \gamma$$

$$\mathrm{Ho}\mathcal{N}(Fm, n) \quad \mathrm{Ho}\mathcal{M}(m, Gn)$$

$$Fq^* \downarrow \qquad\qquad \downarrow Gr_*$$

$$\mathrm{Ho}\mathcal{N}(\mathbf{L}Fm, n) \cong \mathrm{Ho}\mathcal{M}(m, \mathbf{R}Gn)$$

commutes for each pair $m \in \mathcal{M}$, $n \in \mathcal{N}$.

Remark 2.2.12 The proofs of Theorem 2.2.11 and the analogous result for Quillen adjunctions make use of the particular construction of the associated total derived functors. For Quillen adjunctions, one shows that the total left derived functors constructed in the usual way preserve homotopies between fibrant–cofibrant objects [38, 1.3.10]. For deformable adjunctions, one uses the deformations to construct unit and counit maps in the homotopy categories that define a partial adjunction, exhibiting adjoint correspondences between pairs $Fm \to n$ and $n \to Gm$ with $m \in \mathcal{M}_Q$ and $n \in \mathcal{N}_R$. Because the deformations define equivalences of categories $\mathrm{Ho}\mathcal{M} \cong \mathrm{Ho}\mathcal{M}_Q$ and $\mathrm{Ho}\mathcal{N} \cong \mathrm{Ho}\mathcal{N}_R$, this partial adjunction extends to a complete adjunction.

Our point is that both standard arguments are surprisingly fiddly. Hence we were surprised to learn that a formal proof is available. Attempts to show that the total derived functors are adjoints using their defining universal properties alone have not succeeded. However, when these derived functors are constructed via deformations, they turn out to satisfy a stronger universal property, which is enough to prove this result.

Recall that a **pointwise** right Kan extension is one that is constructed as a limit in the target category. From this definition, one would not expect a total left derived functor to be a pointwise Kan extension. Nonetheless:

Proposition 2.2.13 *The total left derived functor of a left deformable functor is a pointwise right Kan extension.*

Proof If $F\colon \mathcal{M} \to \mathcal{N}$ is left deformable, it has a total left derived functor $(\delta FQ, \delta Fq)$ constructed using a left deformation (Q, q) and the localization functor δ. Because Kan extensions are characterized by a universal property, any total derived functor $\mathbf{L}F$ is isomorphic to this one.

Our proof will show that $\mathbf{L}F$ is an **absolute right Kan extension**, meaning a right Kan extension that is preserved by any functor $H\colon \mathrm{Ho}\mathcal{N} \to \mathcal{E}$. We use 2.1.11 to transfer the desired universal property to the subcategory of homotopical functors in $\mathcal{E}^{\mathcal{M}}$. To show that $(H\delta FQ, H\delta Fq)$ is a right Kan extension of $H\delta F$ along $\gamma: \mathcal{M} \to \mathrm{Ho}\mathcal{M}$, consider a homotopical functor $G: \mathcal{M} \to \mathcal{E}$ equipped with a natural transformation $\alpha: G \Rightarrow H\delta F$. Because G sends weak equivalences to isomorphisms, α factors as

$$
\begin{array}{ccc}
G & \overset{\alpha}{\Longrightarrow} & H\delta F \\
{\scriptstyle(Gq)^{-1}}\Big\Downarrow & & \Big\Uparrow{\scriptstyle H\delta Fq} \\
GQ & \underset{\alpha_Q}{\Longrightarrow} & H\delta F Q
\end{array}
$$

Suppose α also factors as $H\delta Fq \cdot \beta$ for some $\beta: G \Rightarrow H\delta FQ$. The components of $H\delta Fq$ at objects of $\mathcal{M}_Q$ are isomorphisms: F preserves weak equivalences between objects in $\mathcal{M}_Q$, δ inverts them, and any functor H preserves the resulting isomorphisms. Hence the components of β are uniquely determined on the objects in $\mathcal{M}_Q$. But both the domain and codomain of β are homotopical functors, and every object in $\mathcal{M}$ is weakly equivalent to one in $\mathcal{M}_Q$, so the isomorphisms $Gq: GQ \cong G$ and $H\delta FQq: H\delta FQ^2 \cong H\delta FQ$ imply that β is unique.   □

Remark 2.2.14   This is the same argument used to prove Theorem 2.2.8, enhanced by the trivial observation that any functor H preserves isomorphisms.

The main theorem from [52] illustrates that what matters is not the details of the construction of the total derived functors but rather that they satisfy the stronger universal property of being absolute Kan extensions. Once the statement is known, the proof is elementary enough to leave to the reader:

Exercise 2.2.15   Suppose $F \dashv G$ is an adjunction between homotopical categories, and suppose also that F has a total left derived functor $\mathbf{L}F$, G has a total right derived functor $\mathbf{R}G$, and both derived functors are absolute Kan extensions. Show that $\mathbf{L}F \dashv \mathbf{R}G$. That is, show that the total derived functors form an adjunction between the homotopy categories, regardless of how these functors may have been constructed.

Remark 2.2.16   A more sophisticated categorical framework allows us to combine Theorem 2.2.9 and its dual, Theorem 2.2.11, and a further result: if a functor is both left and right deformable with respect to a left deformation that preserves fibrant objects or a right deformation that preserves cofibrant objects, then the total left and right derived functors are isomorphic.

To this end, define a double category $\mathbb{D}\mathbf{er}$ whose objects are **deformable categories**: homotopical categories equipped with a left deformation $(Q, \mathcal{M}_Q)$ and a right deformation $(R, \mathcal{M}_R)$ such that either R preserves $\mathcal{M}_Q$ or Q preserves $\mathcal{M}_R$. Horizontal 1-cells are right deformable functors that preserve fibrant objects; vertical 1-cells are left deformable functors that preserve cofibrant objects; and 2-cells are arbitrary natural transformations in the appropriate squares. The results mentioned earlier are subsumed by the following theorem: the map that sends a homotopical category to its homotopy category and the functors to their total derived functors defines a **double pseudofunctor** $\mathbb{D}\mathbf{er} \to \mathbb{C}\mathbf{at}$ landing in the double category of categories, functors, and natural transformations. See [80, §8] for proof.

2.3 Classical derived functors between abelian categories

It is worthwhile to take a moment to explain how this theory fits with the classical definition and construction of derived functors in homological algebra.

The category $\mathbf{Mod}_R$ of left R-modules has the following property: for any R-module A, there exists a projective module P together with an epimorphism $P \twoheadrightarrow A$. By an inductive argument, any module admits a **projective resolution**, an acyclic (except in degree zero) bounded-below chain complex of projectives $P_\bullet$ that maps quasi-isomorphically to the chain complex concentrated in degree zero at the module A. A more refined version of this construction starts with any bounded-below chain complex $A_\bullet$ and produces a chain complex of projectives $P_\bullet$ together with a quasi-isomorphism $P_\bullet \to A_\bullet$.

Making a careful choice of projective resolution, this procedure defines a functor $Q \colon \mathbf{Ch}_{\geq 0}(R) \to \mathbf{Ch}_{\geq 0}(R)$ together with a natural quasi-isomorphism $q \colon Q \Rightarrow 1$; in other words, taking projective resolutions defines a left deformation for the homotopical category of bounded-below chain complexes and quasi-isomorphisms into the subcategory of chain complexes of projective modules. Dually, any R-module embeds into an injective module; iterating this procedure produces an **injective resolution**. When care is taken with the construction, injective resolutions assemble into a right deformation $r \colon 1 \Rightarrow R$ on $\mathbf{Ch}^{\geq 0}(R)$, the category of bounded-below cochain complexes.

Any additive functor $F \colon \mathbf{Mod}_R \to \mathbf{Mod}_S$ induces obvious functors $F_\bullet \colon \mathbf{Ch}_{\geq 0}(R) \to \mathbf{Ch}_{\geq 0}(S)$ and $F^\bullet \colon \mathbf{Ch}^{\geq 0}(R) \to \mathbf{Ch}^{\geq 0}(S)$ that preserve chain homotopies and hence chain homotopy equivalences. Any quasi-isomorphism between non-negatively graded chain complexes of projective objects is a chain homotopy equivalence and so is preserved by $F_\bullet$. Dually, any quasi-isomorphism between cochain complexes of injective objects is a chain homotopy equivalence and so is preserved by $F^\bullet$. Thus the functor $F_\bullet$ has a left derived functor, and $F^\bullet$ has a right derived functor, constructed as in 2.2.8. The functor that is classically referred to as the left derived functor of $F \colon \mathbf{Mod}_R \to \mathbf{Mod}_S$ is the composite

$$\mathbf{Mod}_R \xrightarrow{\deg_0} \mathbf{Ch}_{\geq 0}(R) \xrightarrow{\mathbb{L}F_\bullet} \mathbf{Ch}_{\geq 0}(S) \xrightarrow{H_0} \mathbf{Mod}_S,$$

which takes the projective resolution of an R-module A, applies $F_\bullet$, and then computes the 0th homology.

Remark 2.3.1   The reason one typically only forms left derived functors of right exact functors and right derived functors of left exact functors is that these hypotheses produce long exact sequences from short exact sequences: given a short exact sequence $0 \to A \to B \to C \to 0$ of left R-modules, there is a short exact sequence of projective resolutions $0 \to QA \to QB \to QC \to 0$. Because QC is projective, this short exact sequence is split and hence preserved by any additive functor F. From the short exact sequence $0 \to F_\bullet QA \to F_\bullet QB \to F_\bullet QC \to 0$, we get a long exact sequence of homology groups:

$$\cdots \to H_1(F_\bullet QB) \to H_1(F_\bullet QC) \to H_0(F_\bullet QA)$$

$$\to H_0(F_\bullet QB) \to H_0(F_\bullet QC).$$

When F is right exact, it preserves exactness of $Q_1 A \to Q_0 A \to A \to 0$, which implies that $H_0(F_\bullet QA) = FA$, so the preceding long exact sequence becomes

$$\cdots \to H_2(F_\bullet QC) \to H_1(F_\bullet QA) \to H_1(F_\bullet QB) \to H_1(F_\bullet QC)$$

$$\to FA \to FB \to FC \to 0.$$

Remark 2.3.2   Similar constructions can be given for more general abelian categories, presuming that they have enough projectives or enough injectives. However, it is not always possible to construct *functorial* deformations into chain complexes of projectives or cochain complexes of injectives. In the absence of functorial deformations, this construction defines a total derived functor but not a point-set level functor.

2.4 Preview of homotopy limits and colimits

In closing, we preview a theorem whose statement and proof will motivate much of the next three chapters. Consider diagrams of shape $\mathcal{D}$ in a complete and cocomplete homotopical category $\mathcal{M}$. As illustrated by Example 2.1.16, it is likely that the colimit and limit functors $\mathcal{M}^{\mathcal{D}} \to \mathcal{M}$ will not be homotopical with respect to pointwise weak equivalences in $\mathcal{M}^{\mathcal{D}}$. Because the former functor is cocontinuous and the latter functor is continuous, we hope to be able to construct a left derived functor of colim and a right derived functor of lim, in which case we call the derived functors hocolim and holim. But even if $\mathcal{M}$ is a model category, it is likely that $\mathcal{M}^{\mathcal{D}}$ will not have model structures such that colim or lim are, respectively, left and right Quillen. Hence we need a mechanism for constructing derived functors that does not require a full model structure. This is precisely what the theory of deformations provides.

Note that in the classical literature, a "homotopy colimit" or "homotopy limit" often means something else: it is an object of $\mathcal{M}$ that satisfies an appropriate homotopical universal property, which makes sense only when the hom-sets of $\mathcal{M}$ have some sort of topological structure. To make this precise, we ask at minimum that the category $\mathcal{M}$ is **simplicially enriched**. So that the desired representing object exists, we also ask that $\mathcal{M}$ is **tensored** and **cotensored** over simplicial sets. In the next chapter, we give precise definitions of these concepts.

A priori the simplicial enrichment on $\mathcal{M}$ need not interact meaningfully with the homotopical structure. A common setting that provides the desired compatibility is that of a **simplicial model category**, introduced at the end of Chapter 3. With these hypotheses, it is possible to construct homotopy colimit and limit functors as point-set derived functors of the colimit and limit functors that also satisfy the appropriate homotopical universal properties. Note that our constructions put no restrictions on the category $\mathcal{D}$, except smallness.

More precisely, in Chapter 5, we see how to define a left deformation for colim and a right deformation for lim using the two-sided bar and cobar constructions introduced in Chapter 4. We also show that the bar and cobar constructions defining the homotopy colimit and homotopy limit functors also have an appropriate "local" universal property, representing "homotopy coherent cones" under or over a diagram. The main result of Chapter 5 is a precise statement and proof of this claim, following [79].

First, to explain what we mean by the simplicial enrichment, tensors, and cotensors mentioned earlier, and because this material is needed later, we embark on a brief detour through enriched category theory.

3

Basic concepts of enriched category theory

Most of the categories one encounters in mathematics are **locally small**, meaning that the collection of arrows between any two fixed objects is a set. Frequently in examples, these hom-sets admit additional compatible structures. These are the purview of enriched category theory, which we introduce in this chapter.

Why bother with enrichments? One answer is that it seems silly to forget entirely about structures that are naturally present in many examples, particularly when, as we will see, a fully developed theory is available. Another answer, directed in particular to homotopy theorists, is that ordinary unenriched categories are too coarse to describe all homotopical phenomena.

For instance, a product of a family of objects m_α in a category $\mathcal{M}$ is given by a representation m for the functor $\mathcal{M}^{\mathrm{op}} \to \mathbf{Set}$ displayed on the right:

$$\mathcal{M}(-, m) \stackrel{\cong}{\longrightarrow} \prod_\alpha \mathcal{M}(-, m_\alpha).$$

By the Yoneda lemma, a representation consists of an object $m \in \mathcal{M}$ together with maps $m \to m_\alpha$ for each α that are universal in the sense that for any collection $x \to m_\alpha \in \mathcal{M}$, each of these arrows factors uniquely along a common map $x \to m$. But in certain contexts, we might prefer to require only that the triangles

$$
\begin{array}{ccc}
x & & \\
\vert & \searrow & \\
\exists \vert & \simeq & \\
\vee & & \\
m & \longrightarrow & m_\alpha
\end{array}
\qquad (3.0.1)
$$

commute "up to homotopy." For this to make sense, we are most likely assuming that each hom-set is equipped with a topology; we write $\underline{\mathcal{M}}(x, m_\alpha)$ to distinguish the space from the set. A homotopy between two arrows $x \to m_\alpha$ in the underlying category $\mathcal{M}$ means a path in the space $\underline{\mathcal{M}}(x, m_\alpha)$ between the corresponding points. Now we can define the **homotopy product** to be an object m equipped with a natural (weak) homotopy equivalence

$$\underline{\mathcal{M}}(x, m) \to \prod_\alpha \underline{\mathcal{M}}(x, m_\alpha)$$

for each $x \in \mathcal{M}$. Surjectivity on path components implies the existence and homotopy commutativity of the triangles (3.0.1).

Unusually for homotopy limits (cf. Remark 6.3.1), this homotopy product is a product in the homotopy category $h\mathcal{M}$ of the enriched category $\underline{\mathcal{M}}$. Because we have not specified a class of weak equivalences to invert, this notion of homotopy category is distinct from the one introduced in the previous chapter; we will compare these two notions in Section 10.5. Here the homotopy category $h\mathcal{M}$ of the topological category $\underline{\mathcal{M}}$ has a simple definition: $h\mathcal{M}$ has the same objects as $\underline{\mathcal{M}}$ but hom-sets

$$h\mathcal{M}(x, y) := \pi_0 \underline{\mathcal{M}}(x, y)$$

obtained by applying the path-components functor $\pi_0 \colon \textbf{Top} \to \textbf{Set}$. Because π_0 commutes with products and is homotopical (takes weak equivalences to isomorphisms), it follows that a homotopy product is a product in the homotopy category $h\mathcal{M}$. Similarly, a homotopy coproduct is a coproduct in the homotopy category.

To place the preceding discussion on firm footing, we now describe how to make sense of categorical notions in a way that is compatible with whatever extra structures might be present on the collection of morphisms between two fixed objects.

3.1 A first example

Consider the category $\textbf{Mod}_R$ of left modules over a fixed, not necessarily commutative, ring R. For each pair $A, B \in \textbf{Mod}_R$, the set $\textbf{Mod}_R(A, B)$ of homomorphisms from A to B is itself an abelian group, which we denote $\underline{\textbf{Mod}}_R(A, B)$: the null homomorphism serves as the identity, and the sum of two homomorphisms is defined pointwise using the addition in B. Furthermore, composition in the category $\textbf{Mod}_R$ distributes over this hom-wise addition: in other words, each composition function

$$\textbf{Mod}_R(B, C) \times \textbf{Mod}_R(A, B) \xrightarrow{\ \circ\ } \textbf{Mod}_R(A, C)$$

in **Set** is $\mathbb{Z}$-bilinear, inducing a group homomorphism

$$\underline{\textbf{Mod}}_R(B, C) \otimes_\mathbb{Z} \underline{\textbf{Mod}}_R(A, B) \xrightarrow{\ \circ\ } \underline{\textbf{Mod}}_R(A, C)$$

in **Ab**.

The group $\mathbb{Z}$ plays a special role in **Ab**. First, it represents the forgetful functor **Ab** $\to$ **Set**; homomorphisms $\mathbb{Z} \to A$ correspond to elements of A. In particular, we can represent the identity at A, not to be confused with the unit for the addition law on $\underline{\text{Mod}}_R(A, A)$, by an arrow

$$\mathbb{Z} \overset{\text{id}_A}{\to} \underline{\text{Mod}}_R(A, A)$$

in the category **Ab**. Furthermore, $\mathbb{Z}$ is a unit for the monoidal product

$$- \otimes_\mathbb{Z} - : \mathbf{Ab} \times \mathbf{Ab} \to \mathbf{Ab}$$

in the sense that tensoring on the left or the right with $\mathbb{Z}$ is naturally isomorphic to the identity functor. These structures allow us to express the usual composition axioms **diagrammatically**; that is, we encode the associativity and unit laws by asking that certain diagrams commute inside **Ab**. To summarize these facts, we say the category $\mathbf{Mod}_R$ admits the structure of a category **enriched over** $(\mathbf{Ab}, \otimes_\mathbb{Z}, \mathbb{Z})$.

3.2 The base for enrichment

Before we give the general definition, we must precisely describe the appropriate context for enrichment. As the previous example suggests, the **base** category over which we enrich should be a **symmetric monoidal category** $(\mathcal{V}, \times, *)$. Here $\mathcal{V}$ is an ordinary category, $- \times - : \mathcal{V} \times \mathcal{V} \to \mathcal{V}$ is a bifunctor called the **monoidal product**, and $* \in \mathcal{V}$ is called the **unit object**. We write "$\times$" for the monoidal product because it will be the cartesian product in our main examples: the categories of simplicial sets, spaces, categories, and sets. This notation will also help distinguish the monoidal product from the **tensor**, defined in Section 3.7, of an object of $\mathcal{V}$ with an object in some $\mathcal{V}$-category. But in other examples (such as **Ab**), alternate monoidal products (e.g., $\otimes_\mathbb{Z}$) are preferred, and we stress that we do not mean to suggest that the monoidal product is necessarily the cartesian one.

A symmetric monoidal category is also equipped with specified natural transformations

$$v \times w \cong w \times v, \quad u \times (v \times w) \cong (u \times v) \times w, \quad * \times v \cong v \cong v \times *,$$
$$(3.2.1)$$

expressing symmetry, associativity, and unit conditions on the monoidal product. These must satisfy a handful of coherence conditions, described in [50, XI]. The main theorem is that these natural transformations compose to give a unique isomorphism between any two expressions for a particular product of objects. In practice, this naturality means that we need not concern ourselves with particular parenthesizations or orderings; hence we will not emphasize the role of the structural isomorphisms (3.2.1) here.

Example 3.2.2 The category **sSet** is symmetric monoidal with the cartesian product, computed pointwise, and the terminal simplicial set denoted by Δ^0 or $*$.

Example 3.2.3 Here are a few other symmetric monoidal categories:

- any category with finite products is symmetric monoidal with the terminal object serving as the unit, e.g., **Top**, **Set**, **Cat**, **Gpd**
- $(\mathbf{Mod}_R, \otimes_R, R)$ for any commutative ring R, for example $(\mathbf{Ab}, \otimes_{\mathbb{Z}}, \mathbb{Z})$ or $(\mathbf{Vect}_k, \otimes_k, k)$
- $\mathbf{Ch}_{\geq 0}(R)$ or $\mathbf{Ch}_\bullet(R)$, the category of unbounded chain complexes of modules over a commutative ring; the monoidal product is the tensor product, with the symmetry isomorphism expressing graded commutativity; and the unit is the chain complex consisting of the commutative ring R concentrated in degree zero
- the homotopical categories of S-modules, symmetric spectra, and orthogonal spectra are all symmetric monoidal categories with homotopy categories equivalent to the stable homotopy category [25, 39, 54]
- $(\mathbf{Set}_*, \wedge, * \sqcup *)$ and other examples of this type, as described in 3.3.14

In all of these examples, the category $\mathcal{V}$ is complete and cocomplete. For convenience, let us always suppose without further comment that this is the case.

3.3 Enriched categories

Definition 3.3.1 A $\mathcal{V}$-**category** $\mathcal{D}$ consists of

- a collection of objects $x, y, z \in \mathcal{D}$
- for each pair $x, y \in \mathcal{D}$, a **hom-object** $\underline{\mathcal{D}}(x, y) \in \mathcal{V}$
- for each $x \in \mathcal{D}$, a morphism $\mathrm{id}_x : * \to \underline{\mathcal{D}}(x, x)$ in $\mathcal{V}$
- for each triple $x, y, z \in \mathcal{D}$, a morphism $\circ : \underline{\mathcal{D}}(y, z) \times \underline{\mathcal{D}}(x, y) \to \underline{\mathcal{D}}(x, z)$ in $\mathcal{V}$

such that the following diagrams commute for all $x, y, z, w \in \mathcal{D}$:

$$
\begin{array}{ccc}
\underline{\mathcal{D}}(z, w) \times \underline{\mathcal{D}}(y, z) \times \underline{\mathcal{D}}(x, y) & \xrightarrow{1 \times \circ} & \underline{\mathcal{D}}(z, w) \times \underline{\mathcal{D}}(x, z) \\
{\scriptstyle \circ \times 1} \downarrow & & \downarrow {\scriptstyle \circ} \\
\underline{\mathcal{D}}(y, w) \times \underline{\mathcal{D}}(x, y) & \xrightarrow{\quad \circ \quad} & \underline{\mathcal{D}}(x, w)
\end{array}
$$

$$
\begin{array}{ccccc}
\underline{\mathcal{D}}(x, y) \times * & \xrightarrow{1 \times \mathrm{id}_x} & \underline{\mathcal{D}}(x, y) \times \underline{\mathcal{D}}(x, x) & \quad & \underline{\mathcal{D}}(y, y) \times \underline{\mathcal{D}}(x, y) & \xleftarrow{\mathrm{id}_y \times 1} & * \times \underline{\mathcal{D}}(x, y) \\
& {\scriptstyle \cong} \searrow & \downarrow {\scriptstyle \circ} & & \downarrow {\scriptstyle \circ} & \swarrow {\scriptstyle \cong} & \\
& & \underline{\mathcal{D}}(x, y) & & \underline{\mathcal{D}}(x, y) & &
\end{array}
$$

Here the indicated isomorphisms are the maps (3.2.1) specified by the symmetric monoidal structure on $\mathcal{V}$. An associativity isomorphism is omitted from the first diagram.

Example 3.3.2   A one-object **Ab**-category is a ring with identity. The unique hom-set is the set of endomorphisms of the only object. The composition law defines multiplication and a multiplicative identity. The enrichment defines the addition law and additive identity. The preceding axioms ensure that multiplication distributes over addition, and so on.

Example 3.3.3   Topological spaces are naturally enriched over groupoids. Given spaces X and Y, define the hom-object from X to Y to be the groupoid whose objects are continuous maps $X \to Y$ and whose morphisms are homotopy classes of homotopies between these maps.

Example 3.3.4   When $\mathcal{V}$ has copowers of the unit object that are preserved by the monoidal product in each variable – for example, when $\mathcal{V}$ is cocomplete and **closed** in the sense defined in 3.3.6 – any unenriched category $\mathcal{C}$ has an associated **free $\mathcal{V}$-category**. Its objects are those of $\mathcal{C}$. The hom-object from a to b is the copower $\sqcup_{\mathcal{C}(a,b)} *$. The morphisms identifying the identities are given by including $*$ at the component of the identity arrow in $\sqcup_{\mathcal{C}(a,a)} *$. Because the monoidal product preserves coproducts in each variable, the domains of the composition morphisms have the form

$$\left(\underset{\mathcal{C}(b,c)}{\sqcup} * \right) \times \left(\underset{\mathcal{C}(a,b)}{\sqcup} * \right) \cong \underset{\mathcal{C}(b,c)}{\sqcup} \left(* \times \left(\underset{\mathcal{C}(a,b)}{\sqcup} * \right) \right)$$

$$\cong \underset{\mathcal{C}(b,c)}{\sqcup} \left(\underset{\mathcal{C}(a,b)}{\sqcup} * \times * \right) \cong \underset{\mathcal{C}(b,c) \times \mathcal{C}(a,b)}{\sqcup} *.$$

Hence the composition morphism reindexes the coproduct along the composition function $\mathcal{C}(b, c) \times \mathcal{C}(a, b) \to \mathcal{C}(a, c)$.

When $\mathcal{V}$ is **Cat**, **Top**, or **sSet**, free $\mathcal{V}$-categories have discrete hom-sets in the sense appropriate to each category.

Example 3.3.5   The category **sSet** is enriched over itself using the hom-spaces Y^X defined in Example 1.5.6. The composition map $Z^Y \times Y^X \to Z^X$ is defined as follows: the composite of a pair of n-simplices $f \colon \Delta^n \times X \to Y$ and $g \colon \Delta^n \times Y \to Z$ is the n-simplex

$$\Delta^n \times X \xrightarrow{\Delta \times 1} \Delta^n \times \Delta^n \times X \xrightarrow{1 \times f} \Delta^n \times Y \xrightarrow{g} Z,$$

where the arrow Δ denotes the diagonal map.

This is an instance of a common class of examples worthy of special mention.

Definition 3.3.6 (closed monoidal categories) When each functor $- \times v \colon \mathcal{V} \to \mathcal{V}$ admits a right adjoint $\underline{\mathcal{V}}(v, -)$, the right adjoints in this family of parameterized adjunctions assemble in a unique way into a bifunctor

$$\underline{\mathcal{V}}(-, -) \colon \mathcal{V}^{\mathrm{op}} \times \mathcal{V} \to \mathcal{V}$$

such that there exist isomorphisms

$$\mathcal{V}(u \times v, w) \cong \mathcal{V}(u, \underline{\mathcal{V}}(v, w)) \qquad \forall\, u, v, w \in \mathcal{V} \tag{3.3.7}$$

natural in all three variables [50, IV.7.3].

In this case, $\mathcal{V}$ is enriched over itself using the **internal homs** $\underline{\mathcal{V}}(-, -)$. The required composition law

$$\underline{\mathcal{V}}(v, w) \times \underline{\mathcal{V}}(u, v) \to \underline{\mathcal{V}}(u, w)$$

is adjunct under $- \times u \dashv \underline{\mathcal{V}}(u, -)$ to the composite

$$\underline{\mathcal{V}}(v, w) \times \underline{\mathcal{V}}(u, v) \times u \xrightarrow{\ 1 \times \epsilon\ } \underline{\mathcal{V}}(v, w) \times v \xrightarrow{\ \epsilon\ } w, \tag{3.3.8}$$

where the epsilons are components of the counits of the appropriate adjunctions. The identities are the maps $* \to \underline{\mathcal{V}}(v, v)$ adjunct under $- \times v \dashv \underline{\mathcal{V}}(v, -)$ to the natural isomorphism $* \times v \xrightarrow{\cong} v$.

Triples $(\mathcal{V}, \times, *)$ of this form are called **closed symmetric monoidal categories**. In the special case where the monoidal product is the cartesian product on $\mathcal{V}$, one says that $\mathcal{V}$ is **cartesian closed**.

Remark 3.3.9 In fact, the hom-sets $\mathcal{V}(-, -)$ in the isomorphism (3.3.7) can be replaced with the internal homs to obtain an analogous isomorphism

$$\underline{\mathcal{V}}(u \times v, w) \cong \underline{\mathcal{V}}(u, \underline{\mathcal{V}}(v, w))$$

in $\mathcal{V}$. This follows from the associativity of the monoidal product and the Yoneda lemma, as we will demonstrate in Section 3.7.

Example 3.3.10 In Example 1.5.6, we defined the hom-spaces so that the condition (3.3.7) is satisfied; hence **sSet** is cartesian closed.

Example 3.3.11 The category of modules over a commutative ring is closed. Note that multiplication by $r \in R$ is not R-linear unless r is in the center of the ring. In particular, **Ab** and **Vect**$_k$ are closed symmetric monoidal categories and hence are self-enriched.

Example 3.3.12 The category **Cat** is cartesian closed: functors and natural transformations between two fixed categories themselves assemble into a category. A category, such as **Cat**, that is **Cat**-enriched is called a **2-category**. In a generic 2-category, the morphisms in the hom-categories are called **2-cells**. The term **1-cell** is used for the objects of the hom-categories, that is, the arrows in the **underlying category**, defined in 3.4.5. Objects are also called **0-cells**.

Example 3.3.13 The category of all topological spaces is not cartesian closed. However, there exist several **convenient categories of spaces** that are cartesian closed, complete and cocomplete, and large enough to contain the CW complexes, though not all topological spaces.

We devote Section 6.1 to a thorough exploration of this topic. For now, let **Top** denote some convenient category of spaces.

The closed symmetric monoidal category **Set**$_*$ of based sets, which is isomorphic to the slice category $*/$**Set**, is a special case of a general construction.

Construction 3.3.14 For any cartesian closed symmetric monoidal category $(\mathcal{V}, \times, *)$, which we also suppose to be complete and cocomplete, there is a general procedure for producing a closed symmetric monoidal category $(\mathcal{V}_*, \wedge, S^0)$, where $\mathcal{V}_*$ is the category of **based objects** in $\mathcal{V}$, that is, the slice category under $*$, the terminal object of $\mathcal{V}$. The unit S^0 is defined to be $* \sqcup *$, where the coproduct is taken in $\mathcal{V}$, not in $\mathcal{V}_*$. The basepoints assigned to two objects v and w can be used to define a map $v \sqcup w \to v \times w$. Define the **smash product** to be the pushout

$$
\begin{array}{ccc}
v \sqcup w & \longrightarrow & v \times w \\
\downarrow & & \downarrow \\
* & \longrightarrow & v \wedge w
\end{array}
\tag{3.3.15}
$$

in $\mathcal{V}$. Because $\mathcal{V}$ is cartesian closed, the functors $v \times -$ preserve colimits. It follows that this construction is associative up to isomorphism.[1]

The hom-objects are defined to be the pullbacks

$$
\begin{array}{ccc}
\underline{\mathcal{V}}_*(v, w) & \longrightarrow & \underline{\mathcal{V}}(v, w) \\
\downarrow & & \downarrow \\
\underline{\mathcal{V}}(*, *) & \longrightarrow & \underline{\mathcal{V}}(*, w)
\end{array}
$$

of the maps given by pre- and postcomposing[2] with the basepoint inclusions for v and w. In examples, this internal hom is the object of basepoint-preserving maps from v to w, and its basepoint is the constant map.

---

[1] The smash product is not associative in the (inconvenient) category of *all* based spaces; (cf. Lemma 6.1.3).

[2] We explain precisely what we mean by this in (3.4.14); your intuition is correct.

There is a disjoint basepoint—forgetful adjunction

$$(-)_+ : V \xrightarrow[]{\perp} V_* : U$$

whose left adjoint is defined to be the coproduct with the terminal object.

Lemma 3.3.16 *When V is a cartesian closed symmetric monoidal category, the functor $(-)_+ : V \to V_*$ is **strong monoidal**, that is,*

$$S^0 \cong (*)_+ \qquad v_+ \wedge w_+ \cong (v \times w)_+,$$

and these natural isomorphisms are appropriately associative and unital.

Proof The unit isomorphism is tautologous. By definition, the object $v_+ \wedge w_+$ is the cofiber of

$$v_+ \sqcup w_+ \to (v \sqcup *) \times (w \sqcup *) \cong v \times w \sqcup v \times * \sqcup * \times w \sqcup * \times *$$

$$\cong v \times w \sqcup v \sqcup w \sqcup *.$$

Here the distributivity isomorphisms require that the monoidal structure on V is cartesian closed. The map avoids the first component of the coproduct; hence $v_+ \wedge w_+$ is the composite pushout

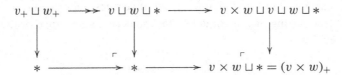

which is seen to have the stated description because the top left arrow is an epimorphism and thus the left-hand pushout is the terminal object. $\qquad\square$

Example 3.3.17 Taking **Top** to be a convenient category of spaces, the category $(\mathbf{Top}_*, \wedge, S^0)$ of based spaces is closed symmetric monoidal with respect to the smash product, as is $(\mathbf{sSet}_*, \wedge, \partial\Delta^1)$. Note these monoidal categories are not cartesian; the smash product is not the categorical product.

3.4 Underlying categories of enriched categories

An example will motivate a general definition. For a fixed discrete group G, the functor category $\mathbf{Top}^G$ of G-spaces and G-equivariant maps has two natural enrichments, which we distinguish by means of special notation. First, the category $\mathbf{Top}^G$ is topologically enriched: the set of G-equivariant maps is a subspace $\underline{\mathbf{Top}}^G(X, Y)$ of the space $\underline{\mathbf{Top}}(X, Y)$ of all continuous maps from X to Y and is topologized as in Remark 6.1.7.

The category $\mathbf{Top}^G$ is also symmetric monoidal: the product of two G-spaces is given the diagonal action, and the one-point space is given the trivial action to serve as the unit object. If X and Y are G-spaces, the space $\underline{\mathbf{Top}}(X, Y)$ of *all* continuous maps has a canonical G-action, by conjugation:

$$g \in G, f \in \underline{\mathbf{Top}}(X, Y), x \in X \qquad \leadsto \qquad g \cdot f(x) := gf(g^{-1}x).$$

We write $\underline{\mathbf{Top}}_G(X, Y)$ for this G-space. With these definitions, $\mathbf{Top}^G$ is cartesian closed.

Morally, the category $\mathbf{Top}^G$ is the underlying (unenriched) category of both $\underline{\mathbf{Top}}^G$ and $\underline{\mathbf{Top}}_G$, but it is unclear whether this intuition can be formalized: the naïve notion of category underlying $\underline{\mathbf{Top}}_G$, obtained by forgetting the topology and the G-action, returns something larger. It turns out that there is a general procedure for extracting an underlying category from an enriched category that gives the desired notion in both examples. To discover the correct definition, it is helpful to generalize.

To define the underlying category of a $\mathcal{V}$-category, we first need some "underlying set" functor $\mathcal{V} \to \mathbf{Set}$. In several examples, in particular, for **Ab**, **sSet**, **Top**, **Top**$_*$, or **Cat**, the unit object for our preferred monoidal structure represents the "underlying set" functor that takes an abelian group, simplicial set, space, or category to its underlying set, set of vertices, set of points, or set of objects. Generalizing, we might define the "underlying set" functor to be the represented functor

$$\mathcal{V}(*, -) \colon \mathcal{V} \to \mathbf{Set}$$

and apply it to the hom-objects of a $\mathcal{V}$-category to define the underlying category. Following your nose leads to the correct definition, but it is helpful to isolate which properties of the functor $\mathcal{V}(*, -)$ facilitate this "change of base" construction because we will use it more generally.

Definition 3.4.1   A functor $F \colon \mathcal{V} \to \mathcal{U}$ between symmetric monoidal categories $(\mathcal{V}, \times, *)$ and $(\mathcal{U}, \otimes, \mathbb{1})$ is **lax monoidal** if there exist associative and unital natural transformations

$$Fv \otimes Fv' \to F(v \times v') \qquad \text{and} \qquad \mathbb{1} \to F(*). \tag{3.4.2}$$

For example, the functor $(-)_+ \colon (\mathcal{V}, \times, *) \to (\mathcal{V}_*, \wedge, S^0)$ of Construction 3.3.14 is **strong monoidal**, meaning lax monoidal with natural isomorphisms (3.4.2).

Lemma 3.4.3 (change of base)   *In the presence of any lax monoidal functor $F \colon \mathcal{V} \to \mathcal{U}$, any $\mathcal{V}$-category has an associated $\mathcal{U}$-category with the same objects.*

Proof   Let $\underline{\mathcal{D}}_{\mathcal{V}}$ be a $\mathcal{V}$-category. Define a $\mathcal{U}$-category $\underline{\mathcal{D}}_{\mathcal{U}}$ with the same objects and with hom-objects $\underline{\mathcal{D}}_{\mathcal{U}}(x, y) = F\underline{\mathcal{D}}_{\mathcal{V}}(x, y)$. The composition and unit maps

$$F\underline{\mathcal{D}}_{\mathcal{V}}(y, z) \otimes F\underline{\mathcal{D}}_{\mathcal{V}}(x, y) \to F(\underline{\mathcal{D}}_{\mathcal{V}}(y, z) \times \underline{\mathcal{D}}_{\mathcal{V}}(x, y)) \xrightarrow{F(\circ)} F\underline{\mathcal{D}}_{\mathcal{V}}(x, z) = \underline{\mathcal{D}}_{\mathcal{U}}(x, z)$$

$$\mathbb{1} \xrightarrow{\quad\quad} F(*) \xrightarrow{F(\mathrm{id}_x)} F\underline{\mathcal{D}}_{\mathcal{V}}(x, x) = \underline{\mathcal{D}}_{\mathcal{U}}(x, x)$$

employ the lax monoidal structure on F.   □

The point, of course, is that $\mathcal{V}(*, -)$ is lax monoidal. For any objects v, $w \in \mathcal{V}$, there is a function

$$\mathcal{V}(*, v) \times \mathcal{V}(*, w) \to \mathcal{V}(*, v \times w), \tag{3.4.4}$$

natural in both v and w, defined by applying the bifunctor $\times$ to the pair of morphisms $* \to v$ and $* \to w$ and precomposing with the isomorphism $* \cong * \times *$. When $\mathcal{V}$ is cartesian monoidal, (3.4.4) is a bijection, but in any case, we can use Lemma 3.4.3 to change the base for the enrichment from $\mathcal{V}$ to **Set**.

Definition 3.4.5   The **underlying category** $\mathcal{C}_0$ of a $\mathcal{V}$-category $\underline{\mathcal{C}}$ has the same objects and has hom-sets

$$\mathcal{C}_0(x, y) := \mathcal{V}(*, \underline{\mathcal{C}}(x, y)).$$

We define the identities $\mathrm{id}_x \in \mathcal{C}_0(x, x)$ to be the specified morphisms $\mathrm{id}_x \in \mathcal{V}(*, \underline{\mathcal{C}}(x, x))$. Composition is defined hom-wise by the following arrow in **Set**:

$$\mathcal{C}_0(y, z) \times \mathcal{C}_0(x, y) \dashrightarrow \mathcal{C}_0(x, z)$$
$$\|$$
$$\mathcal{V}(*, \underline{\mathcal{C}}(y, z)) \times \mathcal{V}(*, \underline{\mathcal{C}}(x, y)) \longrightarrow \mathcal{V}(*, \underline{\mathcal{C}}(y, z) \times \underline{\mathcal{C}}(x, y)) \longrightarrow \mathcal{V}(*, \underline{\mathcal{C}}(x, z))$$
$$\tag{3.4.6}$$

The first arrow is (3.4.4); the second is $\mathcal{V}(*, -)$ applied to the composition morphism for $\underline{\mathcal{C}}$.

Example 3.4.7   Let us determine the underlying categories of the enriched categories $\mathbf{Top}^G$ and $\underline{\mathbf{Top}}_G$. For both, the objects are G-spaces. By Definition 3.4.5, the hom-set from X to Y in the underlying category of the **Top**-enriched category $\mathbf{Top}^G$ is the set $\mathbf{Top}(*, \mathbf{Top}^G(X, Y))$, that is, the set of points of $\mathbf{Top}^G(X, Y)$, that is, the set of G-equivariant maps $X \to Y$. So the underlying category of $\mathbf{Top}^G$ is $\mathbf{Top}^G$. By contrast, in the underlying category of the $\mathbf{Top}^G$-enriched category $\underline{\mathbf{Top}}_G$, the hom-set from X to Y is the set $\mathbf{Top}^G(*, \underline{\mathbf{Top}}_G(X, Y))$. Because $*$ is given the trivial action, G-equivariant maps from $*$ to $\underline{\textbf{Top}}_G(X, Y)$ are precisely G-fixed points. Fixed points for the conjugation action are precisely G-equivariant maps,[3] so the underlying category is again $\textbf{Top}^G$, as desired.

Remark 3.4.8 This discussion generalizes. Suppose $\mathcal{V}$ is a complete and cocomplete closed symmetric monoidal category such as **Set**, **Top**, $\textbf{Top}_*$, **sSet**, $\textbf{sSet}_*$, $\textbf{Vect}_k$, $\textbf{Mod}_R$, **Ab**, and so on. As observed by Ross Street and others, under these hypotheses, $\mathcal{V}^G$ inherits a canonical closed symmetric monoidal structure in such a way that the forgetful functor $U : \mathcal{V}^G \to \mathcal{V}$ preserves the unit, tensor product, and internal homs.[4] As previously, the hom-object $\underline{\mathcal{V}}^G(c, d)$ agrees with $\underline{\mathcal{V}}(c, d)$ and is equipped with a canonical G-action. For formal reasons, the underlying category is the original $\mathcal{V}^G$.

The fact that the underlying category of the $\textbf{Top}^G$-enriched category $\underline{\textbf{Top}}_G$ was $\textbf{Top}^G$ is a special instance of the following result, which says that the underlying category of a closed monoidal category is the category itself.

Lemma 3.4.9 *If $(\mathcal{V}, \times, *)$ is a closed symmetric monoidal category, the underlying category of the $\mathcal{V}$-category $\underline{\mathcal{V}}$ is $\mathcal{V}$.*

Proof From the adjunction (3.3.7) and the unit isomorphism,

$$\mathcal{V}(*, \underline{\mathcal{V}}(x, y)) \cong \mathcal{V}(* \times x, y) \cong \mathcal{V}(x, y), \tag{3.4.10}$$

so the hom-sets of the underlying category of $\underline{\mathcal{V}}$ agree with those of $\mathcal{V}$. By the usual argument, any two putative identities for the same object must coincide, so it remains to show that composition in $\mathcal{V}$ agrees with the definition (3.4.6). Writing this down carefully is somewhat trickier than one might expect.

On account of the isomorphism (3.4.10), we agree to use the same label for corresponding morphisms, for example,

$$* \xrightarrow{f} \underline{\mathcal{V}}(x, y) \qquad \rightsquigarrow \qquad f : x \to y.$$

Writing $\epsilon_y^x : \underline{\mathcal{V}}(x, y) \times x \to y$ for the component at y of the counit of the adjunction for x, the latter f is defined in terms of the former by

$$x \cong * \times x \xrightarrow{\ f \times 1\ } \underline{\mathcal{V}}(x, y) \times x \xrightarrow{\ \epsilon_y^x\ } y. \tag{3.4.11}$$

---

[3] This gives a mnemonic for remembering the notation: $\underline{\textbf{Top}}^G(X, Y) = (\underline{\textbf{Top}}(X, Y))^G = (\underline{\textbf{Top}}_G(X, Y))^G$.

[4] Because U is given by precomposing by the inclusion $e \to G$ of the trivial group into the group G, and $\mathcal{V}$ is complete and cocomplete, U has both left and right adjoints given by the left and right Kan extension.

Now we consider the effect of (3.4.6) on a pair of morphisms $f, g: * \rightrightarrows \underline{\mathcal{V}}(y, z)$. By definition, their composite is

$$* \cong * \times * \xrightarrow{\;g \times f\;} \underline{\mathcal{V}}(y, z) \times \underline{\mathcal{V}}(x, y) \xrightarrow{\;\circ\;} \underline{\mathcal{V}}(x, z) \; . \qquad (3.4.12)$$

Recall from (3.3.8) that $\circ$ was defined to be adjunct to

$$\underline{\mathcal{V}}(y, z) \times \underline{\mathcal{V}}(x, y) \times x \xrightarrow{\;1 \times \epsilon_y^x\;} \underline{\mathcal{V}}(y, z) \times y \xrightarrow{\;\epsilon_z^y\;} z.$$

Employing naturality of the adjunction $- \times x \dashv \underline{\mathcal{V}}(x, -)$, (3.4.12) is adjunct to

$$* \times x \cong * \times * \times x \xrightarrow{\;g \times f \times 1\;} \underline{\mathcal{V}}(y, z) \times \underline{\mathcal{V}}(x, y) \times x \xrightarrow{\;1 \times \epsilon_y^x\;} \underline{\mathcal{V}}(y, z) \times y \xrightarrow{\;\epsilon_z^y\;} z.$$

On precomposing with a unit isomorphism, this arrow becomes

$$x \cong * \times x \xrightarrow{\;g \times 1\;} \mathcal{V}(y, z) \times x \cong \underline{\mathcal{V}}(y, z) \times x \xrightarrow{\;1 \times f\;} \underline{\mathcal{V}}(y, z) \times y \xrightarrow{\;\epsilon_z^y\;} z$$

by functoriality of $\times$ and the definition (3.4.11). By naturality of the unit isomorphism and functoriality of $\times$, this is

$$x \xrightarrow{\;f\;} y \cong * \times y \xrightarrow{\;g \times 1\;} \underline{\mathcal{V}}(y, z) \times y \xrightarrow{\;\epsilon_z^y\;} z \;=\; x \xrightarrow{\;f\;} y \xrightarrow{\;g\;} z$$

by (3.4.11) for g. $\qquad\qquad\qquad\square$

The following exercise is a good way to familiarize oneself with manipulations involving arrows in the underlying category of an enriched category.

Exercise 3.4.13 Let $\underline{\mathcal{D}}$ be a $\mathcal{V}$-category and $* \xrightarrow{g} \underline{\mathcal{D}}(y, z)$ be an arrow in $\mathcal{D}_0$. Using g, define for any $x \in \underline{\mathcal{D}}$

$$g_*: \underline{\mathcal{D}}(x, y) \cong * \times \underline{\mathcal{D}}(x, y) \xrightarrow{\;g \times 1\;} \underline{\mathcal{D}}(y, z) \times \underline{\mathcal{D}}(x, y) \xrightarrow{\;\circ\;} \underline{\mathcal{D}}(x, z).$$
$$(3.4.14)$$

This construction defines an (unenriched) representable functor

$$\underline{\mathcal{D}}(x, -): \mathcal{D}_0 \to \mathcal{V}.$$

Show that the composite of this functor with the underlying set functor $\mathcal{V}(*, -): \mathcal{V} \to \mathbf{Set}$ is the representable functor $\mathcal{D}_0(x, -): \mathcal{D}_0 \to \mathbf{Set}$ for the underlying category $\mathcal{D}_0$.

The upshot is that when $\underline{\mathcal{D}}$ is, say, a topological category, pre- and postcomposition by arrows in the underlying category are continuous. But really more is true. The representable functors $\underline{\mathcal{D}}(x, -)$ defined in Exercise 3.4.13 are in fact $\mathcal{V}$-functors, a notion we now introduce.

3.5 Enriched functors and enriched natural transformations

Small $\mathcal{V}$-categories themselves form a category; in fact, they form a 2-category.[5] For this we need the following notion:

Definition 3.5.1 A $\mathcal{V}$-**functor** $F: \underline{\mathcal{C}} \to \underline{\mathcal{D}}$ between $\mathcal{V}$-categories is given by an object map $\underline{\mathcal{C}} \ni x \mapsto Fx \in \underline{\mathcal{D}}$ together with morphisms

$$\underline{\mathcal{C}}(x, y) \xrightarrow{\;F_{x,y}\;} \underline{\mathcal{D}}(Fx, Fy)$$

in $\mathcal{V}$ for each $x, y \in \underline{\mathcal{C}}$ such that the following diagrams commute for all $x, y, z \in \underline{\mathcal{C}}$:

$$
\begin{array}{ccc}
\underline{\mathcal{C}}(y, z) \times \underline{\mathcal{C}}(x, y) & \xrightarrow{\;\circ\;} & \underline{\mathcal{C}}(x, z) \\
{\scriptstyle F_{y,z} \times F_{x,y}} \downarrow & & \downarrow {\scriptstyle F_{x,z}} \\
\underline{\mathcal{D}}(Fy, Fz) \times \underline{\mathcal{D}}(Fx, Fy) & \xrightarrow{\;\circ\;} & \underline{\mathcal{D}}(Fx, Fz)
\end{array}
\qquad
\begin{array}{ccc}
* & \xrightarrow{\;\mathrm{id}_x\;} & \underline{\mathcal{C}}(x, x) \\
& {\scriptstyle \mathrm{id}_{Fx}} \searrow & \downarrow {\scriptstyle F_{x,x}} \\
& & \underline{\mathcal{D}}(Fx, Fx)
\end{array}
$$

So, in particular, a topological functor between **Top**-categories consists of continuous maps between each hom-space. In the literature, **Top**-enriched functors are frequently called **continuous** functors. A simplicial functor between **sSet**-categories consists of maps between the n-simplices of the appropriate hom-spaces. A 2-**functor**, that is, a **Cat**-functor, consists of maps of objects, morphisms, and 2-cells that preserve domains, codomains, composition, and identities in all dimensions.[6]

Example 3.5.2 A covariant **Ab**-functor from a one-object **Ab**-category R to **Ab** is a left R-module; a contravariant **Ab**-functor is a right R-module.

Write $\mathcal{V}$-**Cat** for the category of $\mathcal{V}$-categories and $\mathcal{V}$-functors.

Exercise 3.5.3 Extend Definition 3.4.5 to define the underlying functor of $\mathcal{V}$-functor and show that your definition is functorial, that is, defines a functor $(-)_0: \mathcal{V}\text{-}\mathbf{Cat} \to \mathbf{Cat}$.

Example 3.5.4 (representable $\mathcal{V}$-functors) When $\underline{\mathcal{V}}$ is a closed $\mathcal{V}$-category, there is a $\mathcal{V}$-functor $\underline{\mathcal{C}}(c, -): \underline{\mathcal{C}} \to \underline{\mathcal{V}}$ whose underlying functor is the

---

[5] Indeed, we see in 7.3.1 that $\mathcal{V}$-**Cat** is a closed symmetric monoidal category.

[6] A **pseudofunctor** is a weakened version of a 2-functor in which the preservation is only up to coherent isomorphism.

representable $\underline{C}(c, -)\colon C_0 \to \mathcal{V}$ of Exercise 3.4.13. The map on objects is obvious; the morphism

$$\underline{C}(x, y) \to \underline{\mathcal{V}}(\underline{C}(c, x), \underline{C}(c, y))$$

is defined to be the transpose of the composition morphism for $\underline{C}$.

Example 3.5.5 Suppose $\mathcal{V}$ is closed, or at least satisfies the hypotheses of Example 3.3.4. A $\mathcal{V}$-functor F from the free $\mathcal{V}$-category on an unenriched category C to a $\mathcal{V}$-category $\underline{D}$ consists of an object function together with maps

$$\underset{C(x,y)}{\sqcup} * \to \underline{D}(Fx, Fy) \qquad \text{or, equivalently,} \qquad C(x, y) \to D_0(Fx, Fy)$$

for all $x, y \in C$. Chasing through the definitions, we see that the free $\mathcal{V}$-category functor is left adjoint to the underlying category functor. Hence a $\mathcal{V}$-functor $C \to \underline{D}$ is a functor $C \to D_0$, and we are content not to introduce notation for free $\mathcal{V}$-categories.

Example 3.5.6 In the unenriched sense, the terminal simplicial set Δ^0 represents the functor $(-)_0\colon \mathbf{sSet} \to \mathbf{Set}$ that takes a simplicial set X to its set X_0 of vertices. In the $\mathbf{sSet}$-enriched sense, the simplicial functor represented by Δ^0, generally written $\underline{\mathbf{sSet}}(\Delta^0, -)$ but which we abbreviate to $(-)^{\Delta^0}$, is the identity on $\mathbf{sSet}$: by definition

$$(X^{\Delta^0})_n = \mathbf{sSet}(\Delta^n \times \Delta^0, X) \cong \mathbf{sSet}(\Delta^n, X) \cong X_n,$$

the last isomorphism by the Yoneda lemma.

Example 3.5.7 When G is a finite group, orthogonal G-spectra are most concisely defined to be $\mathbf{Top}_*^G$-enriched diagrams on a particular $\mathbf{Top}_*^G$-enriched category $\mathcal{I}_G$ taking values in $(\underline{\mathbf{Top}}_*)_G$ [53].

We use arrows in the underlying category to define the notion of a $\mathcal{V}$-natural transformation. While the basic data of a $\mathcal{V}$-natural transformation is unenriched, the naturality condition is stronger than the unenriched one.

Definition 3.5.8 A $\mathcal{V}$-**natural transformation** $\alpha\colon F \Rightarrow G$ between a pair of $\mathcal{V}$-functors $F, G\colon \underline{C} \rightrightarrows \underline{D}$ consists of a morphism $\alpha_x\colon * \to \underline{D}(Fx, Gx)$ in $\mathcal{V}$ for each $x \in \underline{C}$ such that for all $x, y \in \underline{C}$, the following diagram commutes:

$$
\begin{array}{ccc}
\underline{C}(x, y) & \xrightarrow{\ F_{x,y}\ } & \underline{D}(Fx, Fy) \\
{\scriptstyle G_{x,y}}\big\downarrow & & \big\downarrow{\scriptstyle (\alpha_y)_*} \\
\underline{D}(Gx, Gy) & \xrightarrow[\ (\alpha_x)^*\ ]{} & \underline{D}(Fx, Gy)
\end{array}
$$

Here $(\alpha_x)^*$ and $(\alpha_y)_*$ are defined using (3.4.14).

Example 3.5.9   An **Ab**-natural transformation between two **Ab**-functors $R \rightrightarrows \underline{\mathbf{Ab}}$, that is, two left R-modules A and B, is a group homomorphism $A \rightarrow B$ that commutes with scalar multiplication. Hence an **Ab**-natural transformation is exactly a module homomorphism, and the category of **Ab**-functors and **Ab**-natural transformations from R to $\underline{\mathbf{Ab}}$ is precisely the category $\mathbf{Mod}_R$.

$\mathcal{V}$-categories, $\mathcal{V}$-functors, and $\mathcal{V}$-natural transformations assemble into a 2-category $\mathcal{V}$-**Cat**. Extending Exercise 3.5.3, the "underlying set" functor $\mathcal{V}(*, -)\colon \mathcal{V} \rightarrow$ **Set** can be used to define the underlying functor of a $\mathcal{V}$-functor and underlying natural transformation of a $\mathcal{V}$-natural transformation. Indeed:

Proposition 3.5.10   *"Underlying" is a 2-functor* $(-)_0\colon \mathcal{V}$-**Cat** $\rightarrow$ **Cat**.

Remark 3.5.11   We encourage the reader to sketch a proof of Proposition 3.5.10. To organize ideas, it might be helpful to note the following generalization: any lax monoidal functor $F\colon \mathcal{V} \rightarrow \mathcal{U}$ induces a 2-functor $\mathcal{V}$-**Cat** $\rightarrow \mathcal{U}$-**Cat**, extending the construction of Lemma 3.4.3.

Whenever possible, we omit the "$(-)_0$" notation from the underlying notions and instead adopt the following convention: if an unenriched object is denoted with the same letter previous assigned an enriched notion, for example, $F\colon \mathcal{M} \rightarrow \mathcal{N}$ in the presence of a $\mathcal{V}$-functor $F\colon \underline{\mathcal{M}} \rightarrow \underline{\mathcal{N}}$, we mean the former to be the underlying object of the latter.

A priori, there is no notion of isomorphism in a $\mathcal{V}$-category $\underline{\mathcal{C}}$, but the Yoneda lemma implies:

Lemma 3.5.12   *The following are equivalent:*

(i) *$x, y \in \mathcal{C}$ are isomorphic as objects of $\mathcal{C}$.*
(ii) *The representable functors $\mathcal{C}(x, -), \mathcal{C}(y, -)\colon \mathcal{C} \rightrightarrows$ **Set** are naturally isomorphic.*
(iii) *The unenriched representable functors $\underline{\mathcal{C}}(x, -), \underline{\mathcal{C}}(y, -)\colon \mathcal{C} \rightrightarrows \mathcal{V}$ are naturally isomorphic.*
(iv) *The representable $\mathcal{V}$-functors $\underline{\mathcal{C}}(x, -), \underline{\mathcal{C}}(y, -)\colon \underline{\mathcal{C}} \rightrightarrows \underline{\mathcal{V}}$ are $\mathcal{V}$-isomorphic.*

This preliminary version of the $\mathcal{V}$-Yoneda lemma is extended in Lemma 7.3.5 once we have constructed the object of $\mathcal{V}$-natural transformations.

Proof   Applying the underlying category functor $(-)_0\colon \mathcal{V}$-**Cat** $\rightarrow$ **Cat**, the fourth statement implies the third. The third statement implies the second by composing with $\mathcal{V}(*, -)$. The second statement implies the first by the unenriched Yoneda lemma; this is still the main point. Finally, the first implies the last by a direct construction employing the morphisms (3.4.14), left as an exercise. $\square$

Adjunctions and equivalences – which can be encoded by relations between appropriately defined objects, 1-cells, and 2-cells – are definable in any 2-category. Interpreting these definitions in the 2-category $\mathcal{V}$-**Cat** leads to the correct notions. We record these definitions for later use. By an analog of the unenriched argument, this definition of $\mathcal{V}$-equivalence is the same as the standard 2-categorical one.

Definition 3.5.13 A $\mathcal{V}$-**equivalence of categories** is given by a $\mathcal{V}$-functor $F : \underline{\mathcal{C}} \to \underline{\mathcal{D}}$ that is

- **essentially surjective**: every $d \in \underline{\mathcal{D}}$ is isomorphic (in $\mathcal{D}_0$) to some object Fc
- $\mathcal{V}$-**fully faithful**: for each $c, c' \in \underline{\mathcal{C}}$, the map $F_{c,c'} : \underline{\mathcal{C}}(c, c') \to \underline{\mathcal{D}}(Fc, Fc')$ is an isomorphism in $\mathcal{V}$

For example, a **DK-equivalence** (named for William Dwyer and Daniel Kan) of simplicially enriched categories is concisely defined as follows. We see in Chapter 10 that the localization functor **sSet** $\to$ Ho(**sSet**) is lax monoidal; hence simplicial enrichments induce canonical enrichments over the homotopy category of spaces $\mathcal{H} := $ Ho(**sSet**). A simplicial functor is a DK-equivalence just when the resulting $\mathcal{H}$-functor is a $\mathcal{H}$-equivalence.

Definition 3.5.14 A $\mathcal{V}$-**adjunction** consists of $\mathcal{V}$-functors $F : \underline{\mathcal{C}} \to \underline{\mathcal{D}}$, $G : \underline{\mathcal{D}} \to \underline{\mathcal{C}}$ together with

- $\mathcal{V}$-natural isomorphisms $\underline{\mathcal{D}}(Fc, d) \cong \underline{\mathcal{C}}(c, Gd)$ in $\mathcal{V}$

or, equivalently,

- $\mathcal{V}$-natural transformations $\eta : 1 \Rightarrow GF$ and $\epsilon : FG \Rightarrow 1$ satisfying the triangle identities $G\epsilon \cdot \eta_G = 1_G$ and $\epsilon_F \cdot F\eta = 1_F$

We will see a number of examples important to homotopy theory shortly.

3.6 Simplicial categories

Let us pause to explore the meaning of these definitions in our main example: **simplicial categories**, that is, categories enriched over (**sSet**, $\times$, $*$).

Any simplicial category $\underline{\mathcal{C}}$ gives rise to a simplicial object in **Cat**, which we denote $\mathcal{C}_\bullet : \Delta^{\mathrm{op}} \to$ **Cat**:

$$\mathcal{C}_0 \; \underset{\underset{d_1}{\longleftarrow}}{\overset{\overset{d_0}{\longleftarrow}}{\underset{s_0}{\longrightarrow}}} \; \mathcal{C}_1 \; \underset{\underset{d_2}{\longleftarrow}}{\overset{\overset{d_0}{\longleftarrow}}{\underset{\underset{s_1}{\longrightarrow}}{\overset{s_0}{\longrightarrow}}}} \; \underset{\underset{d_1}{\longleftarrow}}{} \mathcal{C}_2 \; \underset{\underset{d_3}{\longleftarrow}}{\overset{\overset{d_0}{\longleftarrow}}{\overset{\overset{s_0}{\longrightarrow}}{\overset{d_1}{\longleftarrow}}}} \mathcal{C}_3 \cdots$$

Each category $\mathcal{C}_n$ has the same objects as $\underline{\mathcal{C}}$. Define $\mathcal{C}_n(x, y) = \underline{\mathcal{C}}(x, y)_n$, that is, arrows in $\mathcal{C}_n$ from x to y are n-simplices in the hom-space $\underline{\mathcal{C}}(x, y)$. The right actions of morphisms in Δ on the hom-spaces of $\underline{C}$ induce identity-on-objects functors between the categories C_n, specifying the simplicial object. Note that there is a fortunate confluence of notation: the category C_0 is the underlying category of $\underline{C}$.

For example, **sSet** is a simplicial object in (large) categories. The objects of each category are simplicial sets X, Y. Morphisms from X to Y in the nth category are maps $\Delta^n \times X \to Y$ of simplicial sets. The face and degeneracy operators act by precomposition on the representable part of the domain.

Conversely, any simplicial object $C_\bullet \colon \Delta^{\mathrm{op}} \to \mathbf{Cat}$ for which each of the constituent functors $d_i \colon C_n \to C_{n-1}$, $s_i \colon C_n \to C_{n+1}$ is the identity on objects corresponds to a simplicially enriched category, where we define $\underline{C}(x, y)_n$ to be $C_n(x, y)$ and use the functors d_i, s_i to specify the simplicial action. This is a convenient mechanism for producing simplicial categories (cf. Example 16.2.3).

A simplicial functor $F \colon \underline{C} \to \underline{D}$ consists of functors $F_n \colon C_n \to D_n$ for each n that commute with the simplicial operator functors. Necessarily, each F_n has the same underlying object function. For example, a simplicial functor $F \colon \underline{C} \to$ **sSet** specifies a simplicial set Fx for each $x \in \underline{C}$ together with a map $\Delta^n \times Fx \to Fy$ for each n-simplex in $\underline{C}(x, y)$ such that the faces and degeneracies of the map $\Delta^n \times Fx \to Fy$ correspond to the faces and degeneracies of the n-simplex.

A simplicial natural transformation between simplicial functors $F, G \colon \underline{C} \rightrightarrows \underline{D}$ is given by arrows in the underlying category of $\underline{D}$ for each object of $\underline{C}$, satisfying an enriched naturality condition. In this case, the data consist of an arrow $\alpha_x \in \mathcal{D}_0(Fx, Gx)$ for each $x \in \underline{C}$. The naturality condition says first that the α_x form a natural transformation between F_0 and G_0 but also that each degenerate image of the vertices α_x forms a natural transformation between the functors F_n and G_n. So the arrows $s_0(\alpha_x) \in \mathcal{D}_1(Fx, Gx)$ should form a natural transformation $F_1 \Rightarrow G_1$; the arrows $s_0 s_0(\alpha_x) = s_1 s_0(\alpha_x) \in \mathcal{D}_2(Fx, Gx)$ form a natural transformation $F_2 \Rightarrow G_2$; and the images of the α_x under the unique degeneracy operator $[n] \twoheadrightarrow [0]$ form a natural transformation $F_n \Rightarrow G_n$.

3.7 Tensors and cotensors

Let us return for a moment to the adjunction defining a closed symmetric monoidal category $(\mathcal{V}, \times, *)$ and prove the claim made in Remark 3.3.9, namely, that the isomorphism (3.3.7) can be interpreted internally to $\mathcal{V}$. To show that $\underline{\mathcal{V}}(u \times v, w)$ and $\underline{\mathcal{V}}(u, \underline{\mathcal{V}}(v, w))$ are isomorphic, we appeal to the Yoneda lemma and prove they represent the same functor $\mathcal{V}^{\mathrm{op}} \to \mathbf{Set}$. For any $x \in \mathcal{V}$, by (3.3.7) and associativity of the monoidal product, we have a sequence of natural isomorphisms

$$\mathcal{V}(x, \underline{\mathcal{V}}(u \times v, w)) \cong \mathcal{V}(x \times u \times v, w) \cong \mathcal{V}(x \times u, \underline{\mathcal{V}}(v, w))$$

$$\cong \mathcal{V}(x, \underline{\mathcal{V}}(u, \underline{\mathcal{V}}(v, w))),$$

proving our claim.

Furthermore, from our discussion of representable functors in Example 3.5.4, we have seen that for each $v \in \mathcal{V}$, the right adjoint $\underline{\mathcal{V}}(v, -)\colon \mathcal{V} \to \mathcal{V}$ is in fact a $\mathcal{V}$-functor. It follows that the left adjoint $- \times v\colon \mathcal{V} \to \mathcal{V}$ canonically inherits the structure of a $\mathcal{V}$-functor: the required maps on hom-objects are defined by

$$\underline{\mathcal{V}}(u, w) \xrightarrow{\;(\eta^v_w)_*\;} \underline{\mathcal{V}}(u, \underline{\mathcal{V}}(v, w \times v)) \cong \underline{\mathcal{V}}(u \times v, w \times v).$$

More generally, suppose $\mathcal{M}$ and $\mathcal{N}$ are $\mathcal{V}$-categories and that we are given an adjunction

$$\mathcal{N}(Fm, n) \cong \mathcal{M}(m, Gn) \tag{3.7.1}$$

between the underlying categories. Inspired by the case just considered, we might ask when it is possible to enrich (3.7.1) to an isomorphism in $\mathcal{V}$.

Attempting to use the same idea and show that $\underline{\mathcal{N}}(Fm, n)$ and $\underline{\mathcal{M}}(m, Gn)$ represent the same functor leads us to consider the hom-set $\mathcal{V}(v, \underline{\mathcal{M}}(m, Gn))$, but already we do not know how to proceed. It would be helpful here if each $\underline{\mathcal{M}}(m, -)\colon \mathcal{M} \to \mathcal{V}$ and each $\underline{\mathcal{N}}(n, -)\colon \mathcal{N} \to \mathcal{V}$ had a left adjoint and these adjoints were preserved by F. Or, dually, we can prove our result if each $\underline{\mathcal{M}}(-, m)$ and $\underline{\mathcal{N}}(-, n)$ has a mutual right adjoint and these are preserved by G.

We return to this situation in Proposition 3.7.10, after presenting the general definitions thus motivated.

Definition 3.7.2   A $\mathcal{V}$-category $\mathcal{M}$ is **tensored** if, for each $v \in \mathcal{V}$ and $m \in \mathcal{M}$, there is an object $v \otimes m \in \mathcal{M}$ together with isomorphisms

$$\underline{\mathcal{M}}(v \otimes m, n) \cong \underline{\mathcal{V}}(v, \underline{\mathcal{M}}(m, n)), \qquad \forall \quad v \in \mathcal{V}, \quad m, n \in \mathcal{M}.$$

By the Yoneda lemma, there is a unique way to make the tensor product into a bifunctor $- \otimes -\colon \mathcal{V} \times \mathcal{M} \to \mathcal{M}$ so that the isomorphism is natural in all three variables. By an argument analogous to the one just given, it follows that each represented $\mathcal{V}$-functor $\underline{\mathcal{M}}(m, -)\colon \underline{\mathcal{M}} \to \underline{\mathcal{V}}$ admits a left $\mathcal{V}$-adjoint $- \otimes m\colon \underline{\mathcal{V}} \to \underline{\mathcal{M}}$. Dually:

Definition 3.7.3   A $\mathcal{V}$-category $\mathcal{M}$ is **cotensored** if, for each $v \in \mathcal{V}$ and $n \in \mathcal{M}$, there is an object $n^v \in \mathcal{M}$ together with isomorphisms

$$\underline{\mathcal{M}}(m, n^v) \cong \underline{\mathcal{V}}(v, \underline{\mathcal{M}}(m, n)) \qquad \forall \quad v \in \mathcal{V} \quad m, n \in \mathcal{M}.$$

By the Yoneda lemma, there is a unique way to make the cotensor product into a bifunctor $(-)^-\colon \mathcal{V}^{\mathrm{op}} \times \mathcal{M} \to \mathcal{M}$ so that the isomorphism is natural in all three variables. As earlier, each represented $\mathcal{V}$-functor $\underline{\mathcal{M}}(-, n)\colon \underline{\mathcal{M}}^{\mathrm{op}} \to \underline{\mathcal{V}}$ admits a mutual right $\mathcal{V}$-adjoint $n^-\colon \underline{\mathcal{V}}^{\mathrm{op}} \to \underline{\mathcal{M}}$.

Remark 3.7.4   If $\underline{\mathcal{M}}$ is tensored and cotensored over $\mathcal{V}$, then the tensor, cotensor, and internal hom form a **two-variable** $\mathcal{V}$-**adjunction**

$$\underline{\mathcal{M}}(v \otimes m, n) \cong \underline{\mathcal{V}}(v, \underline{\mathcal{M}}(m, n)) \cong \underline{\mathcal{M}}(m, n^v).$$

See [79, 14.8] for a definition.

Example 3.7.5   By the discussion at the beginning of this section, any closed symmetric monoidal category is tensored, cotensored, and enriched over itself.

Example 3.7.6   Any locally small category with products and coproducts is enriched, tensored, and cotensored over **Set**. The tensor of a set A with an object $m \in \mathcal{M}$ is the copower $A \cdot m = \coprod_A m$ and the cotensor is the power $m^A = \prod_A m$, introduced in Section 1.2.

More interesting examples appear later, but to produce them, it will help first to develop a bit of the general theory. We begin by observing that the defining $\mathcal{V}$-natural isomorphisms imply that tensors (and dually cotensors) are associative and unital with respect to the monoidal structure on $\mathcal{V}$.

Lemma 3.7.7   *Let* $(\mathcal{V}, \times, *)$ *be a closed symmetric monoidal category, and suppose* $\mathcal{M}$ *is a tensored* $\mathcal{V}$-*category. Then the tensor product is unital and associative, that is, there exist natural isomorphisms*

$$* \otimes m \cong m, \quad (v \times w) \otimes m \cong v \otimes (w \otimes m), \quad \forall v, w \in \mathcal{V} \quad m \in \mathcal{M}.$$

Proof   We use the Yoneda lemma in $\mathcal{V}$ and Lemma 3.5.12. On account of the natural isomorphisms

$$\mathcal{V}(w, \underline{\mathcal{V}}(*, v)) \cong \mathcal{V}(w \times *, v) \cong \mathcal{V}(w, v)$$

for any $v, w \in \mathcal{V}$, we see that $\underline{\mathcal{V}}(*, v) \cong v$, as these objects represent the same functor. Hence

$$\underline{\mathcal{M}}(* \otimes m, n) \cong \underline{\mathcal{V}}(*, \underline{\mathcal{M}}(m, n)) \cong \underline{\mathcal{M}}(m, n), \qquad (3.7.8)$$

which implies that $* \otimes m \cong m$ by Lemma 3.5.12. Naturality of the isomorphisms of (3.7.8) implies that the isomorphism between the representing objects is again natural in m.

Similarly,

$$\mathcal{M}((v \times w) \otimes m, n) \cong \mathcal{V}(v \times w, \underline{\mathcal{M}}(m, n)) \cong \mathcal{V}(v, \underline{\mathcal{V}}(w, \underline{\mathcal{M}}(m, n)))$$

$$\cong \mathcal{V}(v, \underline{\mathcal{M}}(w \otimes m, n)) \cong \mathcal{M}(v \otimes (w \otimes m), n),$$

the isomorphism surrounding the line break being the main point. □

Remark 3.7.9 Proposition 10.1.4 contains the converse to Lemma 3.7.7: such "$\mathcal{V}$-module" structures encode tensors. This is easy to prove now, but we will not need the result until later.

The following lemma answers the question posed at the start of this section and displays some of the utility of tensors and cotensors. In the statement, if an enriched object bears the same name as an unenriched object, we mean to assert that the underlying part of the enriched object is the previously specified unenriched object.

Proposition 3.7.10 *Suppose $\underline{\mathcal{M}}$ and $\underline{\mathcal{N}}$ are tensored and cotensored $\mathcal{V}$-categories and $F: \mathcal{M} \rightleftarrows \mathcal{N}: G$ is an adjunction between the underlying categories. Then the data of any of the following determine the other:*

(i) a $\mathcal{V}$-adjunction $\underline{\mathcal{N}}(Fm, n) \cong \underline{\mathcal{M}}(m, Gn)$
(ii) a $\mathcal{V}$-functor F together with natural isomorphisms $F(v \otimes m) \cong v \otimes Fm$
(iii) a $\mathcal{V}$-functor G together with natural isomorphisms $G(n^v) \cong (Gn)^v$

Proof The proof is similar to the preceding arguments, left as an exercise to ensure that they have been internalized. □

To give examples of $\mathcal{V}$-adjunctions, we must first find a few more tensored, cotensored, and enriched categories. Our search is greatly aided by the following theorem:

Theorem 3.7.11 *Suppose we have an adjunction $F: \mathcal{V} \rightleftarrows \mathcal{U}: G$ between closed symmetric monoidal categories such that the left adjoint F is strong monoidal. Then any tensored, cotensored, and enriched $\mathcal{U}$-category becomes canonically enriched, tensored, and cotensored over $\mathcal{V}$.*

By the theory of doctrinal adjunctions [44] or, alternatively, by the calculus of mates [47], the fact that the left adjoint F is strong monoidal implies that the right adjoint G is lax monoidal. Hence, by Lemma 3.4.3, we can change the base of the enrichment from $\mathcal{U}$ to $\mathcal{V}$ by applying the right adjoint to the hom-objects. By a **strong monoidal adjunction**, we mean an adjunction satisfying the hypotheses of Theorem 3.7.11.

Proof Suppose $\mathcal{M}$ is a tensored and cotensored $\mathcal{U}$-category

$$\underline{\mathcal{M}}_\mathcal{U}(u \otimes m, n) \cong \underline{\mathcal{U}}(u, \underline{\mathcal{M}}_\mathcal{U}(m, n)) \cong \underline{\mathcal{M}}_\mathcal{U}(m, n^u).$$

We define tensor, cotensor, and internal hom over $\mathcal{V}$, using the notation

$$- \star - : \mathcal{V} \times \mathcal{M} \to \mathcal{M}, \quad \{-, -\} : \mathcal{V}^{\mathrm{op}} \times \mathcal{M} \to \mathcal{M},$$

$$\underline{\mathcal{M}}_{\mathcal{V}}(-, -) : \mathcal{M}^{\mathrm{op}} \times \mathcal{M} \to \mathcal{V}$$

by

$$v \star m := Fv \otimes m, \quad \{v, m\} := m^{Fv}, \quad \underline{\mathcal{M}}_{\mathcal{V}}(m, n) := G\underline{\mathcal{M}}_{\mathcal{U}}(m, n).$$

We must show that $\underline{\mathcal{M}}_{\mathcal{V}}(v \star m, n)$, $\underline{\mathcal{V}}(v, \underline{\mathcal{M}}_{\mathcal{V}}(m, n))$, and $\underline{\mathcal{M}}_{\mathcal{V}}(m, \{v, n\})$ represent the same functor $\mathcal{V}^{\mathrm{op}} \to$ **Set**. For the first part, we have natural isomorphisms

$$\mathcal{V}(x, \underline{\mathcal{M}}_{\mathcal{V}}(v \star m, n)) = \mathcal{V}(x, G\underline{\mathcal{M}}_{\mathcal{U}}(Fv \otimes m, n)) \cong \mathcal{U}(Fx, \underline{\mathcal{M}}_{\mathcal{U}}(Fv \otimes m, n))$$

$$\cong \mathcal{M}(Fx \otimes (Fv \otimes m), n) \cong \mathcal{M}((Fx \times Fv) \otimes m, n)$$

$$\cong \mathcal{U}(Fx \times Fv, \underline{\mathcal{M}}_{\mathcal{U}}(m, n))$$

$$\cong \mathcal{U}(F(x \times v), \underline{\mathcal{M}}_{\mathcal{U}}(m, n)) \cong \mathcal{V}(x \times v, G\underline{\mathcal{M}}_{\mathcal{U}}(m, n))$$

$$\cong \mathcal{V}(x, \underline{\mathcal{V}}(v, G\underline{\mathcal{M}}_{\mathcal{U}}(m, n))) = \mathcal{V}(x, \underline{\mathcal{V}}(v, \underline{\mathcal{M}}_{\mathcal{V}}(m, n))).$$

The other half is similar.   □

Corollary 3.7.12   *The strong monoidal adjunction $F \dashv G$ of Theorem 3.7.11 is a $\mathcal{V}$-adjunction with respect to the induced $\mathcal{V}$-category structure on $\mathcal{U}$.*

Proof   By the Yoneda lemma and the definitions given in the proof of Theorem 3.7.11,

$$\mathcal{V}(w, \underline{\mathcal{U}}_{\mathcal{V}}(Fv, u)) \cong \mathcal{V}(w, G\underline{\mathcal{U}}_{\mathcal{U}}(Fv, u)) \cong \mathcal{U}(Fw, \underline{\mathcal{U}}_{\mathcal{U}}(Fv, u))$$

$$\cong \mathcal{U}(Fw \times Fv, u) \cong \mathcal{U}(F(w \times v), u)$$

$$\cong \mathcal{V}(w \times v, Gu) \cong \mathcal{V}(w, \underline{\mathcal{V}}(v, Gu)).$$ □

Example 3.7.13   By 3.3.14, there is a closed monoidal category $(\mathbf{sSet}_*, \wedge, \partial\Delta^1)$ together with an adjunction

$$(-)_+ : \mathbf{sSet} \; \underset{\longleftarrow}{\overset{\perp}{\longrightarrow}} \; \mathbf{sSet}_* : U$$

with strong monoidal left adjoint. It follows that $\mathbf{sSet}_*$ is tensored, cotensored, and enriched over simplicial sets. If K is a simplicial set and $* \to X$ is a based simplicial set, then $K \otimes X = K_+ \wedge X$ is the quotient $K \times X / K \times *$, that is, the coequalizer

$$K \times * \rightrightarrows K \times X$$

of the inclusion at the basepoint component with the constant map at the basepoint. The internal hom $\underline{\textbf{sSet}}_*(X, Y)$ has as n-simplices maps

$$(\Delta^n \times X)/(\Delta^n \times *) \to Y$$

with basepoint the constant map $X \to * \to Y$. The simplicial hom-space just forgets this basepoint. The cotensor Y^K is defined to be $\underline{\textbf{sSet}}_*(K_+, Y)$ with the constant map at the basepoint of Y serving as the basepoint. By Corollary 3.7.12, the adjunction $(-)_+ \dashv U$ is simplicially enriched. It follows that the underlying simplicial set of the cotensor is $(UY)^K$.

Example 3.7.14 A similar analysis shows that the convenient category of based topological spaces is topologically enriched, tensored, and cotensored.

Example 3.7.15 We prove in Lemma 6.1.6 that geometric realization is a strong monoidal left adjoint

$$| - | : \textbf{sSet} \xrightarrow{\;\;\perp\;\;} \textbf{Top} : S$$

This is another instance in which the point-set considerations appearing in the definition of the convenient category of topological spaces play an important role. Because **Top** is a closed monoidal category, it follows that **Top** is tensored, cotensored, and enriched over simplicial sets. Writing $\underline{\textbf{Top}}$ for the internal hom, if $K \in \textbf{sSet}$, $X \in \textbf{Top}$, then $K \otimes X := |K| \times X$ and $\overline{X^K} := \underline{\textbf{Top}}(|K|, X)$. The simplicial enrichment has hom-objects the simplicial sets $S\underline{\textbf{Top}}(X, Y)$. By Corollary 3.7.12, these definitions make $| - | : \textbf{sSet} \rightleftarrows \textbf{Top} : S$ into a simplicially enriched adjunction.

Remark 3.7.16 Because the categories **Top** and **sSet** are both cartesian closed, the right adjoint S, which preserves products, is also strong monoidal. It follows that any simplicially enriched category is also topologically enriched, but tensors and cotensors do not transfer in this direction. This is one reason why we prefer simplicial enrichments to topological ones: they are more general.

Example 3.7.17 ([36, 9.8.7]) Let A be any non-empty topological space, regarded as a functor $\mathbb{1} \to \textbf{Top}$ from the terminal category. Because **Top** is cocomplete, we can form the left Kan extension of this functor along itself. By Theorem 1.2.1, the resulting functor $L : \textbf{Top} \to \textbf{Top}$ is given on objects by

$$X \mapsto \coprod_{\textbf{Top}(A,X)} A.$$

This functor is not simplicial (nor is it continuous). For every pair of spaces X and Y, we would need a map $S\underline{\textbf{Top}}(X, Y) \to S\underline{\textbf{Top}}(LX, LY)$ between the associated total singular complexes. Taking $X = A$ and $Y = A \times I$, the obvious pair of inclusions $i_0, i_1 : A \rightrightarrows A \times I$ defines two vertices in $\underline{\textbf{Top}}(A, A \times I)$

connected by a 1-simplex, namely, the obvious homeomorphism $A \times |\Delta^1| \to A \times I$.

But the images of i_0 and i_1 under L land in distinct components of LY and cannot be so connected. If, however, we replace the Kan extension defining L by its enriched analog, defined in (7.6.7), then this functor is simplicially enriched.

Exercise 3.7.18   Let $\mathcal{M}$ be cocomplete. Show that the category $\mathcal{M}^{\Delta^{op}}$ of simplicial objects in $\mathcal{M}$ is simplicially enriched and tensored, with $(K \otimes X)_n := K_n \cdot X_n$ defined using the copower. Give a formal argument why $\mathcal{M}^{\Delta}$ is simplicially enriched and cotensored if $\mathcal{M}$ is complete.

Digression 3.7.19 (Dold–Kan correspondence)   Exercise 3.7.18 applies to the category $\mathcal{M} = \mathbf{Ab}$. Our particular interest in the category $\mathbf{Ab}^{\Delta^{op}}$ of simplicial abelian groups is on account of the **Dold–Kan correspondence**: there is an (adjoint) equivalence of categories $\mathbf{Ab}^{\Delta^{op}} \rightleftarrows \mathbf{Ch}_{\geq 0}(\mathbb{Z})$.

By the **Ab**-enriched analog of Construction 1.5.1, this adjunction is determined by a cosimplicial object $\Delta \to \mathbf{Ch}_{\bullet}(\mathbb{Z})$ that sends $[n]$ to the normalized Moore complex of the free simplicial abelian group on Δ^n. Explicitly, the simplicial set $\Delta^n : \Delta^{op} \to \mathbf{Set}$ composes with the free abelian group functor $\mathbf{Set} \to \mathbf{Ab}$ to form a simplicial abelian group. The associated chain complex is formed from this graded group with differentials defined to be the alternating sum of the face maps. The normalized chain complex quotients by formal sums of degenerate simplices.

Example 3.7.20   Let $F : \mathcal{M} \rightleftarrows \mathcal{N} : G$ be an unenriched adjunction between cocomplete categories. Levelwise composition defines adjoint functors

$$F_* : \mathcal{M}^{\Delta^{op}} \xrightleftarrows[\bot]{} \mathcal{N}^{\Delta^{op}} : G_* \ .$$

We use Proposition 3.7.10 to prove that $F_* \dashv G_*$ is a simplicial adjunction, using the enrichments of Exercise 3.7.18. Because F is a left adjoint, it commutes with coproducts, and in particular copowers, and hence preserves the tensors defined previously. From the definition of the simplicial enrichment left as an exercise, it is clear from these isomorphisms that F_* is a simplicial functor.[7] It follows from Proposition 3.7.10 that G_* is also simplicially enriched and the adjunction is a simplicial adjunction.

3.8 Simplicial homotopy and simplicial model categories

If $\underline{\mathcal{M}}$ is a simplicially enriched category, then there is a canonical notion of homotopy between maps in the underlying category $\mathcal{M}$. A **homotopy** between

---

[7] Alternatively, we might appeal to Proposition 10.1.5, which characterizes unenriched functors between tensored $\mathcal{V}$-categories that enrich to $\mathcal{V}$-functors.

$f, g : m \rightrightarrows n$ is given by the data of the following left-hand diagram. If $\underline{\mathcal{M}}$ is tensored or cotensored over simplicial sets, the left-hand diagram transposes to one of the right-hand diagrams:

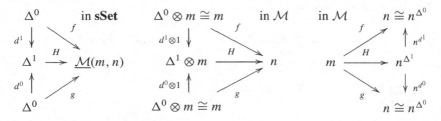

In other words, if $\underline{\mathcal{M}}$ is tensored or cotensored over simplicial sets, we can realize a homotopy as an arrow in the underlying category $\mathcal{M}$. Immediately, we get a notion of simplicial homotopy equivalence in $\mathcal{M}$, as displayed, for example, in (3.8.5).

Definition 3.8.1 If $\mathcal{M}$ is simplicially enriched, tensored, and cocomplete, the **geometric realization** of a simplicial object is

$$|X_\bullet| := \int^{n \in \Delta} \Delta^n \otimes X_n.$$

These coends define a functor $| - | : \mathcal{M}^{\Delta^{\mathrm{op}}} \to \mathcal{M}$.

This construction is an example of a functor tensor product of $\Delta^\bullet : \Delta \to \mathbf{sSet}$ with $X_\bullet : \Delta^{\mathrm{op}} \to \mathcal{M}$. We see more of these in Chapter 4.

Remark 3.8.2 When $\mathcal{M}$ is tensored, there are two reasonable definitions for simplicial tensors on $\mathcal{M}^{\Delta^{\mathrm{op}}}$. One approach, which works equally well for any diagram category $\mathcal{M}^{\mathcal{D}}$, is to define tensors pointwise: given $K \in \mathbf{sSet}$ and $X : \mathcal{D} \to \mathcal{M}$, define $K \otimes X$ to be the functor $d \mapsto K \otimes (Xd)$. The other, particular to $\mathcal{M}^{\Delta^{\mathrm{op}}}$, is the definition suggested by Exercise 3.7.18. This is always the preferred choice because the concordant definition of simplicial homotopy agrees with the older combinatorial definition, parallel to the notion of chain homotopy.

Furthermore:

Lemma 3.8.3 *Suppose $\mathcal{M}$ is simplicially enriched, tensored, and cotensored and admits geometric realizations of simplicial objects. Then geometric realization preserves tensors.*

Proof We prove this in 8.1.6, or see [69, 5.4]. A key observation is that the bifunctor $- \otimes - : \mathbf{Set}^{\Delta^{\mathrm{op}}} \times \mathcal{M} \to \mathcal{M}$, admitting pointwise right adjoints, preserves colimits in both variables. □

Corollary 3.8.4 *If $\mathcal{M}$ is simplicially enriched, tensored, and cotensored, then geometric realization preserves simplicial homotopy equivalences.*

Proof   Because geometric realization preserves tensors, given a simplicial homotopy equivalence in $\mathcal{M}^{\Delta^{op}}$,

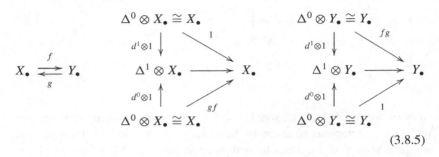

$$(3.8.5)$$

geometric realization produces a simplicial homotopy equivalence in $\mathcal{M}$ between $|X_\bullet|$ and $|Y_\bullet|$. $\square$

In Chapter 7, we show that when $\mathcal{M}$ is cotensored in addition to being tensored and enriched, geometric realization is a simplicial functor, using the simplicial enrichment on $\mathcal{M}^{\Delta^{op}}$ given in Exercise 3.7.18. The notion of simplicial homotopy equivalence, definable in any simplicial category, is obviously preserved by any simplicial functor.

We close this chapter on enriched category theory with the topic that we used to motivate it at the end of Chapter 2. We give a proper definition of a **simplicial model category** in 11.4.4, once we have formally introduced model categories. For present purposes, the following lemma axiomatizes those characteristics that we will need:

Lemma 3.8.6 *A simplicial model category*

(i) is complete and cocomplete

(ii) is tensored, cotensored, and enriched over simplicial sets

*(iii) has subcategories of **cofibrant** and **fibrant** objects preserved, respectively, by tensoring or cotensoring with any simplicial set*

(iv) admits a left deformation Q into the cofibrant objects and a right deformation R into the fibrant objects

(v) is a saturated homotopical category

(vi) has simplicial homotopy equivalences among the weak equivalences

(vii) has the property that the internal hom preserves weak equivalences between cofibrant objects in its first variable, provided the second variable is fibrant, and preserves weak equivalences between fibrant objects in its second variable, provided its first variable is cofibrant.

Proof Properties (i), (iv), and (v) are true for any model category, the latter by the theorem of Quillen mentioned in Remark 2.1.9. Properties (ii), (iii), and (vii) are immediate from the definition of a **simplicial** model category; (iii) and (vii) are consequences of the SM7 axiom [65, §II.2] and Ken Brown's lemma. Property (vi) is non-trivial; see [36, 9.5.16]. □

Examples include **sSet**, **sSet**$_*$, **Top**, and **Top**$_*$ – more details appear in Section 6.2, in preparation for the calculations that appear in that chapter. By work of Daniel Dugger and others, many model categories can be replaced by simplicial model categories with the same homotopy category. For example, one theorem says that any left proper, combinatorial model category is Quillen equivalent to a simplicial model category [17].

Some less technical observations are perhaps more to the point. There are analogous definitions of a **topological model category**, or **sSet**$_*$**-model category**, or $\mathcal{V}$-model category (subject to certain restrictions on $\mathcal{V}$), which are enriched, tensored, and cotensored over **Top**, **sSet**$_*$, or $\mathcal{V}$ in place of **sSet**. A homotopical version of Theorem 3.7.11 says that if the strong monoidal adjunction $F \dashv U$ is additionally a **Quillen adjunction**, then any $\mathcal{U}$-model category is canonically a $\mathcal{V}$-model category.

The adjunctions

$$(-)_+ : \textbf{sSet} \xrightarrow{\;\perp\;} \textbf{sSet}_* : U \qquad |-| : \textbf{sSet} \xrightarrow{\;\perp\;} \textbf{Top} : S$$

are strong monoidal Quillen adjunctions, provided that **Top** is the usual convenient category of spaces. It follows that any based simplicial or topological model category is canonically a simplicial model category. However, extending Remark 3.7.16, the converse is not generally true. For this reason, we submit that simplicial model categories are an appropriate general setting to define homotopy limits and colimits.

4

The unreasonably effective (co)bar construction

The bar construction was first introduced in homological algebra by Samuel Eilenberg and Mac Lane as a way to construct resolutions of algebras over a commutative ring. The name comes from their shorthand use of the character | in place of $\otimes$. It has since been greatly generalized. We will not introduce the most general version of this construction – for this, see, e.g., [79, 23.3–23.5] – but rather one at the right level of generality for present purposes.

One advantage of the bar construction that we do not discuss here but certainly exploit in Part II is that it can easily be extended to enriched contexts. Some of our notation is chosen to ease that generalization. In what follows, $\mathcal{V}$ should typically be read as the category **sSet**, and $\mathcal{M}$ should be be simplicially enriched, tensored, and cotensored. The enrichment will not be used explicitly. Rather, what is needed is a notion of simplicial homotopy; a method for computing the geometric realization of simplicial objects and the totalization of cosimplicial objects; and the ancillary properties of these functors that result from the presence of tensors and cotensors.

Remark 4.0.1 Functors naturally valued in **Set** can be interpreted as taking value in (discrete) simplicial sets, in which case the simplicial tensor is equivalent to the copower, for $A \in$ **Set**, $m \in \mathcal{M}$:

$$A \otimes m = (\sqcup_A \Delta^0) \otimes m \cong \sqcup_A (\Delta^0 \otimes m) \cong \sqcup_A m = A \cdot m. \qquad (4.0.2)$$

In particular, it is occasionally convenient to relax the just-stated conventions and allow $\mathcal{V} =$ **Set** and $\mathcal{M}$ to be arbitrary. On account of isomorphisms such as (4.0.2), there is no real ambiguity when, say, a simplicial set is promoted to a horizontally discrete bisimplicial set.

In Chapter 5, we use the two-sided bar construction and dual two-sided cobar construction introduced here to define homotopy colimit and homotopy limit functors, respectively, for diagrams of any shape taking values in a simplicial model category.

4.1 Functor tensor products

A common form of the coend, introduced in section 1.2, is deserving of a special name. In the presence of a bifunctor $- \otimes -\colon \mathcal{V} \times \mathcal{M} \to \mathcal{M}$, the **functor tensor product** of $F\colon \mathcal{D} \to \mathcal{M}$ with $G\colon \mathcal{D}^{\mathrm{op}} \to \mathcal{V}$ is the coend

$$G \otimes_{\mathcal{D}} F := \int^{\mathcal{D}} G \otimes F = \operatorname{coeq} \left(\coprod_{f\colon d \to d'} Gd' \otimes Fd \overset{f^*}{\underset{f_*}{\rightrightarrows}} \coprod_{d} Gd \otimes Fd \right).$$

(4.1.1)

As our notation might suggest, commonly $\otimes$ is the bifunctor of a tensored $\mathcal{V}$-category, perhaps in the special case given by a monoidal category. Or perhaps $\mathcal{V} = \mathbf{Set}$, $\mathcal{M}$ has coproducts, and $\otimes$ is the copower. In general, its codomain could be some category other than $\mathcal{M}$, but we will not need this extra flexibility here. In the literature, it is fairly common for the notation $G \otimes_{\mathcal{D}} F$ to change to reflect the bifunctor used to define the coend, but we believe the appropriate choice of bifunctor in any context is fairly obvious and hence use uniform notation throughout.

The name and notation are motivated by the following example:

Example 4.1.2 Let R be a ring. A right R-module A is an **Ab**-functor $A\colon R^{\mathrm{op}} \to \underline{\mathbf{Ab}}$ and a left R-module B is an **Ab**-functor $B\colon R \to \underline{\mathbf{Ab}}$. The underlying unenriched functors $A\colon R \to \mathbf{Ab}$, $B\colon R^{\mathrm{op}} \to \mathbf{Ab}$ forget the addition in R and remember only the monoid action on the abelian groups A and B. Nonetheless, the functor tensor product

$$A \otimes_R B = \int^R A \otimes_{\mathbb{Z}} B$$

is the usual tensor product over R of a right and left R-module.

Example 4.1.3 Let $*\colon \mathcal{D}^{\mathrm{op}} \to \mathbf{sSet}$ be the constant functor at the terminal object and let $\mathcal{M}$ be a tensored simplicial category, so that Lemma 3.7.7 applies. Then, by inspection of the formula (4.1.1),

$$* \otimes_{\mathcal{D}} F \cong \operatorname{colim} F.$$

Example 4.1.4 The co-Yoneda lemma 1.4.6 establishes that the covariant and contravariant representables are free modules in the sense that $\mathcal{D}(-, d) \otimes_{\mathcal{D}} F \cong Fd$ and $G \otimes_{\mathcal{D}} \mathcal{D}(d, -) \cong Gd$.

Example 4.1.5 From the formula of 1.2.1, pointwise left Kan extensions are functor tensor products. Given $F\colon \mathcal{C} \to \mathcal{E}$, $K\colon \mathcal{C} \to \mathcal{D}$,

$$\operatorname{Lan}_K F(d) = \mathcal{D}(K-, d) \otimes_{\mathcal{C}} F.$$

In particular, writing $\Delta^\bullet \colon \mathbb{\Delta} \to \mathbf{Top}$ for the functor that sends the nth represented simplicial set to the standard topological n-simplex, for any simplicial set X, by the Yoneda lemma,

$$|X| = \mathbf{sSet}(\Delta^\bullet, X) \otimes_\mathbb{\Delta} \Delta^\bullet \cong X_\bullet \otimes_\mathbb{\Delta} \Delta^\bullet.$$

Hence geometric realization can be defined to be the functor tensor product of $\Delta^\bullet$ with a simplicial set, discrete bisimplicial set, or discrete bisimplicial space defined with respect to the copower, simplicial tensor, or cartesian product of spaces, respectively.

Example 4.1.6   As in Definition 3.8.1, the geometric realization of a simplicial object in a category tensored over simplicial sets is defined to be the functor tensor product of the simplicial object $X_\bullet \colon \mathbb{\Delta}^{\mathrm{op}} \to \mathcal{M}$ with the Yoneda embedding $\Delta^\bullet \colon \mathbb{\Delta} \to \mathbf{sSet}$:

$$|X| := \Delta^\bullet \otimes_{\mathbb{\Delta}^{\mathrm{op}}} X_\bullet. \tag{4.1.7}$$

We use this definition for geometric realization, except when $\mathcal{M} = \mathbf{Set}$. In this case, our conventions dictate that the terms $\Delta^n \otimes X_n$ denote copowers $X_n \cdot \Delta^n$ in $\mathbf{sSet}$, in which case the formula (4.1.7) returns the simplicial set X by the density theorem. Instead, we nearly always prefer to define $|X|$ to be the usual topological space – but this is really a minor point.

Exercise 4.1.8   Let $X \colon \mathbb{\Delta}^{\mathrm{op}} \to \mathbf{sSet}$ be a bisimplicial set. Compute $|X|$.

4.2 The bar construction

The two-sided bar construction is a fattened up version of the functor tensor product in a sense we suggest intuition for here and make precise when we introduce weighted colimits in Chapter 7. Let $\mathcal{M}$ be simplicially enriched, tensored, and cotensored. We define the two-sided bar construction associated to a small category $\mathcal{D}$ and to diagrams $G \colon \mathcal{D}^{\mathrm{op}} \to \mathbf{sSet}$ and $F \colon \mathcal{D} \to \mathcal{M}$.

The first step is to extend the diagram inside the coequalizer (4.1.1) to a simplicial object in $\mathcal{M}$.

Definition 4.2.1   The **two-sided simplicial bar construction** is a simplicial object $B_\bullet(G, \mathcal{D}, F)$ in $\mathcal{M}$ whose n-simplices are defined by the coproduct

$$B_n(G, \mathcal{D}, F) = \coprod_{\vec{d}\,\colon [n] \to \mathcal{D}} Gd_n \otimes Fd_0,$$

writing $\vec{d}$ as shorthand for a sequence $d_0 \to d_1 \to \cdots \to d_n$ of n composable arrows in $\mathcal{D}$.

The diagram $\vec{d}$ is exactly an n-simplex in the nerve of $\mathcal{D}$. The degeneracy and inner face maps in the simplicial object $B_\bullet(G, \mathcal{D}, F)$ reindex the coproducts according to the corresponding maps for the simplicial set $N\mathcal{D}$. The component of the nth face map $B_n(G, \mathcal{D}, F) \to B_{n-1}(G, \mathcal{D}, F)$ at $\vec{d}\colon [n] \to \mathcal{D}$ applies G to $d_{n-1} \to d_n$ to obtain a map $Gd_n \otimes Fd_0 \to Gd_{n-1} \otimes Fd_0$ and includes this at the component that is the restriction of $\vec{d}$ along $d^n\colon [n-1] \to [n]$. The 0th face map is defined similarly.

Example 4.2.2   When $\mathcal{V} = \mathcal{M} = \mathbf{Set}$ and F and G are the constant functors that send everything to the terminal object, $B_\bullet(*, \mathcal{D}, *)$ is the usual nerve $N\mathcal{D}$.

The colimit of the simplicial object $B_\bullet(G, \mathcal{D}, F)$ is the functor tensor product $G \otimes_\mathcal{D} F$, as can be deduced by inspecting the formula (4.1.1).[1] The two-sided bar construction is a "fattened up" colimit, more precisely, a weighted colimit, constructed from the Yoneda embedding $\Delta^\bullet\colon \Delta \to \mathbf{sSet}$ and the tensor structure on $\mathcal{M}$.

Definition 4.2.3   The **bar construction** is the geometric realization of the simplicial bar construction, that is,

$$B(G, \mathcal{D}, F) = |B_\bullet(G, \mathcal{D}, F)| = \Delta^\bullet \otimes_{\Delta^{\mathrm{op}}} B_\bullet(G, \mathcal{D}, F).$$

The unique maps $\Delta^n \to *$ assemble into a natural transformation $\Delta^\bullet \Rightarrow *$. Applying the functor $- \otimes_{\Delta^{\mathrm{op}}} B_\bullet(G, \mathcal{D}, F)$, this induces a map $B(G, \mathcal{D}, F) \to G \otimes_\mathcal{D} F$ from the geometric realization of the simplicial bar construction to its colimit, the functor tensor product. Such comparisons exist between weighted colimits and conical colimits for any cartesian monoidal category $\mathcal{V}$ (cf. Section 7.6).

Example 4.2.4   Regarding a group G as a one-object category, Example 4.2.2 specializes to produce a model $B(*, G, *)$ for the **classifying space** of the group. The classifying space, commonly denoted BG, is defined to be the geometric realization of the simplicial object $B_n(*, G, *) = G^n$. The degeneracy maps in the simplicial object

$$B_\bullet(*, G, *) = \quad \cdots \quad G \times G \times G \; \substack{\longrightarrow \\[-2pt] \longleftarrow \\[-2pt] \longrightarrow \\[-2pt] \longleftarrow \\[-2pt] \longrightarrow} \; G \times G \; \substack{\longrightarrow \\[-2pt] \longleftarrow \\[-2pt] \longrightarrow \\[-2pt] \longleftarrow} \; G \; \substack{\longrightarrow \\[-2pt] \longleftarrow \\[-2pt] \longrightarrow} \; *$$

insert identities. The inner face maps use multiplication in the group. The outer face maps project away from either the first or last copy of G.

Example 4.2.5   As a consequence of the "freeness" mentioned in 4.1.4, we are particularly interested in functor tensor products in the special case where G is a representable functor $\mathcal{D}(-, d)$. Using the Yoneda embedding, the objects $B(\mathcal{D}(-, d), \mathcal{D}, F) \in \mathcal{M}$ depend functorially on $d \in \mathcal{D}$. It is useful to have

---

[1] This has to do with the theory of **final functors** (cf. Example 8.3.8).

notation for the resulting functor, and we follow the common convention and write

$$B(\mathcal{D}, \mathcal{D}, F)\colon \mathcal{D} \to \mathcal{M} \qquad \text{for the functor} \qquad d \mapsto B(\mathcal{D}(-, d), \mathcal{D}, F).$$

Allowing F to vary, we obtain a functor $B(\mathcal{D}, \mathcal{D}, -)\colon \mathcal{M}^{\mathcal{D}} \to \mathcal{M}^{\mathcal{D}}$. We prove in Lemma 5.1.5 that if $\mathcal{M}$ is a simplicial model category, then $B(\mathcal{D}, \mathcal{D}, -)$ is equipped with a natural weak equivalence to the identity functor and hence forms a left deformation on the homotopical category $\mathcal{M}^{\mathcal{D}}$ with pointwise weak equivalences.

Exercise 4.2.6   Taking $\mathcal{M} = \mathbf{Set}$, show that $B_\bullet(\mathcal{D}(-, d), \mathcal{D}, *) \cong N(\mathcal{D}/d)$ so that $B_\bullet(\mathcal{D}, \mathcal{D}, *)$ is naturally isomorphic to the functor $N(\mathcal{D}/-)\colon \mathcal{D} \to \mathbf{sSet}$.

Dually, $B_\bullet(*, \mathcal{D}, \mathcal{D}(d, -)) \cong N(d/\mathcal{D})$, and hence $B_\bullet(*, \mathcal{D}, \mathcal{D}) \cong N(-/\mathcal{D})\colon \mathcal{D}^{\mathrm{op}} \to \mathbf{sSet}$. For example, take $\mathcal{D}$ to be a group G. Then $B(G, G, *)\colon G \to \mathbf{Top}$ is a space with a left G-action, commonly denoted EG, and $B(*, G, G)$ is a space with a right G-action. We will prove in 4.5.5 that these spaces are contractible.

4.3 The cobar construction

There are two ways one might attempt to dualize the bar construction. Replacing $\mathcal{D}$ with $\mathcal{D}^{\mathrm{op}}$ has no meaningful effect; replacing $\mathcal{M}$ with $\mathcal{M}^{\mathrm{op}}$ does. The resulting dual construction is called the **cobar construction**. Of course, citing duality, there is nothing more that needs to be said, but we give a few details nonetheless to ease this introduction.

Let $F, G\colon \mathcal{D} \rightrightarrows \mathcal{M}$. Recall from Exercise 1.2.8 that the set of natural transformations from G to F is computed by the end

$$\mathcal{M}^{\mathcal{D}}(G, F) \cong \int_{d \in \mathcal{D}} \mathcal{M}(Gd, Fd)$$

$$\cong \mathrm{eq}\left(\prod_{d \in \mathcal{D}} \mathcal{M}(Gd, Fd) \overset{f^*}{\underset{f_*}{\rightrightarrows}} \prod_{f\colon d \to d'} \mathcal{M}(Gd, Fd') \right).$$

In the presence of a bifunctor $\{-, -\}\colon \mathcal{V}^{\mathrm{op}} \times \mathcal{M} \to \mathcal{M}$, the **functor cotensor product** or sometimes **functor hom** (for lack of a better name) of $G\colon \mathcal{D} \to \mathcal{V}$ and $F\colon \mathcal{D} \to \mathcal{M}$ is the coend

$$\{G, F\}^{\mathcal{D}} := \int_{d \in \mathcal{D}} \{Gd, Fd\} \cong \mathrm{eq}\left(\prod_{d \in \mathcal{D}} \{Gd, Fd\} \overset{f^*}{\underset{f_*}{\rightrightarrows}} \prod_{f\colon d \to d'} \{Gd, Fd'\} \right).$$

Here it is common to use a cotensor, power, or hom for the bifunctor $\{-, -\}$, the last in the case that F and G have the same codomain. We generally prefer the exponential notation for this sort of bifunctor but here needed some place to put the superscript.

Example 4.3.1   If $\mathcal{M}$ is cotensored over **sSet**, the **totalization** of a cosimplicial object $X^{\bullet} \colon \Delta \to \mathcal{M}$,

$$\text{Tot } X^{\bullet} := \{\Delta^{\bullet}, X^{\bullet}\}^{\Delta},$$

is defined to be the functor cotensor of $X^{\bullet}$ and the Yoneda embedding $\Delta^{\bullet} \colon \Delta \to$ **sSet**.

The **cosimplicial cobar construction** is a "fattened" version of the functor cotensor product. Let $\mathcal{M}$ be complete and cotensored over $\mathcal{V}$. Let $\mathcal{D}$ be small, let $F \colon \mathcal{D} \to \mathcal{M}$, and let $G \colon \mathcal{D} \to \mathcal{V}$. Note that the variance of G parallels that of F, in contrast to the setting for the bar construction.

Definition 4.3.2   The **cosimplicial cobar construction** $C^{\bullet}(G, \mathcal{D}, F)$ is a cosimplicial object in $\mathcal{M}$, defined dually to the simplicial bar construction. Using exponential notation for the cotensor, the n-simplices are

$$C^n(G, \mathcal{D}, F) = \prod_{\vec{d} \colon [n] \to \mathcal{D}} F d_n^{G d_0}.$$

The cosimplicial degeneracy and inner face maps are induced by the corresponding maps of the nerve $N\mathcal{D}$. The dualization is due to the natural variance of the indexing sets of products.

The component of the 0th face map $C^{n-1}(G, \mathcal{D}, F) \to C^n(G, \mathcal{D}, F)$ landing in the component indexed by $\vec{d} \colon [n] \to \mathcal{D}$ uses the arrow $G d_0 \to G d_1$ in $\mathcal{V}$ to define a map $F d_n^{G d_1} \to F d_n^{G d_0}$, by contravariance of the first variable in the cotensor. The face map projects to the component indexed by the restriction of $\vec{d}$ along the 0th face map $[n-1] \to [n]$ and then composes with this map. The nth face map $C^{n-1}(G, \mathcal{D}, F) \to C^n(G, \mathcal{D}, F)$ is defined similarly; see Section 9.1 for more details.

Definition 4.3.3   The **cobar construction** $C(G, \mathcal{D}, F)$ is the totalization of the cosimplicial cobar construction, that is,

$$C(G, \mathcal{D}, F) := \{\Delta^{\bullet}, C^{\bullet}(G, \mathcal{D}, F)\}^{\Delta},$$

where $\Delta^{\bullet} \colon \Delta \to$ **sSet** is the Yoneda embedding.

4.4 Simplicial replacements and colimits

Following [79, §7], we explain the relevance of the bar construction to homotopy theory. Suppose we have a diagram $F : \mathcal{D} \to \mathcal{M}$. We know how to compute its colimit

$$\operatorname{colim}_{\mathcal{D}} F = \operatorname{coeq}\left(\coprod_{f : d \to d'} Fd \rightrightarrows \coprod_{d} Fd\right),$$

but we have seen that the functor $\operatorname{colim}_{\mathcal{D}}$ need not be homotopical. The hope, realized in Corollary 5.1.3, is that a fattened-up version of the colimit functor will preserve pointwise weak equivalence between diagrams.

If $\mathcal{D} = \Delta^{\mathrm{op}}$ and $\mathcal{M}$ is tensored over simplicial sets, we have some idea of what we might do. A fattened version of the colimit of a simplicial object is its geometric realization,[2] as defined in 3.8.1. So one idea would be to try to replace a generic diagram $F : \mathcal{D} \to \mathcal{M}$ by a simplicial object in $\mathcal{M}$ that has the same colimit.

To the small category $\mathcal{D}$ we can associate a simplicial set $N\mathcal{D}$. The data of a simplicial set X can in turn be encoded in a category $\mathbf{el}X$ in a way that retains its higher-dimensional information. This is called the **category of simplices**; the general framework for this construction, which explains the peculiar notation, is described in 7.1.9. The category $\mathbf{el}X$ has simplices $\sigma \in X_n$ as objects. A map from $\sigma \in X_n$ to $\sigma' \in X_m$ is a simplicial operator $\alpha : [n] \to [m]$ such that $\sigma' \cdot \alpha = \sigma$. Note there is a canonical forgetful functor $\mathbf{el}X \to \Delta$.

Example 4.4.1 Objects of $\mathbf{el}N\mathcal{D}$ are finite composable strings of morphisms of $\mathcal{D}$. Face morphisms between such strings either factor some of the arrows in the sequence or introduce new arrows on either end. Degeneracy morphisms forget identities. We denote the forgetful functor by $\Sigma : \mathbf{el}N\mathcal{D} \to \Delta$.

There is a canonical functor $S : \mathbf{el}N\mathcal{D} \to \mathcal{D}^{\mathrm{op}}$ that projects to the source of the string of arrows. To see the contravariance, note that the maps $d^0 : [n-1] \to [n]$ act on objects of $\mathbf{el}N\mathcal{D}$ by precomposing a string $\vec{d} : [n-1] \to \mathcal{D}$ with a new starting arrow $d_{-1} \to d_0$. This arrow defines a map from the source of the latter sequence to the source of the former.

Taking opposite categories, we have a diagram

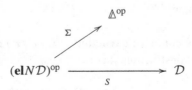

---

[2] Warning: the geometric realization of a simplicial object $X : \Delta^{\mathrm{op}} \to \mathcal{M}$ is not in general its homotopy colimit (cf. 7.7.2 and 14.3.10).

so we can get a simplicial object associated to $F : \mathcal{D} \to \mathcal{M}$ by precomposing with S and then taking the left Kan extension along Σ. But is there any reason why the colimit of $\mathrm{Lan}_\Sigma S^* F$ would have anything to do with the colimit of F? It turns out, yes.

We need two observations. The first is that the precomposition functor $S^* : \mathcal{M}^{\mathcal{D}} \to \mathcal{M}^{(\mathbf{el}N\mathcal{D})^{\mathrm{op}}}$ is full and faithful, which implies that the counit of the adjunction $\mathrm{Lan}_S \dashv S^*$ is an isomorphism $\mathrm{Lan}_S S^* F \xrightarrow{\cong} F$.

The second fact is a general observation about Kan extensions that we prove in even greater generality in 8.1.4. Given $F : \mathcal{D} \to \mathcal{M}$ and $K : \mathcal{D} \to \mathcal{E}$, by the defining universal properties of colimits and Kan extensions, we have a sequence of natural isomorphisms

$$\mathcal{M}(\mathrm{colim}_{\mathcal{E}} \mathrm{Lan}_K F, m) \cong \mathcal{M}^{\mathcal{E}}(\mathrm{Lan}_K F, \Delta m) \cong \mathcal{M}^{\mathcal{D}}(F, \Delta m) \cong \mathcal{M}(\mathrm{colim}_{\mathcal{D}} F, m),$$

where the Δm are constant functors taking values at $m \in \mathcal{M}$. Hence the colimit of a left Kan extension is the same as the colimit of the original functor. In particular,

$$\mathrm{colim}_{\mathcal{D}} F \cong \mathrm{colim}_{\mathcal{D}} \mathrm{Lan}_S S^* F \cong \mathrm{colim}_{(\mathbf{el}N\mathcal{D})^{\mathrm{op}}} S^* F \cong \mathrm{colim}_{\Delta^{\mathrm{op}}} \mathrm{Lan}_\Sigma S^* F.$$

How do we understand this simplicial object $\mathrm{Lan}_\Sigma S^* F$? By Remark 1.2.6,

$$(\mathrm{Lan}_\Sigma S^* F)_n = \mathrm{colim}\left(\Sigma/[n] \xrightarrow{U} (\mathbf{el}N\mathcal{D})^{\mathrm{op}} \xrightarrow{S} \mathcal{D} \xrightarrow{F} \mathcal{M} \right).$$

The slice category $\Sigma/[n]$ is the disjoint union of categories, each of which has a terminal object given by some sequence $\vec{d} = d_0 \to \cdots \to d_n$ together with the identity map $\Sigma \vec{d} \to [n]$. It follows that the colimit is just the coproduct of the images of these objects (cf. 8.3.7). Hence

$$(\mathrm{Lan}_\Sigma S^* F)_n = \coprod_{\vec{d} : [n] \to \mathcal{D}} F d_0 = B_n(*, \mathcal{D}, F),$$

which we recognize as a special case of the two-sided bar construction.

In summary, we have proven:

Lemma 4.4.2 *We can replace a diagram $F : \mathcal{D} \to \mathcal{M}$ by a simplicial object $B_\bullet(*, \mathcal{D}, F)$, or dually by a cosimplicial object $C^\bullet(*, \mathcal{D}, F)$, so that*

$$\mathrm{colim}_{\mathcal{D}} F \cong \mathrm{colim}_{\Delta^{\mathrm{op}}} B_\bullet(*, \mathcal{D}, F) \qquad \text{and} \qquad \lim_{\mathcal{D}} F \cong \lim_{\Delta} C^\bullet(*, \mathcal{D}, F).$$

Before closing, we should note that the proof that the colimit of a left Kan extension is isomorphic to the colimit of the original diagram extends to weighted colimits, in particular to functor tensor products, and in particular to geometric realizations. We state this application as a lemma, which has obvious implications for computing bar constructions.

Write $\Delta^\bullet \colon \mathbb{\Delta} \to \mathbf{sSet}$ for the Yoneda embedding, and suppose that $\mathcal{M}$ is cocomplete and tensored over simplicial sets. Recall the following definition from 1.1.9: a simplicial object $X_\bullet \colon \mathbb{\Delta}^{\mathrm{op}} \to \mathcal{M}$ is n-**skeletal** if it is isomorphic to the left Kan extension of its n-truncation $X_{\leq n}$:

$$\mathcal{M}^{\mathbb{\Delta}^{\mathrm{op}}} \underset{\mathrm{res}_{\leq n}}{\overset{\mathrm{Lan}}{\underset{\perp}{\rightleftarrows}}} \mathcal{M}^{\mathbb{\Delta}^{\mathrm{op}}_{\leq n}}$$

Lemma 4.4.3 *If X is n-skeletal, then*

$$|X| \cong \Delta^\bullet \otimes_{\mathbb{\Delta}^{\mathrm{op}}} X_\bullet \cong \Delta^\bullet \otimes_{\mathbb{\Delta}^{\mathrm{op}}} \mathrm{Lan}\, X_{\leq n} \cong \Delta^\bullet_{\leq n} \otimes_{\mathbb{\Delta}^{\mathrm{op}}_{\leq n}} X_{\leq n}.$$

In words, $|X|$ is isomorphic to the functor tensor product of its n-truncation $X_{\leq n}$ with the restricted Yoneda embedding $\Delta^\bullet_{\leq n} \colon \mathbb{\Delta}_{\leq n} \to \mathbf{Set}^{\mathbb{\Delta}^{\mathrm{op}}_{\leq n}}$.

Proof The proof is left as an exercise for the reader (or see 8.1.4). □

4.5 Augmented simplicial objects and extra degeneracies

We close this chapter by describing a technique for computing the homotopy type of the geometric realization of a simplicial object in a tensored, cotensored, and simplicially enriched category $\mathcal{M}$. We conclude with some examples illustrating its application.

Recall that an augmented simplicial object is a presheaf on the category $\mathbb{\Delta}_+$ of finite ordinals and order-preserving maps.[3] Equivalently, it is a simplicial object X together with an object X_{-1} and a map $d_0 \colon X_0 \to X_{-1}$ such that the two maps $X_1 \rightrightarrows X_0 \to X_{-1}$ agree. Equivalently, an augmentation is specified by a map of simplicial objects $X \to X_{-1}$ whose target is the constant simplicial object on X_{-1}.

In a complete category $\mathcal{M}$, any simplicial object can be augmented by the unique map to the terminal object $*$. It is natural to ask when $X \to *$ is a simplicial homotopy equivalence. The answer presented subsequently makes sense for any category $\mathcal{M}$, though our exploratory discussion, following [18], will tacitly suppose $\mathcal{M} = \mathbf{Set}$.

The intuition is that a contracting homotopy $X \to *$ provides a method to deform each simplex of X down to a point. This data could be given by

- a vertex $* \in X_0$
- for each $x \in X_0$, a 1-simplex from $*$ to x

---

[3] Here are some fun facts: the category $\mathbb{\Delta}_+$ is the free monoidal category (the multiplication given by ordinal sum) containing a monoid (the object $[0]$). Its opposite $\mathbb{\Delta}^{\mathrm{op}}_+$, by duality the free monoidal category containing a comonoid, is also the category of "finite intervals," i.e., finite totally ordered sets with a designated "top" and "bottom" element and maps that preserve these.

- for each $f \in X_1$, a 2-simplex whose initial vertex is $*$, whose 0th face is f, and whose other faces are the previously specified 1-simplices
- for each $\sigma \in X_n$, an $n + 1$-simplex whose initial vertex is $*$, whose 0th face is σ, and whose other faces are the previously specified n-simplices

In other words, a **contracting homotopy** consists of maps $s_{-1} : X_n \to X_{n+1}$ for all $n \geq -1$ that are sections of the face maps $d_0 : X_{n+1} \to X_n$ and satisfy the other simplicial identities that this labeling would suggest [32, III.5].

We can extend this definition to generic augmented simplicial objects by requiring the map $s_{-1} : X_{-1} \to X_0$ to be a section of the augmentation $d_0 : X_0 \to X_{-1}$. The data of the maps s_{-1} might be called a "backward" contracting homotopy; a "forward" contracting homotopy is obtained by replacing the simplicial object X with its opposite, obtained by precomposing with the functor $(-)^{\mathrm{op}} : \Delta \to \Delta$, which reverses the ordering of elements in each set $[n]$. The maps defining the data of a forward contracting homotopy are commonly labelled $s_{n+1} : X_n \to X_{n+1}$.

Contracting homotopies of augmented simplicial objects are sometimes called **extra degeneracies**. The following classical result uses the tensor structure of 3.7.18 to define the notion of simplicial homotopy equivalence.

Lemma 4.5.1 ([62, §6]) *Specifying an augmentation and extra degeneracies for a simplicial object X in a cocomplete category $\mathcal{M}$ is equivalent to specifying an object $X_{-1} \in \mathcal{M}$ and a retract diagram*

$$X_{-1} \overset{s_{-1}}{\to} X \overset{d_0}{\to} X_{-1}$$

whose maps define a simplicial homotopy equivalence between X and the constant simplicial object X_{-1}.

If $\mathcal{M}$ is a tensored, cotensored, and simplicially enriched category, then Corollary 3.8.4 implies that $|X| \to X_{-1}$ is a simplicial homotopy equivalence. If $\mathcal{M}$ is a simplicial model category, 3.8.6.(vi) implies that this map is a weak equivalence. For later reference, we record this corollary.

Corollary 4.5.2 *If an augmented simplicial object $X \to X_{-1}$ in a simplicial model category admits either a forward or backward contracting homotopy, then the natural map $|X| \to X_{-1}$ is a weak equivalence.*

Remark 4.5.3 Exercise 3.7.18 and dual versions of Lemma 3.8.3 and Corollary 3.8.4 imply the dual of Corollary 4.5.2: if a cosimplicial object in a simplicial model category admits an augmentation and a contracting homotopy, then the natural map from the augmentation to its totalization is a simplicial homotopy equivalence and hence a weak equivalence.

Remark 4.5.4   The geometric realization of a simplicial object is not always its homotopy colimit (cf. Remark 5.2.4). However, it is also the case that the homotopy colimit of a simplicial object admitting an augmentation and a contracting homotopy is weakly equivalent to its augmentation. This is proven in Exercise 8.5.14.

Example 4.5.5   Let G be a group and write $EG = B(G, G, *)$. The simplicial set

$$B_\bullet(G, G, *) = \quad \cdots \quad G \times G \times G \times G \rightrightarrows G \times G \times G \rightrightarrows G \times G \rightrightarrows G$$

is augmented by the one-point set $*$ and admits a backward contracting homotopy that inserts identities into the leftmost group in each product. Hence $EG = |B_\bullet(G, G, *)|$ is a contractible G-space. Another proof of contractibility is given in Example 8.5.4.

Example 4.5.6   Suppose a small category $\mathcal{D}$ has a terminal object t. Then $B_\bullet(*, \mathcal{D}, F)$ admits a natural augmentation $\coprod_{d \in \mathcal{D}} Fd \to Ft$ and extra degeneracy: the map

$$\coprod_{\vec{d}: [n] \to \mathcal{D}} Fd_0 \to \coprod_{\vec{d}': [n+1] \to \mathcal{D}} Fd_0'$$

uses the terminal object to extend a sequence of composable arrows $d_0 \to \cdots \to d_n \to t$. Hence $B(*, \mathcal{D}, F)$ is homotopy equivalent to Ft, the image of the terminal object. Recall that $B(*, \mathcal{D}, F)$ is a thickened version of the functor tensor product $* \otimes_\mathcal{D} F \cong \mathrm{colim}_\mathcal{D} F$. In this case, the geometric realization of the simplicial object $B_\bullet(*, \mathcal{D}, F)$ is homotopy equivalent to the ordinary colimit, computed by evaluating at the terminal object by any of the proofs given for Lemma 8.3.1.

A final example will be a key step in the proofs of the next section.

Example 4.5.7   Recall our shorthand notation

$$B(\mathcal{D}, \mathcal{D}, F): \mathcal{D} \to \mathcal{M} \quad \text{for} \quad d \mapsto B(\mathcal{D}(-, d), \mathcal{D}, F).$$

Generalizing Example 4.5.5, the simplicial object $B_\bullet(\mathcal{D}(-, d), \mathcal{D}, F)$ admits an augmentation and extra degeneracy. The augmentation $\coprod_{d'} \mathcal{D}(d', d) \times Fd' \to Fd$ is by composition and is natural in $d \in \mathcal{D}$. The extra degeneracies, which insert the identity at 1_d, are not natural in $d \in \mathcal{D}$. Nonetheless, by Lemma 4.5.1, the natural map $\epsilon: B(\mathcal{D}, \mathcal{D}, F) \Rightarrow F$ produced by the augmentation admits pointwise homotopy inverses; that is, each component ϵ_d is a simplicial homotopy equivalence.

5

Homotopy limits and colimits: The theory

For a fixed small category $\mathcal{D}$ and homotopical category $\mathcal{M}$, the homotopy colimit functor should be a derived functor of colim: $\mathcal{M}^{\mathcal{D}} \to \mathcal{M}$. Assuming that the homotopical category $\mathcal{M}$ is saturated, which it always is in practice, the homotopy type of a homotopy colimit will not depend on which flavor of derived functor is used. Because colimits are left adjoints, we might hope that colim has a left derived functor and, dually, that lim: $\mathcal{M}^{\mathcal{D}} \to \mathcal{M}$ has a right derived functor, defining homotopy limits.

For special types of diagrams (e.g., if $\mathcal{D}$ is a Reedy category), there are simple modifications that produce the right answer. For instance, homotopy pullbacks can be computed by replacing the maps by fibrations between fibrant objects. Or if the homotopical category $\mathcal{M}$ is a model category of a particular sort (cofibrantly generated for colimits or combinatorial for limits), then there exist suitable model structures (the projective and injective, respectively) on $\mathcal{M}^{\mathcal{D}}$ for which Quillen's small object argument can be used to produce deformations for the colimit and limit functors.

These well-documented solutions are likely familiar to those acquainted with model category theory, but they are either quite specialized (those depending on the particular diagram shape) or computationally difficult (those involving the small object argument). By contrast, the theory of derived functors developed in Chapter 2 does not require the presence of model structures on the diagram categories.

In this chapter, we make use of Theorem 2.2.8 to produce point-set derived functors $\mathbb{L}\,\mathrm{colim}, \mathbb{R}\,\mathrm{lim}: \mathcal{M}^{\mathcal{D}} \rightrightarrows \mathcal{M}$ for any small category $\mathcal{D}$ and simplicial model category $\mathcal{M}$ using deformations constructed via the bar and cobar constructions. The hypotheses on $\mathcal{M}$ are more or less unavoidable: although we will not make explicit use of the enrichment, the associated tensor and cotensor structures are essential to the definition of the bar and cobar constructions. The other properties of a simplicial model category, as axiomatized by

Lemma 3.8.6, ensure that the two-sided (co)bar constructions are homotopically well behaved.

This short chapter contains only the statements and proofs of these results. We postpone the discussion of examples to Chapter 6, where we also describe ambient simplicial model categories of interest. Later, in Section 7.7, we see that the homotopy (co)limits constructed here also satisfy a "local" universal property: the homotopy colimit of $F: \mathcal{D} \to \mathcal{M}$ represents a particular simplicial functor of "homotopy coherent cones" under F.

5.1 The homotopy limit and colimit functors

In what follows, we agree to write hocolim: $\mathcal{M}^{\mathcal{D}} \to \mathcal{M}$ for the left derived functor of colim and holim: $\mathcal{M}^{\mathcal{D}} \to \mathcal{M}$ for the right derived functor of lim. Our use of "the" here is misleading: derived functors are not uniquely defined on the point-set level. Nonetheless, it will be convenient to fix a particular construction we will call "the" homotopy (co)limit. Colloquial statements of the form "the homotopy (co)limit is isomorphic to" mean that this particular construction agrees with something or other.

The main goal of this chapter is to show that homotopy colimits and limits can be computed in a simplicial model category by means of a deformation formed using the bar and cobar constructions. This result is a corollary of the following theorem.

Theorem 5.1.1 *Let $\mathcal{M}$ be a simplicial model category with cofibrant replacement Q and fibrant replacement R. The pair*

$$B(\mathcal{D}, \mathcal{D}, Q-): \mathcal{M}^{\mathcal{D}} \to \mathcal{M}^{\mathcal{D}} \qquad B(\mathcal{D}, \mathcal{D}, Q-) \overset{\epsilon_Q}{\Rightarrow} Q \overset{q}{\Rightarrow} 1 \qquad (5.1.2)$$

is a left deformation for colim: $\mathcal{M}^{\mathcal{D}} \to \mathcal{M}$. *Dually,*

$$C(\mathcal{D}, \mathcal{D}, R-): \mathcal{M}^{\mathcal{D}} \to \mathcal{M}^{\mathcal{D}} \qquad 1 \overset{r}{\Rightarrow} R \overset{\eta_R}{\Rightarrow} C(\mathcal{D}, \mathcal{D}, R-)$$

is a right deformation for lim: $\mathcal{M}^{\mathcal{D}} \to \mathcal{M}$.

Corollary 5.1.3 *If $\mathcal{M}$ is a simplicial model category and $\mathcal{D}$ is any small category, then the functors* colim, lim: $\mathcal{M}^{\mathcal{D}} \to \mathcal{M}$ *admit left and right derived functors, which we call* hocolim *and* holim, *defined by*

$$\text{hocolim}_{\mathcal{D}} := \mathbb{L}\,\text{colim}_{\mathcal{D}} \cong B(*, \mathcal{D}, Q-) \quad and$$

$$\text{holim}_{\mathcal{D}} := \mathbb{R}\,\text{lim}_{\mathcal{D}} \cong C(*, \mathcal{D}, R-).$$

Before proving Theorem 5.1.1, let us prove its corollary, after first revisiting the functor tensor product calculations of Examples 4.1.3 and 4.1.4. Recall that $* \otimes_{\mathcal{D}} - : \mathcal{M}^{\mathcal{D}} \to \mathcal{M}$ is isomorphic to the colimit functor and $\mathcal{D}(-, d) \otimes_{\mathcal{D}} - : \mathcal{M}^{\mathcal{D}} \to \mathcal{M}$ is isomorphic to the functor given by evaluation at d. It follows that $\mathcal{D} \otimes_{\mathcal{D}} - : \mathcal{M}^{\mathcal{D}} \to \mathcal{M}^{\mathcal{D}}$ is naturally isomorphic to the identity functor.

Proof of Corollary 5.1.3   By Theorem 5.1.1 and Theorem 2.2.8,

$$\mathbb{L} \operatorname{colim}_{\mathcal{D}}(-) \cong \operatorname*{colim}_{\mathcal{D}} B(\mathcal{D}, \mathcal{D}, Q-) \cong * \otimes_{\mathcal{D}} B(\mathcal{D}, \mathcal{D}, Q-). \qquad (5.1.4)$$

By "Fubini's theorem," iterated coends commute past each other, provided that all the involved colimits exist [50, §IX.8]. This should be regarded as a particular instance of the fact that colimits commute with each other. It follows that functor tensor products commute past either variable of the two-sided bar construction. Hence

$$* \otimes_{\mathcal{D}} B(\mathcal{D}, \mathcal{D}, Q-) \cong B(* \otimes_{\mathcal{D}} \mathcal{D}, \mathcal{D}, Q-) \cong B(*, \mathcal{D}, Q-),$$

the last isomorphism because evaluating a constant functor produces a constant functor. Hence the right-hand side of (5.1.4) is $B(*, \mathcal{D}, Q-)$, as desired.

Dually,

$$\mathbb{R} \lim_{\mathcal{D}}(-) \cong \lim_{\mathcal{D}} C(\mathcal{D}, \mathcal{D}, R-) \cong \{*, C(\mathcal{D}, \mathcal{D}, R-)\}^{\mathcal{D}}.$$

The functor cotensor product is an end, which commutes past the limits defining the cobar construction. But the cobar construction is contravariant in the first variable; hence the limit in **sSet**^{op} is a colimit in **sSet**. Therefore

$$\{*, C(\mathcal{D}, \mathcal{D}, R-)\}^{\mathcal{D}} \cong C(* \otimes_{\mathcal{D}} \mathcal{D}, \mathcal{D}, R-) \cong C(*, \mathcal{D}, R-),$$

as desired. $\square$

Until presenting examples, we do not discuss homotopy limits further as the proofs are entirely dual. The proof of Theorem 5.1.1 occupies the remainder of this section. The first step is to recall that the map $\epsilon : B(\mathcal{D}, \mathcal{D}, F) \Rightarrow F$, as a geometric realization of a pointwise simplicial homotopy equivalence, is itself a pointwise simplicial homotopy equivalence by Corollary 3.8.4. Hence it is a pointwise weak equivalence by 3.8.6.(vi), and it follows that the two-sided bar construction (5.1.2) defines a left deformation on $\mathcal{M}^{\mathcal{D}}$ in the sense of Definition 2.2.1. We record this in the following lemma:

Lemma 5.1.5   *The natural weak equivalence $\epsilon : B(\mathcal{D}, \mathcal{D}, F) \Rightarrow F$ makes $B(\mathcal{D}, \mathcal{D}, -)$ a left deformation for $\mathcal{M}^{\mathcal{D}}$.*

The deformation (5.1.2) is obtained by composing ϵ of the lemma with the left deformation $q : Q \Rightarrow 1$ to the cofibrant objects of the simplicial model category. This certainly defines a left deformation on $\mathcal{M}^{\mathcal{D}}$. It remains to show that (5.1.2) defines a left deformation for the colimit functor, that is, that colim preserves pointwise weak equivalences between diagrams in a full subcategory of $\mathcal{M}^{\mathcal{D}}$ containing the image of $B(\mathcal{D}, \mathcal{D}, Q-)$. The general strategy is presented in the following lemma:

Lemma 5.1.6   *Suppose (Q, q) is a left deformation on $\mathcal{M}$ and $F : \mathcal{M} \to \mathcal{N}$. If*

- *FQ is homotopical*
- *$FqQ : FQ^2 \Rightarrow FQ$ is a natural weak equivalence*

then (Q, q) is a left deformation for F. In other words, F is homotopical on a full subcategory of cofibrant objects if F preserves weak equivalences in the image of Q and also preserves the weak equivalences q_Q.

Proof   This is an easy consequence of the 2-of-3 property. $\square$

By Lemma 5.1.6, it suffices to show that $B(*, \mathcal{D}, Q-) \cong \operatorname{colim}_{\mathcal{D}} B(\mathcal{D}, \mathcal{D}, Q-)$ is homotopical and that the following composite is a natural weak equivalence:

$$\operatorname*{colim}_{\mathcal{D}} B(\mathcal{D}, \mathcal{D}, QB(\mathcal{D}, \mathcal{D}, QF)) \xrightarrow{\operatorname{colim}_{\mathcal{D}} \epsilon} \operatorname*{colim}_{\mathcal{D}} QB(\mathcal{D}, \mathcal{D}, QF)$$

$$\xrightarrow{\operatorname{colim}_{\mathcal{D}} q} \operatorname*{colim}_{\mathcal{D}} B(\mathcal{D}, \mathcal{D}, QF). \quad (5.1.7)$$

Both steps require that we explore the homotopical aspects of the bar construction.

5.2 Homotopical aspects of the bar construction

The good homotopical properties of the two-sided bar construction hinge on the fact that the simplicial bar construction $B_\bullet(G, \mathcal{D}, F)$ is **Reedy cofibrant** when G and F are pointwise cofibrant. Readers conversant in the language of model categories might wish to skip directly to Chapter 14, where this notion is defined. For present purposes, we content ourselves with a conceptual discussion that highlights the main ideas.

In a simplicial model category, there is a class of maps called the **cofibrations**, which have special homotopical properties; an object is **cofibrant**, here meant in the technical sense, just when the unique map from the initial object is a cofibration. A **Reedy cofibrant** simplicial object is a simplicial object in which the inclusion, called the **latching map**, from the degenerate n-simplices, the nth **latching object**, into all n-simplices is a cofibration.

Lemma 5.2.1   *Let $\mathcal{D}$ be a small category and let $\mathcal{M}$ be a simplicial model category. If $F : \mathcal{D} \to \mathcal{M}$ is pointwise cofibrant, then $B_\bullet(*, \mathcal{D}, F)$ is Reedy cofibrant.*

Lemma 5.2.1 explains the role of the functor Q in the left deformation $B(\mathcal{D}, \mathcal{D}, Q-)$. Postcomposition by Q yields a pointwise cofibrant replacement of our diagram F. We cannot resist giving the proof now because it is so simple.

Proof of Lemma 5.2.1   Recall that $B_n(*, \mathcal{D}, F)$ is the coproduct over $N\mathcal{D}_n$ of the image under F of the first object in the sequence of composable arrows. Hence the nth latching object sits inside $B_n(*, \mathcal{D}, F)$ as the coproduct indexed by degenerate simplices in $N\mathcal{D}_n$. By Lemma 11.1.4, the cofibrations, and hence the cofibrant objects, are closed under coproducts. By 11.1.4 again, cofibrations are closed under pushout, and hence coproduct inclusions of cofibrant objects are also cofibrations. By hypothesis, every object in these coproducts is cofibrant, so the preceding observations imply that the nth latching map is a cofibration.   □

Remark 5.2.2   Recall from 3.8.6.(iii) that tensoring with any simplicial set preserves cofibrant objects. Thus the argument just given also shows that $B_\bullet(G, \mathcal{D}, F)$ is Reedy cofibrant for any pointwise cofibrant F taking values in a simplicial model category and any functor $G : \mathcal{D}^{\mathrm{op}} \to \mathbf{sSet}$.

Our proof of Theorem 5.1.1 relies on one major result, a consequence of a standard theorem from Reedy category theory, which we prove as Corollary 14.3.10.

Theorem 5.2.3 ([32, VII.3.6], [36, 18.4.11])   *If $\mathcal{M}$ is a simplicial model category, then*

$$| - | : \mathcal{M}^{\Delta^{\mathrm{op}}} \to \mathcal{M}$$

is left Quillen with respect to the Reedy model structure. In particular, $| - |$ sends Reedy cofibrant simplicial objects to cofibrant objects and preserves pointwise weak equivalences between them.

Remark 5.2.4   At this level of generality, this is the strongest result possible: it is not true that geometric realization preserves all pointwise equivalences (cf. Example 2.2.5). This is why we warned earlier that geometric realization is not a homotopy colimit in all contexts. It is not even homotopical!

Combining 5.2.2 and 5.2.3, we conclude that pointwise cofibrant replacement defines a left deformation for the two-sided bar construction:

Corollary 5.2.5   *The functor $B(G, \mathcal{D}, -)$ preserves weak equivalences between pointwise cofibrant objects. In particular $B(*, \mathcal{D}, Q-)$ is homotopical. Dually, when F is pointwise cofibrant, $B(-, \mathcal{D}, F)$ preserves weak equivalences.*

It remains only to show that the natural map (5.1.7) is a weak equivalence. By naturality of ϵ,

$$
\begin{array}{ccc}
B(\mathcal{D}, \mathcal{D}, QB(\mathcal{D}, \mathcal{D}, QF)) & \xrightarrow{\ \epsilon_{QB}\ } & QB(\mathcal{D}, \mathcal{D}, QF) \\
{\scriptstyle B(\mathcal{D}, \mathcal{D}, q)}\Big\downarrow & & \Big\downarrow{\scriptstyle q} \\
B(\mathcal{D}, \mathcal{D}, B(\mathcal{D}, \mathcal{D}, QF)) & \xrightarrow[\ \epsilon_B\ ]{} & B(\mathcal{D}, \mathcal{D}, QF)
\end{array}
$$

commutes. By 5.2.2, $B_\bullet(\mathcal{D}, \mathcal{D}, QF): \mathcal{D} \to \mathcal{M}^{\Delta^{op}}$ is pointwise Reedy cofibrant; hence, by Theorem 5.2.3, $B(\mathcal{D}, \mathcal{D}, QF)$ is pointwise cofibrant. Applying $\operatorname{colim}_{\mathcal{D}}$ to the left-hand vertical map, we get $B(*, \mathcal{D}, q): B(*, \mathcal{D}, QB(\mathcal{D}, \mathcal{D}, QF)) \to B(*, \mathcal{D}, B(\mathcal{D}, \mathcal{D}, F))$, which is a weak equivalence by Corollary 5.2.5, which says that $B(*, \mathcal{D}, -)$ preserves weak equivalences between pointwise cofibrant diagrams. Hence it remains only to show that the image of the bottom map under $\operatorname{colim}_{\mathcal{D}}$ is a weak equivalence. This is accomplished by the following lemma, which is in some ways the crux of the argument:

Lemma 5.2.6   *When F is pointwise cofibrant, the functor $\operatorname{colim}_{\mathcal{D}}$ preserves the weak equivalence*

$$
\epsilon_{B(\mathcal{D}, \mathcal{D}, F)}: B(\mathcal{D}, \mathcal{D}, B(\mathcal{D}, \mathcal{D}, F)) \to B(\mathcal{D}, \mathcal{D}, F).
$$

Proof   Recall that $\epsilon_F: B(\mathcal{D}, \mathcal{D}, F) \to F$ is the geometric realization of a map between simplicial objects with components at $d \in \mathcal{D}$ determined by the maps

$$
(\mathcal{D}(d_n, d) \times \mathcal{D}(d_{n-1}, d_n) \times \cdots \times \mathcal{D}(d_0, d_1)) \cdot Fd_0 \longrightarrow Fd.
$$

These maps compose the arrows in $\mathcal{D}$ and then evaluate the functor at the composite arrow $d_0 \to d$. Using the fact that functor tensor products, as colimits, commute with both variables of the two-sided bar construction, ϵ_F can be expressed as

$$
B(\mathcal{D}, \mathcal{D}, F) \cong B(\mathcal{D}, \mathcal{D}, \mathcal{D} \otimes_{\mathcal{D}} F) \cong B(\mathcal{D}, \mathcal{D}, \mathcal{D}) \otimes_{\mathcal{D}} F \xrightarrow{\epsilon \otimes_{\mathcal{D}} 1} \mathcal{D} \otimes_{\mathcal{D}} F \cong F,
$$

where this ϵ is the geometric realization of the map of simplicial objects

$$
\epsilon: B_\bullet(\mathcal{D}(-, d), \mathcal{D}, \mathcal{D}(d', -)) \to \mathcal{D}(d', d), \tag{5.2.7}
$$

defined by composing the arrows in $\mathcal{D}$.

Importantly, the augmented simplicial object (5.2.7) admits both backward and forward contracting homotopies. The forward contracting homotopy was used to show that $\epsilon \otimes_{\mathcal{D}} F \colon B(\mathcal{D}, \mathcal{D}, F) \Rightarrow F$ is a natural weak equivalence. Dually, the backward contracting homotopy can be used to show that $G \otimes_{\mathcal{D}} \epsilon \colon B(G, \mathcal{D}, \mathcal{D}) \Rightarrow G$ is a natural weak equivalence.

The argument is now completed by the following commutative diagram, which expresses the "associativity" of the two-sided bar construction, in the sense of natural isomorphisms $B(G, \mathcal{D}, B(\mathcal{D}, \mathcal{D}, F)) \cong B(B(G, \mathcal{D}, \mathcal{D}), \mathcal{D}, F)$. Our map $\mathrm{colim}_{\mathcal{D}}\, \epsilon_B$ is the top map in the following diagram, whose vertical isomorphisms depend on the co-Yoneda lemma and the fact that both variables of the two-sided bar construction commute with colimits:

$$
\begin{array}{ccc}
* \otimes_{\mathcal{D}} B(\mathcal{D}, \mathcal{D}, B(\mathcal{D}, \mathcal{D}, F)) & \xrightarrow{\;*\otimes_{\mathcal{D}}\epsilon\;} & * \otimes_{\mathcal{D}} B(\mathcal{D}, \mathcal{D}, F) \\
\cong \downarrow & & \downarrow \cong \\
* \otimes_{\mathcal{D}} B(\mathcal{D}, \mathcal{D}, \mathcal{D}) \otimes_{\mathcal{D}} B(\mathcal{D}, \mathcal{D}, F) & \xrightarrow{\;*\otimes_{\mathcal{D}}\epsilon\otimes_{\mathcal{D}}B(\mathcal{D},\mathcal{D},F)\;} & * \otimes_{\mathcal{D}} \mathcal{D} \otimes_{\mathcal{D}} B(\mathcal{D}, \mathcal{D}, F) \\
\cong \downarrow & & \downarrow \cong \\
B(*, \mathcal{D}, \mathcal{D}) \otimes_{\mathcal{D}} B(\mathcal{D}, \mathcal{D}, F) & \xrightarrow{\;\epsilon\otimes_{\mathcal{D}}B(\mathcal{D},\mathcal{D},F)\;} & * \otimes_{\mathcal{D}} B(\mathcal{D}, \mathcal{D}, F) \\
\cong \downarrow & & \downarrow \cong \\
B(B(*, \mathcal{D}, \mathcal{D}), \mathcal{D}, F) & \xrightarrow{\;B(\epsilon,\mathcal{D},F)\;} & B(*, \mathcal{D}, F)
\end{array}
$$

Using the backward contracting homotopy mentioned earlier, the map $\epsilon \colon B(*, \mathcal{D}, \mathcal{D}) \to *$ is a weak equivalence and hence, by Corollary 5.2.5, is preserved by the functor $B(-, \mathcal{D}, F)$. Thus the bottom map is a weak equivalence. The result now follows from the 2-of-3 property. $\qquad\square$

6

Homotopy limits and colimits: The practice

Now that we have a general formula for homotopy limit and colimit functors in any simplicial model category, we should take a moment to see what these objects look like in particular examples, in particular, for traditional topological spaces. This demands that we delve into a topic that is perhaps overdue. In addition to the category of simplicial sets, several choices are available for a simplicial model category of spaces. In Section 6.1, we discuss the point-set topological considerations that support the definitions of two of the aforementioned convenient categories of spaces: k-**spaces** and **compactly generated spaces**.[1] In Section 6.2, we use these results to list several simplicial model categories of spaces, paying particular attention to their fibrant and cofibrant objects, which feature in the construction of homotopy limits and colimits. The short Section 6.3 concludes this background segment with some important preparatory remarks.

Finally, in Sections 6.4 and 6.5, we turn to examples, describing the spaces produced by the formulae of Corollary 5.1.3 and considering what simplified models might be available in certain cases. A particularly intuitive presentation of the theory of homotopy limits and colimits, including several of the examples discussed in those sections, can be found in the unfinished yet extremely clear notes on this topic by [18]. We close this chapter in Section 6.6 with a preview of Part II. We show that our preferred formulae for homotopy limits and colimits are *isomorphic* to certain functor cotensor and functor tensor products. We give an immediate application that illustrates the power of this perspective, proving that the homotopy colimit of a diagram of based spaces has a certain relationship to the homotopy colimit of the underlying unbased diagram.

---

[1] Note that neither of these models is the one described in the paper [81], whose title introduced the phrase "A convenient category of topological spaces."

6.1 Convenient categories of spaces

For this section only, let us revert to the old naïve notation and write **Top** for the category of all topological spaces and continuous maps. At issue is that the category **Top** of all spaces is not particularly well behaved categorically. The following results illustrate some pathologies:

Theorem 6.1.1 ([9, 7.1.1-2]) *The product, unit, and internal hom for any symmetric monoidal closed structure on the category* **Top** *necessarily have the cartesian product, one-point set, and set of all continuous functions as underlying sets. But there is no such structure in which the monoidal product is the cartesian product of spaces. Hence* **Top** *is not cartesian closed.*

In other words, it is not possible to topologize the set of continuous maps from X to Y in such a way that there exists a continuous "evaluation" function

$$\mathbf{Top}(X, Y) \times X \to Y$$

satisfying the desired universal property: $\mathbf{Top}(X, Y)$ should be terminal in the category of spaces W equipped with a continuous map $W \times X \to Y$. For pathological spaces X, the problem, familiar from undergraduate point-set topology, is that the functor $- \times X \colon \mathbf{Top} \to \mathbf{Top}$ does not preserve all colimits [63, §22 Example 7].

Remark 6.1.2 By a theorem of Maria Cristina Pedicchio and Sergio Solimini, there is a unique closed symmetric monoidal structure on **Top** [64]; the underlying sets of the product, unit, and hom-spaces are determined by Theorem 6.1.1. This closed monoidal structure topologizes the hom-spaces using the topology of pointwise convergence: a subbasis is given by sets of functions that take a specified point in the domain into a specified open subset of the codomain [9, 7.1.6]. Note that this is not the most useful topology on hom-spaces: even for metric spaces, a sequence of continuous functions that converges pointwise need not have a continuous limit. The monoidal product $X \otimes Y$, whose underlying set is the cartesian product $X \times Y$, is topologized so that a continuous function $X \otimes Y \to Z$ is exactly a function $X \times Y \to Z$ that is separately continuous in each variable.

For based spaces, the situation is even worse. From our experience with CW complexes, we might hope that an internal hom would be right adjoint to the smash product. However:

Lemma 6.1.3 (Kathleen Lewis [59, 1.7.1]) *The smash product in the category* **Top**$_*$ *of based spaces and basepoint-preserving maps is not associative. In particular, if $\mathbb{Q}$ and $\mathbb{N}$ are topologized as subspaces of the real line with basepoint 0, then $(\mathbb{Q} \wedge \mathbb{Q}) \wedge \mathbb{N}$ and $\mathbb{Q} \wedge (\mathbb{Q} \wedge \mathbb{N})$ are not homeomorphic.*

We suggest two alternatives to **Top**, which require the following definitions from point-set topology. Historically, the terminology has been somewhat variable. Our terminology and notation follow [60, §2] and [57, §5], but our presentation owes more to [87] and [9], which helped us separate the benefits of the two alternatives presented here.

Definition 6.1.4   A subspace $A \subset X$ is **compactly closed** if its restriction along any continuous function $K \to X$ with compact Hausdorff domain is closed in K. A space X is a k-**space** if every compactly closed subspace of X is closed.

Examples of k-spaces include CW complexes, compact spaces, locally compact spaces, topological manifolds, and first-countable spaces, including metric spaces. Write **kTop** for the full subcategory of k-spaces; **kTop** forms a **coreflective** subcategory of **Top**, meaning the inclusion

$$\mathbf{kTop} \xrightarrow{} \mathbf{Top}$$
$$\underset{k}{\underset{\perp}{\longleftarrow}}$$

has a right adjoint, called k-**ification**. The functor k maps a space X to a space with the same underlying set topologized so that the closed subsets of kX are precisely the compactly closed subsets of X. The identity function $kX \to X$ is continuous and has a universal property: kX has the finest topology so that any map $K \to X$, with K compact Hausdorff, factors through $kX \to X$. This map is the counit of the coreflection; the unit is the identity. One readily checks that compactly closed subsets of kX are closed – hence, $kkX = kX$ – and that a continuous map $f : X \to Y$ gives rise to a continuous map $kf : kX \to kY$.

Theorem 6.1.5   *The category* **kTop** *is complete and cocomplete. Colimits are formed as in* **Top**, *and limits are formed by applying the functor k to the corresponding limit in* **Top**. *The category* **kTop** *is cartesian closed.*

Proof   See [87, §3] for a proof of cartesian closure. The assertions about limits and colimits follow formally from 6.1.9.   □

In particular, the product of two k-spaces is the k-ification of the usual product topology. If one of the spaces is locally compact, these two notions coincide, but in general the k-space product topology will be finer. This modification is desirable: the functor $| - | : \mathbf{sSet} \to \mathbf{Top}$ does not preserve products, but $| - | : \mathbf{sSet} \to \mathbf{kTop}$ does. One interpretation of this result is that the k-ification of the product topology is the topology of the product of CW complexes.

Lemma 6.1.6   *If X and Y are simplicial sets, then the geometric realization of their product is homeomorphic to the k-ification of the product of their geometric realizations.*

Proof　The canonical map $|\Delta^m \times \Delta^n| \to |\Delta^m| \times |\Delta^n|$ is a continuous bijection between compact Hausdorff spaces and hence a homeomorphism. Note that $|\Delta^n|$ is locally compact, so the right-hand product agrees with its k-ification. By the density theorem, generic simplicial sets X and Y are colimits, indexed by their categories of simplices, of representable simplicial sets (see 7.2.7). Because **kTop** is cartesian closed, the k-space product preserves colimits in both variables, and the result follows. □

A function $f \colon X \to Y$ is k-**continuous** if and only if its composite with any continuous function $K \to X$ whose domain is a compact Hausdorff space K is continuous. A continuous function is clearly k-continuous. Conversely, a k-continuous function whose domain is a k-space is also continuous; this is the universal property that characterizes the k-space topology.

Remark 6.1.7 ([9, §7.2])　In the cartesian closed structure on **kTop**, the topology on the space of maps $X \to Y$ is the k-ification of the compact-open topology. This differs from the usual compact-open topology if X happens to have non-Hausdorff compact subspaces.

Note that k-continuity is not a local property unless compact subsets of X are locally compact, which is automatic if X is Hausdorff but not true in general. Hence we might prefer to introduce a separation condition on k-spaces so that the topology assigned to the hom-spaces gives a better notion of convergence for sequences of continuous functions.

Definition 6.1.8　A space X is **weak Hausdorff** if the image of any continuous map with compact Hausdorff domain is closed in X. A weak Hausdorff k-space is called **compactly generated**.

CW complexes, metric spaces, and topological manifolds are compactly generated. The weak Hausdorff condition should be thought of as a separation axiom: Hausdorff spaces (T_2-spaces) are weak Hausdorff and weak Hausdorff spaces are T_1. In a weak Hausdorff space, the image of a map $K \to X$ with compact Hausdorff domain is a compact Hausdorff subspace and, in particular, is locally compact. A subspace of a weak Hausdorff space is compactly closed if and only if its intersection with compact Hausdorff subspaces is closed. The slogan is that the compact Hausdorff subspaces of a compactly generated space determine its topology. In particular, a map $X \to Y$ between compactly generated spaces is continuous if and only if its restriction to any compact Hausdorff subspace of X is continuous.

Write **CGTop** for the category of compactly generated spaces. It is a **reflective** subcategory of **kTop**; that is, the inclusion functor

$$\mathbf{CGTop} \underset{\longrightarrow}{\overset{w}{\underset{\perp}{\longleftarrow}}} \mathbf{kTop}$$

has a left adjoint. The reflector w takes a k-space X to its maximal weak Hausdorff quotient, which is again a k-space.

Digression 6.1.9 (reflective subcategories)   Let us pause a moment to review the general features of this situation. A category C is a **reflective subcategory** of $\mathcal{D}$ if there is a full inclusion

$$C \xrightarrow[\substack{R \\ \leftarrow \; \mathtt{I} \; \searrow}]{} \mathcal{D} \tag{6.1.10}$$

that admits a left adjoint R. Because the right adjoint is full and faithful, the counit of the adjunction is an isomorphism. The unit $d \to Rd$ is universal among maps from an object $d \in \mathcal{D}$ to an object, its **reflection**,[2] in the subcategory C.

An important point is that the adjunction (6.1.10) is **monadic**, meaning C is canonically equivalent to the category of algebras for the monad R on $\mathcal{D}$. The proof is elementary: the essential point is that R is an **idempotent monad**, meaning its multiplication is a natural isomorphism. In particular, the functor $C \to \mathcal{D}$ not only preserves but creates all limits. So if $\mathcal{D}$ is complete, then C is as well, with its limits formed as in the larger category.

If $\mathcal{D}$ is also cocomplete, then C is too, but colimits are computed by applying the functor R to the corresponding colimit in $\mathcal{D}$. This situation is familiar from algebraic geometry: limits of sheaves on a space X are formed pointwise, as they are for presheaves, whereas colimits have to pass through the "sheafification" functor that reflects presheaves into the subcategory of sheaves.

Exercise 6.1.11   Verify that the procedure just described produces colimits in C.

These general remarks prove the initially unjustified assertions in Theorem 6.1.5. Colimits, formed in **Top**, of k-spaces are necessarily k-spaces. The category **kTop** is also complete, with limits formed by applying k-ification to the corresponding limit in **Top**. Similarly, limits of weak Hausdorff spaces are necessarily weak Hausdorff. Hence limits in **CGTop** are formed as in **kTop** by applying k to the ordinary limit. In particular, geometric realization can also be regarded as a strong monoidal functor $| - | :$ **sSet** $\to$ **CGTop**.

Certain colimits of compactly generated spaces happen to be weak Hausdorff already, in which case the functor w has no effect. For instance, the category **CGTop** is closed under colimits of sequences of closed inclusions, pushouts along closed inclusions, and quotients by closed equivalence relations [57, §5.2]. However, generic colimits constructed by applying w are poorly behaved and, in particular, change the underlying set of the resulting topological space.

---

[2] The mnemonic is that an object looks at its reflection.

Thus one tends to work only in the category **CGTop** when the desired colimit constructions avoid weak Hausdorffication.

Like **kTop**, the category **CGTop** is cartesian closed with internal homs as described in 6.1.7. Hence it follows from Construction 3.3.14 that the categories of based k-spaces and based compactly generated spaces are also closed monoidal. To summarize, we state the theorem:

Theorem 6.1.12 *The categories* **kTop** *and* **CGTop** *are cartesian closed. The categories* **kTop**$_*$ *and* **CGTop**$_*$ *are closed symmetric monoidal with respect to the smash product.*

6.2 Simplicial model categories of spaces

We can now use the extension of Theorem 3.7.11 mentioned at the end of Chapter 3 to produce several simplicial model categories of spaces in which to compute homotopy limits and homotopy colimits.

Example 6.2.1 ([38, 4.2.8]) The primary example, due to Quillen, is **sSet**. All objects are cofibrant. Fibrant objects, called **Kan complexes**, are simplicial sets X in which any horn $\Lambda_k^n \to X$ admits an extension along the canonical inclusion $\Lambda_k^n \to \Delta^n$. See Example 11.3.5.

Example 6.2.2 By Lemma 6.1.6, the usual geometric realization–total singular complex adjunction is strong monoidal, at least when we interpret its target as landing in **kTop** or **CGTop**. It is a left Quillen functor by a classical theorem of Quillen [65]. Hence the fact that Quillen's model structures on **kTop** and **CGTop** form monoidal model categories [38, 4.2.11] implies that they are simplicial model categories. For both model structures, all objects are fibrant. The cofibrant spaces are the cell complexes. See Example 11.3.6.

Example 6.2.3 By Construction 3.3.14, there is a strong monoidal Quillen adjunction with left adjoint $(-)_+ : $ **sSet** $\to$ **sSet**$_*$ [38, 4.2.9]. This makes the monoidal model categories of based simplicial sets into a simplicial model category. All objects are cofibrant. Fibrant objects are based Kan complexes.

A final pair of examples makes use of the following result:

Proposition 6.2.4 *Geometric realization extends to a strong monoidal left Quillen functor* $|-| : $ **sSet**$_*$ $\to$ **CGTop**$_*$.

Proof See [38, 4.2.17] for a proof that geometric realization is a left Quillen functor. Here we show that it is strong monoidal, that is, that it commutes with smash products. Recall that the smash product of a pair of based spaces was defined by the pushout (3.3.15). Because geometric realization is a left adjoint, it preserves the pushout. By Lemma 6.1.6, it also commutes with the constituent coproduct and product. The conclusion follows.   □

Example 6.2.5   Applying Proposition 6.2.4, it follows that the closed monoidal categories of based k-spaces and based compactly generated spaces are simplicial model categories. All objects are fibrant. Cofibrant objects are non-degenerately based cell complexes.

In the sequel, we write **Top** to mean either **CGTop** or **kTop**. Because the weak Hausdorff property is stable under the sorts of colimits we need, including those involved in the bar construction, the distinction between these categories is immaterial. We do not discuss the inner workings of the convenient category of spaces in the future and have no further use for the category of *all* spaces.

6.3 Warnings and simplifications

Before presenting examples, we should issue a warning.

Remark 6.3.1   We always take the weak equivalences in the category $\mathcal{M}^{\mathcal{D}}$ of diagrams of shape $\mathcal{D}$ in a homotopical category $\mathcal{M}$ to be defined pointwise. By the universal property of localization, there is a canonical map

$$\mathcal{M}^{\mathcal{D}} \xrightarrow{\ \gamma^{\mathcal{D}}\ } (\mathrm{Ho}\mathcal{M})^{\mathcal{D}}$$
$$\gamma \downarrow \qquad \nearrow$$
$$\mathrm{Ho}(\mathcal{M}^{\mathcal{D}})$$

but it is not typically an equivalence of categories.[3] Indeed, some of the pioneering forays into abstract homotopy theory were motivated by attempts to understand the essential image of the functor $\mathrm{Ho}(\mathcal{M}^{\mathcal{D}}) \to (\mathrm{Ho}\mathcal{M})^{\mathcal{D}}$.

The diagonal functor $\Delta\colon \mathcal{M} \to \mathcal{M}^{\mathcal{D}}$ is homotopical and hence acts as its own derived functors. By Exercise 2.2.15, the total derived functor $\mathbf{L}\mathrm{colim}\colon \mathrm{Ho}(\mathcal{M}^{\mathcal{D}}) \to \mathrm{Ho}\mathcal{M}$ is left adjoint to $\Delta\colon \mathrm{Ho}\mathcal{M} \to \mathrm{Ho}(\mathcal{M}^{\mathcal{D}})$, but unless the indicated comparison is an equivalence, this is not the same as the diagonal functor $\Delta\colon \mathrm{Ho}\mathcal{M} \to \mathrm{Ho}(\mathcal{M})^{\mathcal{D}}$. Hence homotopy colimits are not typically colimits in the homotopy category:

---

[3] This is only true in very special cases, such as when $\mathcal{D}$ is discrete. This is the reason why homotopy products and homotopy coproducts *are* products and coproducts in the homotopy category, as discussed in the introduction to Chapter 3.

Example 6.3.2   Fix a field k. A chain complex $A_\bullet \in \mathbf{Ch}_\bullet(k)$ is quasi-isomorphic to its homology, interpreted as a chain complex with zero differentials.[4] Among chain complexes with zero differentials, quasi-isomorphisms are just pointwise isomorphisms. It follows that the category $\mathrm{Ho}(\mathbf{Ch}_\bullet(k))$ is equivalent to the category of graded vector spaces. The latter category is complete and cocomplete, but these (co)limits are not homotopy (co)limits. A high-level way to see this is the following: $\mathbf{Ch}_\bullet(k)$ is a stable $(\infty, 1)$-category, so homotopy pullbacks are also homotopy pushouts, which, mutatis mutandis, is not true for graded vector spaces.

Unlike the situation in the previous example, more often homotopy categories have few actual limits and colimits. We learned about the following example of a non-existent colimit in the homotopy category of spaces from Michael Andrews and Markus Hausmann.

Example 6.3.3   The identity and reflection about some axis of symmetry define a parallel pair of maps $S^1 \rightrightarrows S^1$ of degree 1 and -1. Suppose there existed a coequalizer X in $\mathcal{H} \cong \mathrm{Ho}(\mathbf{Top})$. Recall that cohomology is a represented functor on $\mathcal{H}$, that is,

$$H^n(X; G) = \mathcal{H}(X, K(G, n)).$$

Because represented functors take colimits to limits, it follows that

$$H^n(X; G) \longrightarrow H^n(S^1; G) \overset{1}{\underset{-1}{\rightrightarrows}} H^n(S^1; G)$$

is an equalizer diagram. By a routine calculation, $\tilde{H}^n(X; G) = 0$, unless $n = 1$, $H^1(X; \mathbb{Z}) = 0$, and $H^1(X; \mathbb{Z}/2) = \mathbb{Z}/2$. For any abelian group G, the universal coefficient theorem gives us short exact sequences:

$$0 \to \mathrm{Ext}(H_{i-1}(X; \mathbb{Z}), G) \to H^i(X; G) \to \mathrm{Hom}(H_i(X; Z), G) \to 0.$$

Writing A for $H_1(X; \mathbb{Z})$, we easily compute that $\mathrm{Hom}(A, \mathbb{Z}/2) = \mathbb{Z}/2$ and $\mathrm{Ext}(A, \mathbb{Z}/2) = \mathrm{Hom}(A, \mathbb{Z}) = \mathrm{Ext}(A, \mathbb{Z}) = 0$. But these calculations contradict the long exact sequence

$$0 \to \mathrm{Hom}(A, \mathbb{Z}) \overset{2}{\to} \mathrm{Hom}(A, \mathbb{Z}) \to \mathrm{Hom}(A, \mathbb{Z}/2)$$

$$\to \mathrm{Ext}(A, \mathbb{Z}) \overset{2}{\to} \mathrm{Ext}(A, \mathbb{Z}) \to \mathrm{Ext}(A, \mathbb{Z}/2).$$

These warnings dispatched, the following simplification is frequently applied. By Corollary 5.1.3, given any $F : \mathcal{D} \to \mathcal{M}$ taking values in a simplicial

---

[4] Construction of the requisite quasi-isomorphism makes use of the fact that a vector space decomposes into a direct sum involving any subspace.

model category, the object $B(*, \mathcal{D}, QF)$ has the right homotopy type for hocolim F. But if the diagram happens to be pointwise cofibrant already, then Corollary 5.2.5 implies that $B(*, \mathcal{D}, F)$ and $B(*, \mathcal{D}, QF)$ are weakly equivalent and the "correction" given by applying Q is not necessary. Put another way, our homotopy colimit functor $B(*, \mathcal{D}, Q-)$ defines which objects have the correct homotopy type to be the homotopy colimit. But in computing examples, one typically prefers a simpler construction, here omitting the pointwise cofibrant replacement, that gives the right homotopy type.

Remark 6.3.4   In the simplicial model category **Top**, not all objects are cofibrant. Nonetheless, by a folklore result, the homotopy colimit of a diagram $F : \mathcal{D} \to$ **Top** can always be computed by $B(*, \mathcal{D}, F)$. In other words, it is not necessary that the objects in the diagram are pointwise cofibrant for the bar construction to give the correct homotopy type.

The proof relies on the fact that in the category of spaces, Theorem 5.2.3 holds under weaker hypotheses than in the general case, and the simplicial bar construction satisfies these conditions even if the diagram F is not pointwise cofibrant. We were delighted to discover the beautiful proof of this fact in [19, §A]. For the reader's convenience, it is reproduced in Section 14.5.

6.4 Sample homotopy colimits

These preliminaries completed, we are now ready to deploy Corollary 5.1.3 in the field. Let us warm up with a very simple example.

Example 6.4.1   Consider a functor $f : 2 \to$ **Top** whose image is the arrow $X \xrightarrow{f} Y$. Because the category 2 has only a single non-identity arrow, $N2$ is 1-skeletal. Because degenerate simplices in the simplicial bar construction correspond to degenerate simplices in the nerve of the diagram shape (the fact that enabled the proof of Lemma 5.2.1), $B_\bullet(*, 2, f)$ is 1-skeletal. Hence Lemma 4.4.3 implies that hocolim$(f) = B(*, 2, f)$ can be computed as the functor tensor product of the 1-truncated simplicial object.

By definition, the first two spaces in the simplicial bar construction $B_\bullet(*, 2, f)$ are

$$B_0 = X \sqcup Y \quad \text{and} \quad B_1 = X^0 \sqcup X^f \sqcup Y^1,$$

where the superscripts indicate the arrow that indexes each component of B_1. The first X in B_1 corresponds to the domain of the image of the identity at 0, the second X corresponds to the domain of f, and the Y corresponds to the domain of the image of the identity at 1.

By 5.1.3, the homotopy colimit is the (1-truncated) geometric realization, which is[5]

$$\text{hocolim}(f) = \text{colim} \left(\begin{array}{c} \Delta^1 \otimes B_0 \xrightarrow{\ s_0\ } \Delta^1 \otimes B_1 \\ \\ \Delta^0 \otimes B_1 \rightrightarrows \Delta^0 \otimes B_0 \end{array} \right)$$

with maps s^0, d^0, d^1, d_0, d_1

$$= \text{colim} \left(\begin{array}{c} X \times I \sqcup Y \times I \xrightarrow{\ s_0\ } (X^0 \times I) \sqcup (X^f \times I) \sqcup (Y^1 \times I) \\ \\ X^0 \sqcup X^f \sqcup Y^1 \rightrightarrows X \sqcup Y \end{array} \right)$$

with maps s^0, d^0, d^1, d_0, d_1

(6.4.2)

writing $I = |\Delta^1| \cong [0,1] \subset \mathbb{R}$. This colimit is computed by first taking the disjoint union

$$(X \sqcup Y) \sqcup ((X^0 \times I) \sqcup (X^f \times I) \sqcup (Y^1 \times I))$$

of the two objects on the right of (6.4.2) and then identifying any two points that appear in the images of any pair of corresponding maps (s_0 and s^0 or d_i and d^i). In particular, s_0 includes $X^0 \times I$ into the 0 component of the top coproduct, and s^0 projects onto X in the bottom coproduct. Hence the cylinders $X^0 \times I$ and $Y^1 \times I$ are identified with their projections X and Y. The intermediate result is

$$X \sqcup (X^f \times I) \sqcup Y, \tag{6.4.3}$$

but it remains to identify the images of the d_i and d^i.

The images of X^0 and Y^1 under d_0 and d^0 and d_1 and d^1 were already identified when we quotiented using the degeneracy operator. The image of X^f under d^1 is X and the map restricts to the identity. The image of X^f under d_0 is $X \times \{0\} \subset X^f \times I$, so the zeroth face of this cylinder gets glued to the X in (6.4.3). The image of X^f under d^0 is $f(X) \subset Y$, whereas the image of X_f

---

[5] Recall that the colimit under consideration is a coend. In particular, the parallel maps d^0 and d^1 are not coequalized.

under d_0 is $X \times \{1\} \subset X^f \times I$. So the first face of this cylinder gets identified with Y by gluing $(x, 1)$ to $f(x)$. The result is

$$\text{hocolim}(f) = X \times I \sqcup Y/_{(x,1)\sim f(x)}.$$

This is the usual **mapping cylinder** Mf.

Remark 6.4.4 The mapping cylinder deformation retracts onto Y. In 8.5.9, as a consequence of the theory of homotopy final functors, we see that $\text{hocolim}(X \to Y)$ necessarily has the same homotopy type as Y.

Example 6.4.5 A similar calculation reveals that the homotopy colimit of a diagram of spaces $X \xleftarrow{f} A \xrightarrow{g} Y$ is the space

$$X \sqcup (A \times I) \sqcup A \sqcup (A \times I) \sqcup Y/_{\langle f(a)\sim(a,0),\ (a,1)\sim a\sim(a,0),\ (a,1)\sim g(a)\rangle}$$

that identifies one end of the left cylinder with its image in X, one end of the right cylinder with its image in Y, and the other two ends of the cylinders together. This space is homeomorphic to the **double mapping cylinder**

$$X \sqcup (A \times I) \sqcup Y/_{\langle f(a)\sim(a,0),\ (a,1)\sim g(a)\rangle},$$

a familiar model for the **homotopy pushout**.

Example 6.4.6 A special case has a special name. The homotopy pushout of a diagram $* \leftarrow X \to *$ in **Top** is the (unreduced) **suspension** of X. The homotopy pushout in **Top**$_*$ is the reduced suspension of X. The relation between the reduced suspension and the unreduced suspension is governed by a general result, Theorem 6.6.5, which is proven shortly.

Reconsidering Example 2.1.16, the homotopy pushout of $D^n \leftarrow S^{n-1} \to D^n$ is homeomorphic to S^n, which was also the ordinary pushout. The homotopy pushout of $* \leftarrow S^{n-1} \to *$ is the unreduced suspension of S^{n-1}, also homeomorphic to S^n.

In contrast to the ordinary pushout, the homotopy pushout is homotopy invariant, that is, given a commutative diagram

$$
\begin{array}{ccccc}
X & \longleftarrow & A & \longrightarrow & Y \\
\wr \downarrow & & \wr \downarrow & & \downarrow \wr \\
X' & \longleftarrow & A' & \longrightarrow & Y'
\end{array}
$$

with vertical maps weak homotopy equivalences, the induced map $P \to P'$ between the homotopy pushouts is a weak homotopy equivalence. This is, of course, a consequence of our definition of the homotopy pushout as a derived functor, but let us check this directly. After identifying $|\Delta^1|$ with the interval $[0, 1]$, we obtain an open cover of P by $U = X \cup (A \times [0, 2/3))$ and $V = (A \times (1/3, 1]) \cup Y$ such that the maps

$$a \mapsto (a, 1/2) \colon A \to U \cap V, \qquad X \hookrightarrow U, \qquad Y \hookrightarrow V$$

are all a deformation retracts. Define analogous spaces U' and V' for X', A', Y' and note that $P \to P'$ restricts to weak equivalences $U \to U'$, $V \to V'$, and $U \cap V \to U' \cap V'$, because the displayed inclusions are compatible with the deformation retracts onto the weak equivalences $X \to X'$, $Y \to Y'$, $A \to A'$. It follows from a classical topological theorem, recorded as Lemma 14.5.9, that $P \to P'$ is a weak equivalence, as desired.

Exercise 6.4.7   Show that the homotopy colimit of $Y \xleftarrow{f} X \to *$ in **Top**$_*$ is the reduced mapping cone C_f of the based map f, also called the **homotopy cofiber**. Writing $\mathcal{H}_*$ for the homotopy category of based spaces, show using either formal or classical arguments that the cofiber sequence $X \to Y \to C_f$ induces an exact sequence of based sets $\mathcal{H}_*(C_f, Z) \to \mathcal{H}_*(Y, Z) \to \mathcal{H}_*(X, Z)$; that is, show that the image of the first map is the kernel of the second. Homotopy invariance says that if we replace f by a homotopy equivalent diagram $X' \xrightarrow{f'} Y'$, then the induced map between the homotopy cofibers is a homotopy equivalence.

Example 6.4.8   Consider a diagram $X \xrightarrow{f} Y \xrightarrow{g} Z$ given by a functor $F \colon \mathbf{3} \to \mathbf{Top}$. Extending 6.4.1, the geometric realization of the 1-truncation of $B_\bullet(*, \mathbf{3}, F)$ is formed by gluing three mapping cylinders Mf, Mg, and Mgf along the spaces X, Y, and Z. But, as mentioned in Remark 6.4.4, 8.5.9 implies that the homotopy colimit should have the homotopy type of Z, which this space does not. From this we see that the "higher homotopies" in the bar construction play an essential role.

Indeed, the simplicial set $N\mathbf{3}$ is 2-skeletal, so $B(*, \mathbf{3}, F)$ is homeomorphic to the geometric realization of the 2-truncation of $B_\bullet(*, \mathbf{3}, F)$. The non-degenerate 2-simplex of $N\mathbf{3}$ glues in the space $X \times |\Delta^2|$ along the three cylinders in the obvious way.

A simpler construction also has the correct homotopy type:

$$(X \times I) \sqcup (Y \times I) \sqcup Z / {\langle (x,1) \sim (f(x),0), (y,1) \sim g(y) \rangle} .$$

The inductive version of this construction is called the **mapping telescope** and is the homotopy colimit of a sequence

$$X_1 \to X_2 \to X_3 \to \cdots .$$

The proof that the mapping telescope computes the homotopy colimit of this diagram is given in Example 11.5.11.

Example 6.4.9   The homotopy colimit of the constant diagram of shape $\mathcal{D}$ at the terminal object is $B(*, \mathcal{D}, *)$, which, by Example 4.2.2, is isomorphic to the geometric realization of the nerve of $\mathcal{D}$. In the case where $\mathcal{D}$ is a 1-object groupoid, that is, a group G, this space

$$BG := B(*, G, *) \cong B(* \otimes_G G, G, *) \cong * \otimes_G B(G, G, *) \cong \operatorname*{colim}_G EG$$

$$(6.4.10)$$

is called the classifying space of G. More generally, we refer to $B(*, \mathcal{D}, *)$ as the **classifying space** of the category.

The formula (6.4.10) gives a concrete model of BG as a CW complex. For instance, let $G = \mathbb{Z}/2$. By definition, $B_n(*, \mathbb{Z}/2, *)$ is the discrete space on 2^n vertices: ordered n-tuples of 0s and 1s. Geometric realization produces a CW complex with the following description. Start with a single 0-cell $*$. A priori, there are two 1-cells because $B_1(*, \mathbb{Z}/2, *)$ has two vertices 0 and 1. But the first of these is degenerate; degeneracy maps in $B_\bullet(*, G, *)$, the nerve of a category, insert identities, here 0s. So the remaining 1-cell 1 is glued at both endpoints to $*$. For convenience, we remember the fact that the other 1-cell 0 was collapsed to the point $*$.

Continuing, a priori, we have four 2-cells labelled 00, 01, 10, 11. The first three are degenerate, so the only one contributing to the homotopy type is the fourth. Its three faces are, in order, 1, 0, and 1; recall that the outer face maps are projections and the inner face maps are compositions. Hence the associated 2-simplex $|\Delta^2|$ is attached to the loop at $*$ using an attaching map $S^1 \to S^1$ of degree 2.

Inductively, there is a single non-degenerate n-cell $11 \cdots 1$ whose outer two faces are attached to the non-degenerate $(n - 1)$-cell and whose remaining faces are degenerate. Careful examination of these attaching maps reveals that we have just described the standard CW structure for $\mathbb{R}P^\infty$, the classifying space for $\mathbb{Z}/2$.

Example 6.4.11 (Borel construction)   Let $X: G \to \mathbf{Top}$ be a G-space. The ordinary colimit of this diagram is the **orbit space**. By Corollary 5.1.3,

$$\operatorname*{hocolim}_G X = \operatorname*{colim}_G B(G, G, X) \cong \operatorname*{colim}_G B(G, G, G \otimes_G X)$$

$$\cong \operatorname*{colim}_G (B(G, G, G) \otimes_G X).$$

The action on the right of $B(G, G, G)$ is free and the product $\otimes$ is the cartesian product on **Top**. Hence the G-space $B(G, G, G) \otimes_G X$ is homeomorphic to $B(G, G, *) \times X = EG \times X$ endowed with diagonal action. In Example 4.5.5, we saw that EG was contractible. So, if the G-action on X is free, then

$$\operatorname*{hocolim}_{G} X = \operatorname*{colim}_{G}(EG \times X) \cong \operatorname*{colim}_{G} X \cong X/G.$$

For instance, the antipodal $\mathbb{Z}/2$ action on S^n produces $\mathbb{R}P^n$ as its orbit space, and this action is free, so $\mathbb{R}P^n$ is also the homotopy orbit space. If the action on X is trivial,

$$\operatorname*{hocolim}_{G} X \cong (\operatorname*{colim}_{G} EG) \times X \cong EG/G \times X \cong BG \times X.$$

The construction of the space $\operatorname{hocolim}_G X$ is called the **Borel construction**.

For instance, $\mathbb{Z}/2$ acts freely on S^n by the antipodal action, and these actions are compatible with the inclusions

$$S^1 \hookrightarrow S^2 \hookrightarrow \cdots \hookrightarrow \cup_n S^n =: S^\infty.$$

Here the colimit and the homotopy colimit coincide. Indeed, extending the computation of $B\mathbb{Z}/2 = \mathbb{R}P^\infty$, we see that $E\mathbb{Z}/2 = B(\mathbb{Z}/2, \mathbb{Z}/2, *)$ admits precisely this CW decomposition: we start with two 0-cells, to which we attach a parallel pair of 1-cells, to which we attach a parallel pair of 2-cells, and so on. The free $\mathbb{Z}/2$-action is the antipodal map, exchanging the cells in each degree.

Because the maps are equivariant, we can quotient by $\mathbb{Z}/2$ before forming the colimit, returning our previous calculation: $B\mathbb{Z}/2 \cong (E\mathbb{Z}/2)/(\mathbb{Z}/2) \cong \cup_n S^n/_{\langle \pm 1 \rangle} \cong \cup_n \mathbb{R}P^n = \mathbb{R}P^\infty$.

6.5 Sample homotopy limits

Homotopy limits, being somewhat harder to visualize, are less frequently expounded on in the literature. Therefore let us dualize and compute a few homotopy limits.

Example 6.5.1 Let X_α be a collection of simplicial sets indexed by some set I. We use the cobar construction to compute their homotopy product. The nerve of I, regarded as a discrete category, is 0-skeletal, so the cobar construction is isomorphic to its 0-truncation, which produces the product of fibrant replacements of these objects, that is,

$$\operatorname*{holim}_{\alpha \in I} X_\alpha \cong \prod_{\alpha \in I} RX_\alpha.$$

Now that we know this object has the correct homotopy type, we might ask whether a simpler construction is available. Tautologically, there is a weak equivalence $X_\alpha \to RX_\alpha$ from each simplicial set to its fibrant replacement. Iterating Lemma 6.1.6, geometric realization preserves finite products. A finite product of weak equivalences of topological spaces is again a weak equivalence;

one way to see this is to note that binary products of spaces are homotopy pull-backs because all spaces are fibrant. So if the index set is finite, then the ordinary product $\prod_\alpha X_\alpha$ is weakly equivalent to the homotopy product. For infinite products, however, the "correction" provided by the fibrant replacements is necessary.

Example 6.5.2   Dualizing Example 6.4.1, the homotopy limit of $X \xrightarrow{f} Y$ is the **mapping path space**. Because the nerve of the walking arrow category is 1-skeletal, it suffices, by the dual of Lemma 4.4.3, to compute the totalization of the 1-truncated cosimplicial object

$$C^0 = X \times Y \xrightarrow[\substack{d^0 \\ d^1}]{\substack{\longrightarrow \\ \longleftarrow s^0 \\ \longrightarrow}} X \times Y \times Y = C^1.$$

To improve the aesthetics, let us write X^I for $\underline{\mathbf{Top}}(|\Delta^1|, X)$, the cotensor X^{Δ^1} defined by 3.7.15, 6.1.7, and 6.2.2. Recall that this means that X^I is the space of continuous maps $|\Delta^1| \to X$ topologized with the k-ification of the compact open topology; that is, X^I is the k-ification of the path space of X.

Observing that the right adjoint $(-)^I$ preserves products, Corollary 5.1.3 implies that

$$\text{holim } f = \lim \left(\begin{array}{c} X \times Y \xrightarrow[\substack{d^0 \\ d^1 \; d^0 \\ s^0 \\ s^0}]{} X \times Y \times Y \\ \\ X^I \times Y^I \times Y^I \xrightarrow{\hspace{2cm}} X^I \times Y^I \end{array} \right)$$

Using ad hoc methods, justified by the fact that limits commute with each other, we can simplify this expression. The first step might be to pull back along the degeneracy maps, restricting to the subspace of the lower left-hand product where the paths in the first and last coordinates are constant:

$$= \lim \left(\begin{array}{c} X \times Y^I \times Y \\ \\ d^1 \; \Big\Vert \; d^0 \\ \\ X \times Y \xrightarrow[\substack{d^0 \\ d^1}]{} X \times Y \times Y \end{array} \right)$$

The face map $d^0 \colon X \times Y \to X \times Y \times Y$ projects and then acts by the identity, whereas $d^0 \colon X \times Y^I \times Y \to X \times Y \times Y$ evaluates at the endpoint of each path. The effect of the limit condition imposed by these maps is to restrict to the subspace $X \times Y^I \subset X \times Y^I \times Y$ in which the third coordinate is identified with the endpoint of the path in Y. We are now left with the pullback

$$
\begin{array}{ccc}
\operatorname{holim} f \cong Nf & \longrightarrow & Y^I \\
\downarrow & \lrcorner & \downarrow \; {}^{d^1 = \mathrm{ev}_0} \\
X & \xrightarrow{\;\; d^1 = f \;\;} & Y
\end{array}
$$

which defines the mapping path space.

Example 6.5.3 Consider a diagram $* \overset{p}{\underset{q}{\rightrightarrows}} X$ in **Top**, picking out a pair of points in X. If $p \neq q$, the limit of this diagram is empty; if $p = q$, it is a singleton. Let us compute the homotopy limit. Again, the nerve of the category $\bullet \rightrightarrows \bullet$ is 1-skeletal, so it suffices to compute the totalization of a 1-truncated cosimplicial object. As we saw in the previous example, the resulting homotopy limit will be a subspace of $C^0 \times (C^1)^{\Delta^1}$. Using subscripts to indicate which objects are indexed by which arrows, the relevant spaces are

$$
C^0 = * \times X \qquad (C^1)^{\Delta^1} = *_0^{\Delta^1} \times X_p^{\Delta^1} \times X_q^{\Delta^1} \times X_1^{\Delta^1},
$$

where we have used the fact that the cotensor with Δ^1, as a right adjoint, preserves the categorical product.

The limit condition imposed by the degeneracy map $s^0 \colon \Delta^1 \to \Delta^0$ replaces the space $(C^1)^{\Delta^1}$ by a subspace isomorphic to $X_p^{\Delta^1} \times X_q^{\Delta^1} \times X$. The limit condition imposed by $d^1 \colon \Delta^0 \to \Delta^1$ replaces two copies of X^{Δ^1} in the product by the respective pullbacks

$$
\begin{array}{ccc}
\Gamma(X, p) & \longrightarrow & X^{\Delta^1} \\
\downarrow & \lrcorner & \downarrow \; {}^{d^1} \\
* & \xrightarrow{\;\; p \;\;} & X \cong X^{\Delta^0}
\end{array}
\qquad \text{and} \qquad
\begin{array}{ccc}
\Gamma(X, q) & \longrightarrow & X^{\Delta^1} \\
\downarrow & \lrcorner & \downarrow \; {}^{d^1} \\
* & \xrightarrow{\;\; q \;\;} & X \cong X^{\Delta^0}
\end{array}
$$

The pullback $\Gamma(X, p)$ is the space of paths in X starting at p. Finally, the limit condition imposed by $d^0 \colon \Delta^0 \to \Delta^1$ demands that the codomains of the paths in $\Gamma(X, p)$ and $\Gamma(X, q)$ coincide with some specified $x \in X$. It follows that the homotopy limit is homeomorphic to the subspace of X^{Δ^1} of paths from p to q in X. Taking $p = q$, the homotopy limit is the **loop space** $\Omega(X, p)$.

Exercise 6.5.4 Dualize Example 6.4.5 to show that the homotopy limit of $p \colon Y \to X, q \colon Z \to X$ is the subspace of $X^I \times Y \times Z$ of triples (f, y, z) with $p(y) = f(0), q(z) = f(1)$. This space is called the **homotopy pullback**.

The homotopy limit of Example 6.5.3 can also be constructed as a homotopy pullback in based spaces. Note that the underlying space of this homotopy pullback is isomorphic to the homotopy pullback of the underlying spaces. This is an instance of a general phenomenon that can assist with the computation of homotopy limits in $\mathbf{sSet}_*$ or $\mathbf{Top}_*$.

Lemma 6.5.5   *Suppose given a diagram* $F : \mathcal{D} \to \mathcal{M}_*$, *where* $\mathcal{M}$ *is either simplicial sets or spaces. Then the underlying space of its homotopy limit is the homotopy limit of the composite of* F *with the forgetful functor* $\mathcal{M}_* \to \mathcal{M}$.

Proof   The proof is left to the reader (or see Corollary 8.1.2).   □

Example 6.5.6   The limit of

$$\cdots X_n \hookrightarrow \cdots \hookrightarrow X_3 \hookrightarrow X_2 \hookrightarrow X_1 \qquad\qquad (6.5.7)$$

is $\cap_n X_n$. Dualizing Example 6.4.8 and making use of the topological properties of the standard interval, the homotopy limit has the same homotopy type as a space with the following description: the homotopy limit is the space of maps $[1, \infty) \to X_1$ such that $f([n, \infty)) \subset X_n$ for all $n > 0$. For example, let X_n be the ball of radius $1/n$ in $\mathbb{R}^2$ about the origin, but with the origin removed. Forgetting the distance to the origin but remembering the angle defines a natural homotopy equivalence from the sequence of inclusions (6.5.7) to the constant sequence at S^1. The limit of the former is empty and the limit of the latter is, of course, S^1. By the homotopy invariance of $C(*, \mathcal{D}, -)$, the homotopy limit, the space of paths in the punctured disk that approach the origin, has the homotopy type of S^1.

Example 6.5.8   The limit of $X : G \to \mathbf{Top}$ is X^G the space of fixed points. The homotopy limit is the **homotopy fixed point** space X^{hG} computed by $\mathbf{Top}^G(EG, X)$. Precomposition with the contraction $EG \to *$ gives a canonical map $X^G \to X^{hG}$, which is not generally an equivalence. We prove that this homotopy colimit has this form in Section 7.7.

6.6 Homotopy colimits as weighted colimits

Historically, the first definitions of homotopy limits and colimits, such as the definition of Pete Bousfield and Kan [10], employed functor tensor and cotensor products. Our facility with the two-sided bar construction makes the following theorem easy to prove:

Theorem 6.6.1   *Let* $F : \mathcal{D} \to \mathcal{M}$ *be any diagram in a complete and cocomplete, tensored, cotensored, and simplicially enriched category* $\mathcal{M}$. *There are* *natural isomorphisms*

$$B(*, \mathcal{D}, F) \cong N(-/\mathcal{D}) \otimes_\mathcal{D} F \quad \text{and} \quad C(*, \mathcal{D}, F) \cong \{N(\mathcal{D}/-), F\}^\mathcal{D}.$$

In particular, the homotopy colimit of a pointwise cofibrant diagram F can be computed by the functor tensor product with $N(-/\mathcal{D})$. Dually, the homotopy limit of a pointwise fibrant diagram can be computed by the functor cotensor product with $N(\mathcal{D}/-)$.

Proof Extending previous notation, here we regard the contravariant Yoneda embedding as a functor $\mathcal{D} \colon \mathcal{D}^\text{op} \to \mathbf{sSet}^\mathcal{D}$, taking values in discrete simplicial sets. By Exercise 4.2.6, $B_\bullet(*, \mathcal{D}, \mathcal{D})$ is the horizontally discrete bisimplicial set $N(-/\mathcal{D})$. By the co-Yoneda lemma, which is used to prove Exercise 4.1.8, the geometric realization of a bisimplicial set is isomorphic to its diagonal; hence $B(*, \mathcal{D}, \mathcal{D}) \cong N(-/\mathcal{D})$, and the claimed isomorphism is an easy calculation:

$$N(-/\mathcal{D}) \otimes_\mathcal{D} F \cong B(*, \mathcal{D}, \mathcal{D}) \otimes_\mathcal{D} F \cong B(*, \mathcal{D}, \mathcal{D} \otimes_\mathcal{D} F) \cong B(*, \mathcal{D}, F). \quad \square$$

The result of Theorem 6.6.1 is important enough to merit a more explicit demonstration. By Fubini's theorem for iterated coends and cocontinuity of simplicial tensors,

$$N(-/\mathcal{D}) \otimes_\mathcal{D} F \cong B(*, \mathcal{D}, \mathcal{D}) \otimes_\mathcal{D} F$$

$$\cong \int^{d \in \mathcal{D}} |B_\bullet(*, \mathcal{D}, \mathcal{D}(d, -))| \otimes Fd$$

$$\cong \int^{d \in \mathcal{D}} \left(\int^{n \in \Delta} \Delta^n \times B_n(*, \mathcal{D}, \mathcal{D}(d, -)) \right) \otimes Fd$$

$$\cong \int^{d \in \mathcal{D}} \int^{n \in \Delta} \Delta^n \otimes (B_n(*, \mathcal{D}, \mathcal{D}(d, -)) \otimes Fd)$$

$$\cong \int^{n \in \Delta} \Delta^n \otimes \left(\int^{d \in \mathcal{D}} B_n(*, \mathcal{D}, \mathcal{D}(d, -)) \otimes Fd \right). \quad (6.6.2)$$

Similarly,

$$B(*, \mathcal{D}, F) \cong \int^{n \in \Delta} \Delta^n \otimes B_n(*, \mathcal{D}, F)$$

$$\cong \int^{n \in \Delta} \Delta^n \otimes \left(\coprod_{\vec{d}\colon [n] \to \mathcal{D}} Fd_0 \right). \quad (6.6.3)$$

So, if we can show that

$$\int^{d \in \mathcal{D}} B_n(*, \mathcal{D}, \mathcal{D}(d, -)) \otimes Fd \cong \coprod_{\vec{d}\colon [n] \to \mathcal{D}} Fd_0,$$

then we may conclude that (6.6.2) and (6.6.3) are isomorphic. The tensor inside the left-hand coend is indexed by the discrete simplicial set $N(d/\mathcal{D})_n$, so it may be rewritten as

$$\int^{d\in\mathcal{D}} \coprod_{N(d/\mathcal{D})_n} Fd.$$

Elements of $N(d/\mathcal{D})_n$ are strings $\vec{d}\colon [n] \to \mathcal{D}$ of n composable arrows in $\mathcal{D}$ together with an arrow $d \to d_0$ in $\mathcal{D}$. Thus this coend is isomorphic to

$$\int^{d\in\mathcal{D}} \coprod_{\vec{d}\colon [n]\to\mathcal{D}} \mathcal{D}(d,d_0)\cdot Fd \cong \coprod_{\vec{d}\colon [n]\to\mathcal{D}} \int^{d\in\mathcal{D}} \mathcal{D}(d,d_0)\cdot Fd \cong \coprod_{\vec{d}\colon [n]\to\mathcal{D}} Fd_0,$$

the last step by the co-Yoneda lemma.

Example 6.6.4   Let $X\colon \Delta^{\mathrm{op}} \to \mathbf{sSet}$ be a bisimplicial set. By Theorem 6.6.1,

$$\operatorname*{hocolim}_{\Delta^{\mathrm{op}}} X \cong N(\Delta/-)^{\mathrm{op}} \otimes_{\Delta^{\mathrm{op}}} X$$

because $d/(\mathcal{D}^{\mathrm{op}}) \cong (\mathcal{D}/d)^{\mathrm{op}}$. This homotopy colimit is larger than the geometric realization of $|X|$, though we see in Example 7.7.2 that there is a canonical comparison map $\operatorname{hocolim}_{\Delta^{\mathrm{op}}} X \to |X|$ that is a weak equivalence of simplicial sets. Comparison results like this motivate our discussion of enriched homotopy theory in Part II.

The isomorphism of Theorem 6.6.1 simplifies the proof of the following comparison, the colimit counterpart to Lemma 6.5.5, between homotopy colimits in based and unbased spaces.

Theorem 6.6.5 (Dror Farjoun [36, 18.8.4])   *Given a diagram $F\colon \mathcal{D} \to \mathbf{sSet}_*$, writing $U\colon \mathbf{sSet}_* \to \mathbf{sSet}$ for the forgetful functor, there is a cofiber sequence*

$$B(*, \mathcal{D}, *) \to \operatorname{hocolim} UF \to U \operatorname{hocolim} F.$$

Similarly, given $F\colon \mathcal{D} \to \mathbf{Top}_*$, whose constituent spaces are non-degenerately based, there is a cofiber sequence

$$B(*, \mathcal{D}, *) \to \operatorname{hocolim} UF \to U \operatorname{hocolim} F.$$

Theorem 6.6.5 says that the space underlying the homotopy colimit of a pointwise cofibrant diagram of based spaces is the quotient of the homotopy colimit of the underlying spaces by the classifying space of the diagram category. For example, the classifying space of the category $\bullet \leftarrow \bullet \rightarrow \bullet$ indexing a pushout diagram is an interval. This interval includes into the unreduced suspension at the basepoint component of the product $X \times I$. The quotient of this map is the usual construction of the reduced suspension.

To isolate the main point, we prove Theorem 6.6.5 in a slightly different form:

Theorem 6.6.6 *Let $F : \mathcal{D} \to \mathbf{Top}_*$ be any diagram and let $U : \mathbf{Top}_* \to \mathbf{Top}$. There is a pushout diagram in $\mathbf{Top}$,*

$$
\begin{array}{ccc}
N(-/\mathcal{D}) \otimes_{\mathcal{D}} * & \longrightarrow & N(-/\mathcal{D}) \otimes_{\mathcal{D}} U F \\
\downarrow & & \downarrow \\
* & \longrightarrow & U(N(-/\mathcal{D}) \otimes_{\mathcal{D}} F)
\end{array}
$$

The cofibrancy hypothesis on the functor F in the topological case of Theorem 6.6.5 is to ensure that the first map of the claimed cofiber sequence is a cofibration, which implies that the associated pushout is a homotopy pushout (see Corollary 14.3.2). But even without this hypothesis, it remains the case that those maps form two sides of a pushout square whose other vertex is the point. Even though these pushouts might not have good homotopical properties, we submit that our proof of the variation described by Theorem 6.6.6 illuminates the reason why Theorem 6.6.5 is true.

Proof of Theorem 6.6.6 The functor tensor products are defined as coequalizers:

$$
\begin{array}{ccccc}
\coprod_{d \to d'} |N(d'/\mathcal{D})| \times * & \rightrightarrows & \coprod_{d} |N(d/\mathcal{D})| \times * & \longrightarrow & N(-/\mathcal{D}) \otimes_{\mathcal{D}} * \cong B(*, \mathcal{D}, *) \\
\downarrow & & \downarrow & & \downarrow \\
\coprod_{d \to d'} |N(d'/\mathcal{D})| \times U F d & \rightrightarrows & \coprod_{d} |N(d/\mathcal{D})| \times U F d & \longrightarrow & N(-/\mathcal{D}) \otimes_{\mathcal{D}} U F \cong B(*, \mathcal{D}, U F) \\
\downarrow & & \downarrow & & \downarrow \\
U(\bigvee_{d \to d'} |N(d'/\mathcal{D})|_+ \wedge F d) & \rightrightarrows & U(\bigvee_{d} |N(d/\mathcal{D})|_+ \wedge F d) & \longrightarrow & U(N(-/\mathcal{D}) \otimes_{\mathcal{D}} F) \cong U B(*, \mathcal{D}, F)
\end{array}
$$

The forgetful functor $U : \mathbf{Top}_* \to \mathbf{Top}$ preserves connected colimits, including coequalizers. (This is the case for the forgetful functor associated to any slice category under an object.) The summands of the various coproducts displayed here encode the appropriate notions of simplicial tensor in $\mathbf{Top}$ and $\mathbf{Top}_*$, as described in 3.3.14. From the definition given there, if $K_\alpha \in \mathbf{sSet}$ and

$X_\alpha \in \mathbf{Top}_*$ we have pushout squares

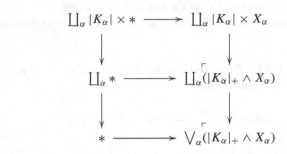

It follows that each of the first two columns forms two sides of a pushout square over the point. Hence the dotted arrows bear the same relationship.   □

In effect, using language we are finally ready to introduce, Theorem 6.6.1 asserts that homotopy limits and homotopy colimits are special cases of **weighted limits and colimits**, at least when we do not have to worry about the "corrections" provided by the deformations Q and R. This observation has a lot of mileage because the general theory of weighted limits and colimits is well developed and allows us to give simple proofs of a number of facts, which we have been postponing. For instance, our proof of Theorem 6.6.6 is a "calculation with the weights," which, when combined with the appropriate homotopical input (e.g., Theorem 11.5.1), proves Theorem 6.6.5.

In particular, our focus on enriched category theory helps us complement the "global" (i.e., derived functor) perspective on homotopy colimits from the "local" one, which defines homotopy colimits to be objects that represent "homotopy coherent cones." With this motivation, let us turn our focus once more to enriched category theory.

Part II

Enriched homotopy theory

7

Weighted limits and colimits

"Weighted" notions of limits and colimits are necessary to describe a comprehensive theory of limits and colimits in enriched categories. Weighted limits extend classical limits in two ways. First, the universal property of the representing object is enriched: an isomorphism of sets is replaced by an isomorphism in the base category for the enrichment. Second, and most interestingly, the shapes of cones that limits and colimits represent are vastly generalized. To build intuition, we begin our journey by examining the role played by weighted limits and colimits in unenriched category theory. In the unenriched context, weighted limits reduce to classical ones. Despite this fact, this perspective can be conceptually clarifying, as we see in the examples presented in this chapter and in Chapter 14.

After first gaining familiarity in the unenriched case, we define internal hom-objects for $\mathcal{V}$-**Cat**, which will be necessary to encode the enriched universal properties of weighted limits and colimits. We then introduce the general theory of weighted limits and colimits in a $\mathcal{V}$-category, also called $\mathcal{V}$-**limits** and $\mathcal{V}$-**colimits**; describe an important special case; and discuss $\mathcal{V}$-completeness and $\mathcal{V}$-cocompleteness. We close this chapter with applications to homotopy limits and colimits, establishing their previously advertised local universal property.

7.1 Weighted limits in unenriched category theory

The limit of a diagram $F \colon \mathcal{C} \to \mathcal{M}$ is an object of $\mathcal{M}$ that represents the functor $\mathcal{M}^{\mathrm{op}} \to \mathbf{Set}$ that maps $m \in \mathcal{M}$ to the set of cones over F with summit m. One way to encode the set of cones over F with summit m is by the set $\mathbf{Set}^{\mathcal{C}}(*, \mathcal{M}(m, F))$ of natural transformations from the constant functor at the terminal object to $\mathcal{M}(m, F) \colon \mathcal{C} \to \mathbf{Set}$. In this presentation, we think of the constant functor $* \colon \mathcal{C} \to \mathbf{Set}$ as specifying the nature of the cone over a diagram of shape $\mathcal{C}$: for each $c \in \mathcal{C}$, the component $* \to \mathcal{M}(m, Fc)$ picks out an arrow

$m \to Fc$ in $\mathcal{M}$. Naturality says that these arrows assemble to form a cone over F with summit m. By definition, a limit is a representation

$$\mathcal{M}(m, \lim F) \cong \mathbf{Set}^{\mathcal{C}}(*, \mathcal{M}(m, F)).$$

Dually, the colimit of F is a representation

$$\mathcal{M}(\mathrm{colim}\, F, m) \cong \mathbf{Set}^{\mathcal{C}^{\mathrm{op}}}(*, \mathcal{M}(F, m)).$$

Here we reversed the variance of the constant functor $*: \mathcal{C}^{\mathrm{op}} \to \mathbf{Set}$ to match that of $\mathcal{M}(F, m): \mathcal{C}^{\mathrm{op}} \to \mathbf{Set}$.

Suppose instead $*$ were replaced by a functor $* \sqcup *: \mathcal{C} \to \mathbf{Set}$ constant at the two-element set. A natural transformation from this functor to $\mathcal{M}(m, F)$ gives rise to two cones over F with summit m. For arbitrary $W: \mathcal{C} \to \mathbf{Set}$, referred to in this context as a **weight**, the data of a natural transformation $W \Rightarrow \mathcal{M}(m, F)$ consists of arrows $m \to Fc$ indexed by the elements of the set Wc, for each $c \in \mathcal{C}$. If $Wf: Wc \to Wc'$ maps $x \in Wc$ to $x' \in Wc'$, naturality implies that the triangle under m whose legs are indexed by x and x' and whose base is Ff must commute.

Definition 7.1.1   A **limit** of $F: \mathcal{C} \to \mathcal{M}$ **weighted by** $W: \mathcal{C} \to \mathbf{Set}$ is an object $\lim^{W} F \in \mathcal{M}$ together together with a representation

$$\mathcal{M}(m, \lim^{W} F) \cong \mathbf{Set}^{\mathcal{C}}(W, \mathcal{M}(m, F)).$$

Example 7.1.2   Let $f: 2 \to \mathcal{M}$ be the functor with image $f: a \to b$ and let $W: 2 \to \mathbf{Set}$ have image $* \sqcup * \to *$, the function from a two element set to a one element set. A natural transformation $W \Rightarrow \mathcal{M}(m, f)$ consists of two arrows $h: m \to a$, $k: m \to a$ which have a common composite with f. From the defining universal property, we see that $\lim^{W} f$ is the pullback

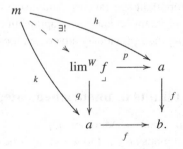

The pair of maps p, q forming the limit cone of shape W is called the **kernel pair** of f.

There is a reason that weighted limits have not attracted more notice in unenriched category theory. Suppose $\mathcal{M}$ is complete, or at least that the limits appearing below exist. Then, by Exercise 1.2.8, the fact that $\mathcal{M}$ is cotensored over **Set**, and the fact that representables preserve limits,

$$\mathbf{Set}^{\mathcal{C}}(W, \mathcal{M}(m, F)) \cong \int_{c \in \mathcal{C}} \mathbf{Set}(Wc, \mathcal{M}(m, Fc)) \cong \int_{c \in \mathcal{C}} \mathcal{M}(m, Fc^{Wc})$$

$$\cong \mathcal{M}(m, \int_{c \in \mathcal{C}} Fc^{Wc}).$$

The end inside the right-hand hom is the functor cotensor product of $W \colon \mathcal{C} \to$ **Set** with $F \colon \mathcal{C} \to \mathcal{M}$. By this calculation, it has the defining universal property of the weighted limit. Hence

$$\lim{}^{W} F \cong \int_{c \in \mathcal{C}} Fc^{Wc}. \qquad (7.1.3)$$

In particular, when $\mathcal{M} = \mathbf{Set}$, Exercise 1.2.8 and (7.1.3) imply that $\lim^{W} F$ is just the set of natural transformations from W to F.

Example 7.1.4 In particular, by the Yoneda lemma, the limit of F weighted by the representable functor $\mathcal{C}(c, -)$ is Fc.

Example 7.1.5 If $\mathcal{M}$ is complete and $\mathcal{C}$ is small, then for any $F \colon \mathcal{C} \to \mathcal{M}$, $K \colon \mathcal{C} \to \mathcal{D}$, there exists a pointwise right Kan extension of F along K defined by

$$\mathrm{Ran}_{K} F(d) = \int_{c \in \mathcal{C}} Fc^{\mathcal{D}(d, Kc)}.$$

Comparing with (7.1.3), we see that $\mathrm{Ran}_{K} F(d)$ is the limit of F weighted by the functor $\mathcal{D}(d, K-) \colon \mathcal{C} \to \mathbf{Set}$.

We can translate the end on the right-hand side of the formula (7.1.3) into a conventional limit:

$$\int_{c \in \mathcal{C}} Fc^{Wc} \cong \mathrm{eq}\left(\prod_{c \in \mathcal{C}} Fc^{Wc} \rightrightarrows \prod_{c \to c' \in \mathcal{C}} Fc'^{Wc}\right)$$

$$\cong \mathrm{eq}\left(\prod_{c} \prod_{Wc} Fc \rightrightarrows \prod_{c \to c'} \prod_{Wc} Fc'\right). \qquad (7.1.6)$$

Recall the limit of a diagram $G \colon \mathcal{D} \to \mathcal{M}$ can be constructed as the equalizer of the two natural maps between the products indexed by the objects and morphisms of $\mathcal{D}$:

$$\lim_{\mathcal{D}} G \cong \mathrm{eq}\left(\prod_{d \in \mathcal{D}} Gd \rightrightarrows \prod_{d \to d' \in \mathcal{D}} Gd'\right). \qquad (7.1.7)$$

This is quite similar to (7.1.6). Indeed, let $\mathcal{D}$ be the **category of elements** of W, typically denoted $\mathbf{el}\,W$. Objects are pairs $(c \in \mathcal{C}, x \in Wc)$ and morphisms $(c, x) \to (c', x')$ are given by arrows $f\colon c \to c'$ in $\mathcal{C}$ such that $Wf(x) = x'$. There is a canonical forgetful functor $\Sigma\colon \mathbf{el}\,W \to \mathcal{C}$. Comparing (7.1.6) with (7.1.7), we see that

$$\lim{}^{W} F \cong \int_{c \in \mathcal{C}} Fc^{Wc} \cong \lim_{\mathbf{el}\,W} F\Sigma. \tag{7.1.8}$$

The procedure that translated the functor $W\colon \mathcal{C} \to \mathbf{Set}$ into the functor $\Sigma\colon \mathbf{el}\,W \to \mathcal{C}$ is called the **Grothendieck construction**. The slogan is that weighted limits in unenriched category theory are ordinary limits indexed by the category of elements of the weight.[1]

We make use of the dual form as well.

Construction 7.1.9 (Grothendieck construction) Given a presheaf $W\colon \mathcal{C}^{\mathrm{op}} \to \mathbf{Set}$, the **contravariant Grothendieck construction** produces a functor $\mathbf{el}\,W \to \mathcal{C}$. By definition, objects in the category $\mathbf{el}\,W$ are pairs $(c \in \mathcal{C}, x \in Wc)$ and morphisms $(c, x) \to (c', x')$ are arrows $f\colon c \to c'$ in $\mathcal{C}$ such that $Wf\colon Wc' \to Wc$ takes x' to x.

The functor $\Sigma\colon \mathbf{el}\,W \to \mathcal{C}$ so constructed is called a **discrete right fibration**, a special type of functor characterized by a certain lifting property: given any arrow $f\colon c \to c'$ in $\mathcal{C}$ and element x' in the fiber over c', there is a unique lift of f that has codomain x'.

Given a discrete right fibration $\Sigma\colon \mathcal{B} \to \mathcal{C}$, define a functor $W\colon \mathcal{C}^{\mathrm{op}} \to \mathbf{Set}$ by taking Wc to be the fiber $\Sigma^{-1}(c)$ and $Wf\colon Wc' \to Wc$ by defining the image of $x' \in Wc' = \Sigma^{-1}(c)$ to be the domain of the unique lift of f with codomain x'. We leave it to the reader to verify that these constructions define an equivalence between the category $\mathbf{Set}^{\mathcal{C}^{\mathrm{op}}}$ and the full subcategory of $\mathbf{Cat}/\mathcal{C}$ of discrete right fibrations over $\mathcal{C}$.

Example 7.1.10 The category of elements of a simplicial set $X\colon \Delta^{\mathrm{op}} \to \mathbf{Set}$ is called the **category of simplices**, introduced in section 4.4. By 7.1.9, its objects are n-simplices in X, for some n, and its morphisms $\alpha\colon \sigma \to \sigma'$ are simplicial operators $\alpha\colon [n] \to [m]$, where $\sigma \in X_n$ and $\sigma' \in X_m$, such that $\sigma' \cdot \alpha = \sigma$. The functor $\Sigma\colon \mathbf{el}\,X \to \Delta$ is a discrete right fibration: given $\alpha\colon [n] \to [m]$ and $\sigma' \in X_m$, there is a unique lift $\sigma' \cdot \alpha \to \sigma'$.

Example 7.1.11 Given $F\colon \mathcal{C} \to \mathcal{M}$, $K\colon \mathcal{C} \to \mathcal{D}$, as in Example 7.1.5, the category of elements of $\mathcal{D}(d, K-)\colon \mathcal{C} \to \mathbf{Set}$ has arrows $d \to Kc$ as objects;

---

[1] Note the distinction between the meaning of "indexed by" and "weighted over" – however, some authors, notably [46], called weighted limits "indexed limits."

morphisms are arrows $f : c \to c'$ in C such that the triangle

commutes. Thus the category of elements of $\mathcal{D}(d, K-)$ is the slice category d/K, and the functor $d/K \to C$ produced by the Grothendieck construction is just the usual forgetful functor. Hence (7.1.8) gives us the formula for pointwise right Kan extensions stated without proof in 1.2.6:

$$\operatorname{Ran}_K F(d) \cong \lim{}^{\mathcal{D}(d,K-)} F \cong \lim(d/K \to C \xrightarrow{F} \mathcal{M}).$$

The category of elements of a functor $W : C \to \mathbf{Set}$ can detect whether the functor is representable.

Lemma 7.1.12 *The functor $W : C \to \mathbf{Set}$ is representable if and only if its category of elements has an initial object.*

Dually, a contravariant functor is representable if and only if its category of elements has a terminal object. This characterization of representability is fundamentally **Set** based and will not generalize to the enriched context.

Proof By the Yoneda lemma, natural transformations $\alpha : C(c, -) \Rightarrow W$ correspond bijectively to objects $(c, \alpha_c(1_c))$ in the category of elements of W. This α is a natural isomorphism just when the corresponding element $x = \alpha_c(1_c) \in Wc$ has the following universal property: for any $x' \in Wc'$, there is a unique arrow $f : c \to c'$ such that $Wf(x) = x'$. But this says precisely that x is initial in the category of elements for W. □

Example 7.1.13 The category of elements of a represented functor $C(-, c) : C^{\mathrm{op}} \to \mathbf{Set}$ is isomorphic to the slice category C/c, which has a terminal object: the identity at c. Indeed, the identity at c defines the canonical representation of $C(-, c)$ by the object c. Note that any isomorphism $c' \cong c$ is also terminal in C/c and also defines a representation $C(-, c') \cong C(-, c)$.

Example 7.1.14 Because the standard n-simplex Δ^n is a represented simplicial set, we know its category of simplices has a terminal object: the unique non-degenerate n-simplex. The fact that this object is terminal in the category of simplices says that every other simplex in Δ^n is uniquely a face or degeneracy of the standard n-simplex.

Example 7.1.15 Writing $\Delta\colon \mathcal{M} \to \mathcal{M}^{\mathcal{C}}$ for the diagonal functor, a cone with summit m is precisely a natural transformation $\Delta m \Rightarrow F$. A representation for the contravariant functor

$$\mathcal{M}^{\mathcal{C}}(\Delta-, F)\colon \mathcal{M}^{\mathrm{op}} \to \mathbf{Set}$$

is precisely a limit for F. By Lemma 7.1.12, this exists just when the category of elements has a terminal object. By Example 7.1.11, the category of elements is the slice category Δ/F. Hence a limit is precisely a terminal object in the slice category Δ/F.[2]

Assuming the weighted limits exist, they assemble into a bifunctor

$$\lim{}^{-} -\colon (\mathbf{Set}^{\mathcal{C}})^{\mathrm{op}} \times \mathcal{M}^{\mathcal{C}} \to \mathcal{M}$$

that is covariant in the diagram but contravariant in the weight. In particular, for any weight W, there is a unique natural transformation $W \Rightarrow *$. Hence, for any $F\colon \mathcal{C} \to \mathcal{M}$, there is a canonical comparison map

$$\lim F \to \lim{}^{W} F$$

from the ordinary limit to the W-weighted limit.

Example 7.1.16 Revisiting Example 7.1.2, the ordinary limit of $f\colon a \to b$ is just a. The comparison map $a \to \lim^{W} f$ is the canonical arrow induced by the universal property of the pullback

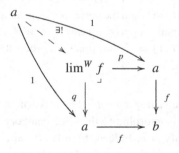

Example 7.1.17 The map induced by $\mathcal{C}(c, -) \Rightarrow *$ between the weighted limits of a diagram $F\colon \mathcal{C} \to \mathcal{M}$ is the leg $\lim F \to Fc$ of the limit cone. For any $f\colon c \to c'$ in $\mathcal{C}$, the unique map from the representable at c' to the terminal functor factors through $f^*\colon \mathcal{C}(c', -) \Rightarrow \mathcal{C}(c, -)$. Functoriality of the weighted

---

[2] Indeed, this was the first definition of a limit encountered by the author, in Peter Johnstone's Part III Category Theory course.

limit bifunctor then tells us that the legs of the limit cone commute with Ff:

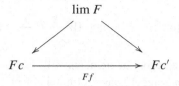

7.2 Weighted colimits in unenriched category theory

The dual notion of weighted colimit is defined by replacing the categories $\mathcal{M}$ and $\mathcal{C}$ of Definition 7.1.1 with $\mathcal{M}^{\mathrm{op}}$ and $\mathcal{C}^{\mathrm{op}}$.

Definition 7.2.1 A **colimit** of $F : \mathcal{C} \to \mathcal{M}$ **weighted by** $W : \mathcal{C}^{\mathrm{op}} \to \mathbf{Set}$ is an object $\mathrm{colim}^W F \in \mathcal{M}$ together with a representation

$$\mathcal{M}(\mathrm{colim}^W F, m) \cong \mathbf{Set}^{\mathcal{C}^{\mathrm{op}}}(W, \mathcal{M}(F, m)).$$

Note the variance of the weight is now opposite that of the diagram.

Exercise 7.2.2 Express a **cokernel pair**, that is, a pushout of $f : a \to b$ along itself as a weighted colimit.

Suppose $\mathcal{M}$ is cocomplete, or at least that the following colimits exist. By 1.2.8,

$$\mathbf{Set}^{\mathcal{C}^{\mathrm{op}}}(W, \mathcal{M}(F, m)) \cong \int_{c \in \mathcal{C}} \mathbf{Set}(Wc, \mathcal{M}(Fc, m))$$

$$\cong \int_{c \in \mathcal{C}} \mathcal{M}(Wc \cdot Fc, m) \cong \mathcal{M}(\int^{c \in \mathcal{C}} Wc \cdot Fc, m).$$

Hence the Yoneda lemma provides an isomorphism

$$\mathrm{colim}^W F \cong \int^{c \in \mathcal{C}} Wc \cdot Fc$$

from which we conclude that weighted colimits are computed as functor tensor products formed using the copower $- \cdot - : \mathbf{Set} \times \mathcal{M} \to \mathcal{M}$.

Example 7.2.3 The value of a pointwise left Kan extension of $F : \mathcal{C} \to \mathcal{M}$ along $K : \mathcal{C} \to \mathcal{D}$ can be computed at an object $d \in \mathcal{D}$ as the colimit of F weighted by $\mathcal{D}(K-, d) : \mathcal{C}^{\mathrm{op}} \to \mathbf{Set}$:

$$\mathrm{Lan}_K F(d) \cong \int^{c \in \mathcal{C}} \mathcal{D}(Kc, d) \cdot Fc \cong \mathrm{colim}^{\mathcal{D}(K-, d)} F.$$

By the duals of the remarks made in Section 7.1, the weighted colimit $\mathrm{colim}^W F$ is isomorphic to the colimit of F reindexed along the functor

$\Sigma\colon \mathbf{el}W \to \mathcal{C}$ produced by the Grothendieck construction. In other words, a weighted colimit is a colimit indexed over the category of elements for W:

$$\mathrm{colim}^W F \cong \mathrm{colim}(\mathbf{el}W \xrightarrow{\Sigma} \mathcal{C} \xrightarrow{F} \mathcal{M}). \qquad (7.2.4)$$

Example 7.2.5 The category of elements of $W = \mathcal{C}(-, c)\colon \mathcal{C}^{\mathrm{op}} \to \mathbf{Set}$ is the slice category $\mathcal{C}/c$. This category has a terminal object. Hence the colimit over any diagram of this shape is just given by evaluating at 1_c. This gives another proof of the co-Yoneda lemma:

$$Fc \cong \mathrm{colim}^{\mathcal{C}(-,c)} F \cong \int^{x \in \mathcal{C}} \mathcal{C}(x, c) \cdot Fx. \qquad (7.2.6)$$

Example 7.2.7 When $F\colon \mathcal{C} \to \mathbf{Set}$, the formula (7.2.6) is symmetric; hence

$$Fc \cong \int^{x \in \mathcal{C}} Fx \cdot \mathcal{C}(x, c) \cong \mathrm{colim}^F \mathcal{C}(-, c).$$

Applying the Grothendieck construction to F, we obtain a functor $\mathbf{el}F \to \mathcal{C}$. Letting $c \in \mathcal{C}$ vary, (7.2.4) tells us that

$$F \cong \mathrm{colim}_{\mathbf{el}F} \mathcal{C}(c, -),$$

that is, that F is the colimit, indexed over its category of elements, of representable functors. This is the precise statement of the density theorem promised in Example 1.4.6.

Example 7.2.8 The density theorem tells us that any simplicial set is canonically a colimit, indexed by its category of simplices, of the represented simplicial sets. Working backward through the preceding analysis, we could also say that any $X \in \mathbf{sSet}$ is canonically the colimit of the Yoneda embedding $\Delta^\bullet\colon \Delta \to \mathbf{sSet}$ weighted by X itself, that is, that

$$X \cong \mathrm{colim}^X \Delta^\bullet.$$

This description fits well with our intuitive understanding of the role played by the weight. For each element of X_n, here thought of as the nth object of the weight, there is a corresponding arrow with domain Δ^n to the weighted colimit X, which, by the Yoneda lemma, corresponds to an n-simplex in the weighted colimit. The face and degeneracy maps in the weight assign faces and degeneracies to the corresponding elements in the weighted colimit.

Example 7.2.9 Given $H\colon \mathcal{C}^{\mathrm{op}} \times \mathcal{C} \to \mathbf{Set}$, the coend $\int^{\mathcal{C}} H$ is the colimit of H weighted by the hom functor $\mathcal{C}(-, -)\colon \mathcal{C}^{\mathrm{op}} \times \mathcal{C} \to \mathbf{Set}$. This follows from the Fubini theorem and the co-Yoneda lemma:

$$\mathrm{colim}^{\mathcal{C}(-,-)}\, H \cong \int^{(x,y)\in\mathcal{C}^{\mathrm{op}}\times\mathcal{C}} \mathcal{C}(x,y) \times H(y,x)$$

$$\cong \int^x \int^y \mathcal{C}(x,y) \times H(y,x) \cong \int^x H(x,x).$$

We can apply the Grothendieck construction to $H\colon \mathcal{C}^{\mathrm{op}} \times \mathcal{C} \to \mathbf{Set}$ regarded as a covariant functor to express coends as ordinary colimits. The result $\mathrm{tw}\mathcal{C} \to \mathcal{C}^{\mathrm{op}} \times \mathcal{C}$ is called the **twisted arrow category**: its objects are arrows in $\mathcal{C}$ and morphisms $f \to g$ are factorizations of g through f, that is, diagrams

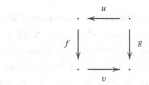

The projections to $\mathcal{C}^{\mathrm{op}}$ and to $\mathcal{C}$ are, respectively, the domain and codomain functors. The functor $\mathbf{tw}\mathcal{C} \to \mathcal{C}^{\mathrm{op}} \times \mathcal{C}$ is a discrete left fibration: given f, u, v, as displayed, there is a unique choice for g.

Remark 7.2.10 There is an alternative method for realizing coends as colimits via the category [50, §IX.5] denotes $\mathcal{C}^{\S}$. The obvious functor $\mathcal{C}^{\S} \to \mathrm{tw}\mathcal{C}$ is final (cf. Example 8.3.9), so any colimit indexed over the twisted arrow category is isomorphic to the colimit of the diagram reindexed along $\mathcal{C}^{\S}$.

When $\mathcal{M}$ is cocomplete, weighted colimits define a bifunctor

$$\mathrm{colim}^- -\colon \mathbf{Set}^{\mathcal{C}^{\mathrm{op}}} \times \mathcal{M}^{\mathcal{C}} \to \mathcal{M}$$

that is covariant in both the weight and the diagram. The unique natural transformation from a weight to the constant functor $*$ produces a canonical comparison from weighted colimits to ordinary colimits.

The practical utility of weighted notions of limit and colimit for defining objects via universal properties has a lot to do with the "cocontinuity of the weight," which we invite the reader to explore on his or her own in the following exercise. This feature, an immediate consequence of the definition, extends to the enriched context.

Exercise 7.2.11 Let $W\colon \mathcal{D} \to \mathbf{Set}$ be the colimit of a diagram of functors $j \mapsto W^j\colon \mathcal{J} \to \mathbf{Set}^{\mathcal{D}}$. Give formulas that express limits and colimits weighted by W in terms of limits and colimits weighted by the W^j. Use the former to compute the limit of a simplicial object in a complete category $\mathcal{E}$ weighted by the simplicial set $\partial\Delta^n$.

7.3 Enriched natural transformations and enriched ends

The point of introducing weighted limits and colimits, of course, is not for **Set** but for categories enriched in some general closed symmetric monoidal category $(\mathcal{V}, \times, *)$. Here closure of the monoidal category $\mathcal{V}$ is essential: weights are $\mathcal{V}$-functors whose target is $\underline{\mathcal{V}}$.

The definition essentially parallels Definitions 7.1.1 and 7.2.1, except, of course, we enrich to $\mathcal{V}$-functors, $\mathcal{V}$-natural transformations, and isomorphisms taking place in $\mathcal{V}$. For this last point, we need a way to encode the set of $\mathcal{V}$-natural transformations between a pair of $\mathcal{V}$-functors as an object in $\mathcal{V}$. Doing so requires a brief digression.

Digression 7.3.1 (the closed symmetric monoidal category $\mathcal{V}$-Cat) The category $\mathcal{V}$-**Cat** of $\mathcal{V}$-categories and $\mathcal{V}$-functors admits a natural monoidal product, but it is not the cartesian product, unless $\mathcal{V}$ itself is cartesian monoidal. Instead, define the tensor product $\underline{\mathcal{C}} \otimes \underline{\mathcal{D}}$ of two $\mathcal{V}$-categories to be the $\mathcal{V}$-category whose objects are pairs (c, d) and whose hom-objects are

$$\underline{\mathcal{C}} \otimes \underline{\mathcal{D}}((c, d), (c', d')) := \underline{\mathcal{C}}(c, c') \otimes \underline{\mathcal{D}}(d, d').$$

Composition and identities are defined in the obvious way; note that these definitions make explicit use of the symmetry isomorphism in $\mathcal{V}$. The tensor, cotensor, and internal hom bifunctors introduced in section 3.7 are $\mathcal{V}$-functors whose domain is the tensor product of the obvious $\mathcal{V}$-categories (cf. Remark 3.7.4).

Modulo size issues, the category $\mathcal{V}$-**Cat** is closed: given $\mathcal{V}$-categories $\underline{\mathcal{D}}$ and $\underline{\mathcal{M}}$, where $\underline{\mathcal{D}}$ is small, define a $\mathcal{V}$-category $\underline{\mathcal{M}}^{\underline{\mathcal{D}}}$ whose objects are $\mathcal{V}$-functors $F, G: \underline{\mathcal{D}} \rightrightarrows \underline{\mathcal{M}}$ and whose hom-objects, taking inspiration from 1.2.8, are defined by the formula

$$\underline{\mathcal{M}}^{\underline{\mathcal{D}}}(F, G) = \int_{d \in \underline{\mathcal{D}}} \underline{\mathcal{M}}(Fd, Gd). \tag{7.3.2}$$

Here the end (7.3.2) is meant in the enriched sense, indicated by the use of $\int_{\underline{\mathcal{D}}}$ in place of $\int_{\mathcal{D}}$. Recall that unenriched ends are computed as equalizers, for example,

$$\int_{\mathcal{D}} \mathcal{M}(Fd, Gd) \cong \mathrm{eq}\left(\prod_d \mathcal{M}(Fd, Gd) \rightrightarrows \prod_{d,d'} \prod_{\mathcal{D}(d,d')} \mathcal{M}(Fd, Gd') \right),$$

where we have taken the liberty of decomposing the right-hand product over the arrows in $\mathcal{D}$ as a nested product first over pairs of objects and then over the corresponding hom-set. In the enriched context, the innermost product over the hom-set $\mathcal{D}(d, d')$ should be replaced by a cotensor with the hom-object $\underline{\mathcal{D}}(d, d')$. In the particular case of (7.3.2), the ambient category is $\mathcal{V}$, so for this example, the cotensor is just the internal hom. Hence the formula for the enriched end defining the hom-object of $\mathcal{V}$-natural transformations is the equalizer

$$\int_{\underline{\mathcal{D}}} \underline{\mathcal{M}}(Fd, Gd) := \text{eq}\left(\prod_d \underline{\mathcal{M}}(Fd, Gd) \rightrightarrows \prod_{d,d'} \underline{\mathcal{V}}(\underline{\mathcal{D}}(d, d'), \underline{\mathcal{M}}(Fd, Gd'))\right).$$

(7.3.3)

The component of the top arrow indexed by the ordered pair d, d' projects to $\underline{\mathcal{M}}(Fd, Gd)$ and composes in the second coordinate with $\underline{\mathcal{D}}(d, d')$. The analogous component of the bottom arrow projects to $\underline{\mathcal{M}}(Fd', Gd')$ and pre-composes in the first coordinate with $\underline{\mathcal{D}}(d, d')$.

Exercise 7.3.4 Describe the underlying set of $\underline{\mathcal{M}}^{\underline{\mathcal{D}}}(F, G)$. Show that if $\underline{\mathcal{D}}$ is the free $\mathcal{V}$-category on an unenriched category $\mathcal{D}$, the enriched end $\int_{\underline{\mathcal{D}}} \underline{\mathcal{M}}$ reduces to the unenriched end $\int_{\mathcal{D}} \mathcal{M}$. Hence the enriched end formula strictly generalizes the unenriched formula.

The object of $\mathcal{V}$-natural transformations allows us to state the $\mathcal{V}$-Yoneda lemma.

Lemma 7.3.5 ($\mathcal{V}$-Yoneda lemma) *Given a small $\mathcal{V}$-category $\underline{\mathcal{D}}$, and object $d \in \underline{\mathcal{D}}$, and a $\mathcal{V}$-functor $F : \underline{\mathcal{D}} \to \underline{\mathcal{V}}$, the canonical map is a $\mathcal{V}$-natural isomorphism*

$$Fd \xrightarrow{\cong} \underline{\mathcal{V}}^{\underline{\mathcal{D}}}(\underline{\mathcal{D}}(d, -), F).$$

Proof Using Lemma 3.5.12, one verifies directly that Fd is the appropriate end. □

7.4 Weighted limits and colimits

Our main purpose is to define and study weighted limits and colimits in a category $\mathcal{M}$ enriched over a closed symmetric monoidal category $\mathcal{V}$.

Definition 7.4.1 Given a $\mathcal{V}$-functor $F : \underline{\mathcal{D}} \to \underline{\mathcal{M}}$ and a $\mathcal{V}$-functor $W : \underline{\mathcal{D}} \to \underline{\mathcal{V}}$, the **weighted limit** of F by W, if it exists, is an object $\lim^W F$ of $\mathcal{M}$ together with a $\mathcal{V}$-natural isomorphism

$$\underline{\mathcal{M}}(m, \lim^W F) \cong \underline{\mathcal{V}}^{\underline{\mathcal{D}}}(W, \underline{\mathcal{M}}(m, F)).$$

Dually, given $F : \underline{\mathcal{D}} \to \underline{\mathcal{M}}$ and $W : \underline{\mathcal{D}}^{\text{op}} \to \underline{\mathcal{V}}$, the **weighted colimit** of F by W, if it exists, is an object $\text{colim}^W F$ of $\mathcal{M}$ together with a $\mathcal{V}$-natural isomorphism

$$\underline{\mathcal{M}}(\text{colim}^W F, m) \cong \underline{\mathcal{V}}^{\underline{\mathcal{D}}^{\text{op}}}(W, \underline{\mathcal{M}}(F, m)).$$

Remark 7.4.2 Note that the weight for a limit of a diagram has the same variance but the weight for a colimit of a diagram has contrasting variance. We like using the letter "W" for the weight for both limits and colimits, but of course this means that when we discuss limits and colimits in parallel, the functor W cannot be the same in both cases.

In practice, the author prefers the notation

$$W \star F := \text{colim}^W F \qquad \text{and} \qquad \{W, F\} := \lim^W F$$

used in [46] and standard within the categorical community because the former evokes a tensor product and the latter a cotensor product (cf. Theorem 7.6.3). However, we suspect the clunkier notation used here will prove easier to digest while these concepts remain unfamiliar.

We are already well acquainted with a pair of simple, but important, examples.

Example 7.4.3 Let $\mathbf{1}$ be the free $\mathcal{V}$-category on the terminal category, that is, the $\mathcal{V}$-category with a single object with hom-object the monoidal unit $*$. A $\mathcal{V}$-functor $\mathbf{1} \to \underline{\mathcal{M}}$ is just an object of $\mathcal{M}$; the map on hom-objects necessarily picks out the designated identity. The $\mathcal{V}$-category $\underline{\mathcal{V}}^{\mathbf{1}}$ of $\mathcal{V}$-functors and $\mathcal{V}$-natural transformations is just $\underline{\mathcal{V}}$. Hence the limit of $n \colon \mathbf{1} \to \underline{\mathcal{M}}$ weighted by $v \colon \mathbf{1} \to \underline{\mathcal{V}}$ is defined by the universal property

$$\underline{\mathcal{M}}(m, \lim^v n) \cong \underline{\mathcal{V}}(v, \underline{\mathcal{M}}(m, n)),$$

which characterizes the **cotensor** of n with v. Dually, the colimit of $m \colon \mathbf{1} \to \underline{\mathcal{M}}$ weighted by $v \colon \mathbf{1}^{\text{op}} \to \underline{\mathcal{V}}$ is defined by the universal property

$$\underline{\mathcal{M}}(\text{colim}^v m, n) \cong \underline{\mathcal{V}}(v, \underline{\mathcal{M}}(m, n)),$$

which characterizes the **tensor** of v and m.

Example 7.4.4 By the $\mathcal{V}$-Yoneda lemma 7.3.5, limits and colimits weighted by representable $\mathcal{V}$-functors are computed by evaluating the diagram at the representing object: the isomorphisms

$$\underline{\mathcal{M}}(m, \lim^{\underline{\mathcal{D}}(d,-)} F) \cong \underline{\mathcal{V}}^{\underline{\mathcal{D}}}(\underline{\mathcal{D}}(d, -), \underline{\mathcal{M}}(m, F-)) \cong \underline{\mathcal{M}}(m, Fd)$$

imply that $\lim^{\underline{\mathcal{D}}(d,-)} F \cong Fd$.

Example 7.4.5 Let $\mathcal{V} = \mathbf{Cat}$ and let $\mathcal{D}$ be the unenriched category that indexes pullback diagrams $\bullet \to \bullet \leftarrow \bullet$. Define the weight $W \colon \mathcal{D} \to \mathbf{Cat}$ and diagram $F : \mathcal{D} \to \mathbf{Cat}$ to be the functors whose respective images are

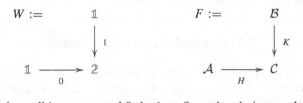

where 2 is the walking arrow and $0, 1 : \mathbb{1} \rightrightarrows 2$ are the obvious endpoint inclusions. Applying the defining universal property of the W-weighted limit of F to the unit object $\mathbb{1} \in \mathbf{Cat}$, which represents the identity functor, the category $\lim^W F$ is characterized by

$$\lim^W F \cong \underline{\mathbf{Cat}}(\mathbb{1}, \lim^W F) \cong \underline{\mathbf{Cat}}^{\mathcal{D}}(W, \underline{\mathbf{Cat}}(\mathbb{1}, F)) \cong \underline{\mathbf{Cat}}^{\mathcal{D}}(W, F).$$

Unpacking the definition of the right-hand category, we see that an object of $\lim^W F$ is a natural transformation $W \Rightarrow F$, that is, a triple ($a \in \mathcal{A}, b \in \mathcal{B}, Ha \to Kb \in \mathcal{C}$). A morphism from $Ha \to Kb$ to $Ha' \to Kb'$ is a pair of morphisms $a \to a'$ in $\mathcal{A}$ and $b \to b'$ in $\mathcal{B}$ such that the obvious square commutes. In this way, we see that $\lim^W F$ is the **comma category** H/K, which generalizes the notions of slice category encountered previously.

Example 7.4.6 The trick we just employed generalizes. Consider $W, F :$ $\mathcal{D} \rightrightarrows \mathcal{V}$. Recall that the unit object in a closed symmetric monoidal category represents the identity $\mathcal{V}$-functor. In particular, we see from the defining universal property that

$$\lim^W F \cong \underline{\mathcal{V}}(*, \lim^W F) \cong \underline{\mathcal{V}}^{\mathcal{D}}(W, \underline{\mathcal{V}}(*, F)) \cong \underline{\mathcal{V}}^{\mathcal{D}}(W, F).$$

In other words, the weighted limit of a $\mathcal{V}$-valued $\mathcal{V}$-functor is the object of $\mathcal{V}$-natural transformations from the weight to the diagram.

Remark 7.4.7 Using Example 7.4.6, we can reexpress the defining universal properties of weighted (co)limits in terms of weighted limits in the $\mathcal{V}$-category $\underline{\mathcal{V}}$, much like ordinary limits and colimits are defined in terms of limits in **Set**:

$$\underline{\mathcal{M}}(m, \lim^W F) \cong \underline{\mathcal{V}}^{\mathcal{D}}(W, \underline{\mathcal{M}}(m, F)) \cong \lim^W \underline{\mathcal{M}}(m, F)$$

$$\underline{\mathcal{M}}(\operatorname{colim}^W F, m) \cong \underline{\mathcal{V}}^{\mathcal{D}^{\mathrm{op}}}(W, \underline{\mathcal{M}}(F, m)) \cong \lim^W \underline{\mathcal{M}}(F, m).$$

We draw an important conclusion from these isomorphisms: enriched representable functors preserve weighted limits in the codomain variable and colimits in the domain one. The reason is that this is essentially tautologous: weighted colimits and limits in $\underline{\mathcal{M}}$ are defined representably as weighted limits in $\underline{\mathcal{V}}$, and hence in terms of ordinary limits in $\mathcal{V}$, and thus ultimately in terms of ordinary limits in **Set** (which are just sets of cones).

7.5 Conical limits and colimits

As always, we suppose that the base for enrichment $\mathcal{V}$ is a closed symmetric monoidal category that is complete and cocomplete. Recall that the functor assigning the free $\mathcal{V}$-category to an unenriched category defines a left adjoint to the underlying category functor: a $\mathcal{V}$-functor $F: \mathcal{D} \to \underline{\mathcal{M}}$ corresponds to an unenriched functor $F: \mathcal{D} \to \mathcal{M}$ taking values in the underlying category (cf. 3.3.4 and 3.4.5). In this special case, where the domain of our diagram is a free $\mathcal{V}$-category, we might choose our weight to be the functor $*: \mathcal{D} \to \mathcal{V}$ that is constant at the monoidal unit. The resulting weighted limit is called a **conical limit**.[3]

The conical limit $\lim^* F$ has defining universal property

$$\underline{\mathcal{M}}(m, \lim^* F) \cong \underline{\mathcal{V}}^{\mathcal{D}}(*, \underline{\mathcal{M}}(m, F)) \cong \lim^* \underline{\mathcal{M}}(m, F),$$

the last isomorphism by 7.4.7. The underlying set functor $\mathcal{V}(*, -): \mathcal{V} \to$ **Set** yields an isomorphism

$$\mathcal{M}(m, \lim^* F) \cong \mathcal{V}^{\mathcal{D}}(*, \underline{\mathcal{M}}(m, F)).$$

By Exercise 7.3.4, the right-hand side is the set of natural transformations $* \Rightarrow \underline{\mathcal{M}}(m, F)$. By the definition of arrows in the underlying category $\mathcal{M}$, a natural transformation $* \Rightarrow \underline{\mathcal{M}}(m, F)$ is just a cone over F with summit m in $\mathcal{M}$. In particular, conical limits in $\underline{\mathcal{M}}$ are ordinary limits in $\mathcal{M}$, though the defining universal property of the conical limit is stronger. Conical colimits are, of course, defined dually:

$$\underline{\mathcal{M}}(\operatorname{colim}^* F, m) \cong \lim^* \underline{\mathcal{M}}(F, m).$$

In passing from ordinary limits to weighted limits, we are permitted more flexibility in specifying both the shape of cones and the universal property satisfied by representing objects. For conical limits, the cones are unchanged but the universal property is enriched. To say that the disjoint union of simplicial sets is a coproduct in the unenriched sense is to say that maps $X \sqcup Y \to Z$ correspond to pairs of maps $X \to Z$ and $Y \to Z$. To say $X \sqcup Y$ is the conical coproduct is to say that the spaces of maps $Z^{X \sqcup Y} \cong Z^X \times Z^Y$ are isomorphic. The constituent isomorphism on vertices, obtained by applying the underlying

---

[3] Note that when $\mathcal{D}$ is not a free $\mathcal{V}$-category, there might not exist a "constant" $\mathcal{V}$-functor $*: \mathcal{D} \to \mathcal{V}$. (For instance, consider $\mathcal{V} = \mathbf{Mod}_R$.) A special case where constant $\mathcal{V}$-functors exist for any enriched category $\mathcal{D}$ is when $\mathcal{V}$ is cartesian closed, in which case the monoidal unit $*$ is the terminal object. See [46, §3.9] – which, however, is titled "The inadequacy of conical limits" – for a discussion.

set functor $(-)_0 \colon$ **sSet** $\to$ **Set**, is precisely the unenriched statement; the space-level isomorphism also posits natural bijections between the sets of maps

$$\{(X \sqcup Y) \times \Delta^n \to Z\} \cong \{(X \times \Delta^n) \sqcup (Y \times \Delta^n) \to Z\}$$

$$\cong \{X \times \Delta^n \to Z\} \times \{Y \times \Delta^n \to Z\}.$$

As this example illustrates, the universal property of a conical limit or colimit is much stronger than that asserted by the corresponding classical limit or colimit. Even so, in common settings, ordinary limits and colimits, assuming they exist, automatically satisfy an enriched universal property. For instance:

Lemma 7.5.1 *Consider* $F \colon \mathcal{D} \to \mathcal{V}$ *taking values in a closed symmetric monoidal category* $\mathcal{V}$. *Then its ordinary limit is a conical limit.*

Proof By (7.3.3) and 7.4.6 and

$$\lim{}^* F \cong \underline{\mathcal{V}}^{\mathcal{D}}(*, F) \cong \mathrm{eq}\left(\prod_{d \in \mathcal{D}} \underline{\mathcal{V}}(*, Fd) \rightrightarrows \prod_{d,d' \in \mathcal{D}} \underline{\mathcal{V}}(*, Fd')^{\mathcal{D}(d,d')}\right)$$

$$\cong \mathrm{eq}\left(\prod_{d \in \mathcal{D}} Fd \rightrightarrows \prod_{d \to d' \in \mathcal{D}} Fd'\right) \cong \lim_{\mathcal{D}} F \qquad \square$$

Example 7.5.2 Let $\mathcal{V} = $ **Cat** and let $\underline{\mathcal{M}}$ be the 2-category freely generated by a 2-cell endomorphism, as depicted:

$$a \underset{f}{\overset{f}{\Longrightarrow}}{\Downarrow\alpha} b$$

The underlying category of $\underline{\mathcal{M}}$, forgetting the 2-cells, is isomorphic to the category 2. In the underlying category, the object b is the product $b \times b$, but this limit is not conical. The conical limit $b \times b$ must have an isomorphism

$$\underline{\mathcal{M}}(a, b \times b) \cong \underline{\mathbf{Cat}}(*, \underline{\mathcal{M}}(a, b)) \times \underline{\mathbf{Cat}}(*, \underline{\mathcal{M}}(a, b)) \cong \underline{\mathcal{M}}(a, b) \times \underline{\mathcal{M}}(a, b).$$

Here $\underline{\mathcal{M}}(a, b)$ is the category corresponding to the monoid $\mathbb{N}$, which is not isomorphic to its square.

There is a good reason why this example was somewhat contrived:

Theorem 7.5.3 *When* $\underline{\mathcal{M}}$ *is tensored over* $\mathcal{V}$, *all limits in* $\mathcal{M}$ *are conical* $\mathcal{V}$-*limits in* $\underline{\mathcal{M}}$.

Proof If $\mathcal{M}$ is tensored, then each $\underline{\mathcal{M}}(m, -)$ is a right adjoint and hence sends limits in $\mathcal{M}$ to limits in $\mathcal{V}$. By Lemma 7.5.1, ordinary limits in $\mathcal{V}$ are conical, and the result follows.

More slowly, by the Yoneda lemma, to show that the natural map

$$\mathcal{M}(m, \lim_{\mathcal{D}} F) \to \lim_{\mathcal{D}} \mathcal{M}(m, F)$$

is an isomorphism in $\mathcal{V}$, it suffices to show that its image under $\mathcal{V}(v, -)$ is an isomorphism for all $v \in \mathcal{V}$. This follows from the postulated adjunctions $- \otimes m \dashv \mathcal{M}(m, -)$ and the fact that ordinary representables preserve ordinary limits:

$$\mathcal{V}(v, \mathcal{M}(m, \lim_{\mathcal{D}} F)) \cong \mathcal{M}(v \otimes m, \lim_{\mathcal{D}} F) \cong \lim_{\mathcal{D}} \mathcal{M}(v \otimes m, F)$$

$$\cong \lim_{\mathcal{D}} \mathcal{V}(v, \mathcal{M}(m, F)) \cong \mathcal{V}(v, \lim_{\mathcal{D}} \mathcal{M}(m, F)).$$

By Lemma 7.5.1, the ordinary limit over $\mathcal{D}$ of $\mathcal{M}(m, F) \colon \mathcal{D} \to \mathcal{V}$ is conical. By Remark 7.4.7, this conical limit in $\mathcal{V}$ encodes the desired universal property. $\quad\square$

The dual of Theorem 7.5.3 says that the presence of cotensors, a $\mathcal{V}$-limit, implies that any ordinary colimits are $\mathcal{V}$-colimits. For instance, by the axiomatization of Lemma 3.8.6, all the limits and colimits in a simplicial model category are conical, meaning they satisfy a simplicially enriched universal property.

7.6 Enriched completeness and cocompleteness

A classical theorem says that ordinary limits can be constructed out of products and equalizers via the formula displayed in (7.1.7). In particular, if a category has equalizers and small products, then it has all small limits. We would like an analogous result that characterizes $\mathcal{V}$-categories that admit weighted limits for any small $\mathcal{V}$-category $\mathcal{D}$ and any weight $W \colon \mathcal{D} \to \mathcal{V}$. Such $\mathcal{V}$-categories are called $\mathcal{V}$-**complete**.

For the sake of contrast, let us dualize and explore conditions for $\mathcal{V}$-**cocompleteness**. As a preliminary, we should note that in a tensored $\mathcal{V}$-category $\mathcal{M}$, there is a notion of enriched functor tensor product of $\mathcal{V}$-functors $G \colon \mathcal{D}^{\mathrm{op}} \to \mathcal{V}$ and $F \colon \mathcal{D} \to \mathcal{M}$ defined using the enriched coend $\int^{\mathcal{D}}$ in place of the ordinary coend $\int^{\mathcal{D}}$:

$$G \otimes_{\mathcal{D}} F := \int^{d \in \mathcal{D}} Gd \otimes Fd$$

$$\cong \mathrm{coeq} \left(\coprod_{d, d'} \mathcal{D}(d, d') \otimes (Gd' \otimes Fd) \rightrightarrows \coprod_{d} Gd \otimes Fd \right). \quad (7.6.1)$$

Example 7.6.2   The enriched versions of the isomorphisms of 4.1.4 also hold – that is, $\mathcal{D}(-, d) \otimes_{\mathcal{D}} F \cong Fd$ and $G \otimes_{\mathcal{D}} \mathcal{D}(d, -) \cong Gd$ – as can economically be verified by one of our favorite methods for computing colimits: guess and check. In other words, write down the obvious cone and show that it satisfies the universal property to be the desired colimit.

A priori, the formula (7.6.1) uses a mix of weighted colimits (the tensors) and ordinary unenriched colimits (the coproducts and coequalizer). But when $\underline{\mathcal{M}}$ is also cotensored over $\mathcal{V}$, the dual of Theorem 7.5.3 says that these ordinary colimits are conical colimits, in which case the right-hand side of (7.6.1) is a weighted colimit.

To calculate its universal property, we use the fact that enriched representables preserve weighted colimits to see that

$$\underline{\mathcal{M}}(G \otimes_{\underline{\mathcal{D}}} F, m) \cong \underline{\mathcal{M}}\left(\int^{\underline{\mathcal{D}}} Gd \otimes Fd, m\right) \cong \int_{\underline{\mathcal{D}}} \underline{\mathcal{M}}(Gd \otimes Fd, m)$$

$$\cong \int_{\underline{\mathcal{D}}} \underline{\mathcal{V}}(Gd, \underline{\mathcal{M}}(Fd, m))$$

$$\cong \underline{\mathcal{V}}^{\underline{\mathcal{D}}^{\mathrm{op}}}(G, \underline{\mathcal{M}}(F, m)) \cong \lim{}^{G} \underline{\mathcal{M}}(F, m).$$

By 7.4.7, this says that $G \otimes_{\underline{\mathcal{D}}} F$ is the limit of F weighted by G.

With this simple calculation, we have just proven two theorems:

Theorem 7.6.3   *When $\underline{\mathcal{M}}$ is tensored and cotensored, then there are natural isomorphisms*

$$\mathrm{colim}^{W} F \cong W \otimes_{\underline{\mathcal{D}}} F \qquad \text{and} \qquad \lim{}^{W} F \cong \{W, F\}^{\underline{\mathcal{D}}}$$

whenever these weighted limits and colimits exist.

Corollary 7.6.4   *A $\mathcal{V}$-category $\underline{\mathcal{M}}$ is $\mathcal{V}$-complete and $\mathcal{V}$-cocomplete if it is tensored and cotensored and if its underlying category is complete and cocomplete.*

A category satisfying the hypotheses of Corollary 7.6.4 is called $\mathcal{V}$-**bicomplete**.

Exercise 7.6.5   Prove, using the defining universal property, that in a cotensored $\mathcal{V}$-category, colimits with arbitrary weights preserve (pointwise) tensors. Note that this implies a complementary result to Lemma 3.8.3: in a tensored and cotensored simplicial category $\mathcal{M}$, geometric realization preserves both pointwise tensors and the tensors defined in 3.8.2.

In a $\mathcal{V}$-complete and $\mathcal{V}$-cocomplete category $\underline{\mathcal{M}}$, weighted limits and colimits define $\mathcal{V}$-functors

$$\lim{}^{-} - : (\underline{\mathcal{V}}^{\underline{\mathcal{D}}})^{\mathrm{op}} \otimes \underline{\mathcal{M}}^{\underline{\mathcal{D}}} \to \underline{\mathcal{M}} \qquad \mathrm{colim}^{-} - : \underline{\mathcal{V}}^{\underline{\mathcal{D}}^{\mathrm{op}}} \otimes \underline{\mathcal{M}}^{\underline{\mathcal{D}}} \to \underline{\mathcal{M}}.$$

Example 7.6.6   The geometric realization of a simplicial object $X \colon \Delta^{\mathrm{op}} \to \mathcal{M}$ in a tensored and cotensored simplicial category is the colimit weighted by the Yoneda embedding. Note that the unique natural transformation $\Delta^{\bullet} \Rightarrow *$

induces a canonical map

$$\text{colim}^{\Delta^\bullet} X \cong |X| \longrightarrow \pi_0 X \cong \text{coeq}(X_1 \rightrightarrows X_0).$$

For instance, this gives rise to the comparison between the two-sided bar construction $B(G, \mathcal{D}, F)$ and the functor tensor product $G \otimes_\mathcal{D} F$ mentioned in Section 4.2.

Now consider a diagram of $\mathcal{V}$-functors

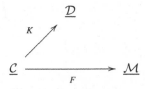

where $\mathcal{C}$ is small. Taking inspiration from 7.2.3, we define the left and right $\mathcal{V}$-enriched Kan extensions of $F : \mathcal{C} \to \mathcal{M}$ along $K : \mathcal{C} \to \mathcal{D}$ to be the weighted colimit and weighted limit

$$\text{Lan}_K F(d) := \text{colim}^{\mathcal{D}(K, d)} F \cong \mathcal{D}(K, d) \otimes_{\mathcal{C}} F \tag{7.6.7}$$

$$\text{Ran}_K F(d) := \lim^{\mathcal{D}(d, K)} F \cong \{\mathcal{D}(d, K), F\}^{\mathcal{C}}.$$

This definition adopts the preferred terminology of Kelly, who posits that the only Kan extensions worthy of the name are those defined pointwise (cf. Section 1.3).

Exercise 7.6.8   Suppose $\mathcal{M}$ is $\mathcal{V}$-bicomplete. Describe the universal property of left and right Kan extensions. Show conversely that a $\mathcal{V}$-functor satisfying the appropriate universal property is a pointwise Kan extension in the sense defined in (7.6.7).

Example 7.6.9 (extension of scalars)   Recall that an **Ab**-functor $f : R \to S$ between one-object **Ab**-categories is precisely a ring homomorphism, and an **Ab**-functor $M : R \to \underline{\textbf{Ab}}$ is a left R-module. The left **Ab**-enriched Kan extension of M along f is the left S-module $S \otimes_R M : S \to \underline{\textbf{Ab}}$ obtained via **extension of scalars**.

7.7 Homotopy (co)limits as weighted (co)limits

Recognizing homotopy colimits as weighted colimits will allow us to understand their local universal property. Suppose $\mathcal{M}$ is a simplicial model category with all objects cofibrant and let $F : \mathcal{D} \to \mathcal{M}$. By Theorem 6.6.1 and Theorem 7.6.3, our preferred model of the homotopy colimit of F is just the colimit of

F weighted by $N(-/\mathcal{D})\colon \mathcal{D}^{\mathrm{op}} \to \mathbf{sSet}$. From the defining universal property of the weighted colimit

$$\mathcal{M}(\mathrm{hocolim}_\mathcal{D} F, m) \cong \underline{\mathbf{sSet}}^{\mathcal{D}^{\mathrm{op}}}(N(-/\mathcal{D}), \underline{\mathcal{M}}(F, m)),$$

we see that the homotopy colimit is an object $\mathrm{hocolim}_\mathcal{D} F \in \mathcal{M}$ equipped with a universal simplicial natural transformation $N(d/\mathcal{D}) \Rightarrow \underline{\mathcal{M}}(Fd, \mathrm{hocolim}_\mathcal{D} F)$.

Example 7.7.1 For instance, consider the case where $\mathcal{D}$ is the pushout diagram $b \leftarrow a \to c$ and $\mathcal{M} = \mathbf{Top}$. The slice categories $b/\mathcal{D}$ and $c/\mathcal{D}$ are isomorphic to the terminal category, so the corresponding components of the simplicial natural transformation pick out points in the spaces $\underline{\mathcal{M}}(Fb, \mathrm{hocolim}_\mathcal{D} F)$ and $\underline{\mathcal{M}}(Fc, \mathrm{hocolim}_\mathcal{D} F)$. The space $N(a/\mathcal{D})$ is the wedge of two intervals, picking out a path in $\underline{\mathcal{M}}(Fa, \mathrm{hocolim}_\mathcal{D} F)$. By naturality, this path defines a homotopy between the chosen maps $Fa \to Fb \to \mathrm{hocolim}_\mathcal{D} F$ and $Fa \to Fc \to \mathrm{hocolim}_\mathcal{D} F$. This describes the local universal property of the homotopy pushout.

Example 7.7.2 The weight for the homotopy colimit of a simplicial object is the opposite of the cosimplicial simplicial set $[n] \mapsto N(\Delta/[n])$. There is a natural transformation of cosimplicial simplicial sets from $N(\Delta/-)$ to the Yoneda embedding $\Delta^\bullet$ called the "last vertex" map. Its components $N(\Delta/[n])_m \to \Delta^n_m$ take $[n_0] \to [n_1] \to \cdots \to [n_m] \to [n]$ to the map $\alpha\colon [m] \to [n]$, where $\alpha(i)$ is defined to be the image of $n_i \in [n_i]$ in $[n]$. Note that the represented simplicial sets Δ^n are each isomorphic to their opposites. Hence the "last vertex" map equally defines a natural transformation $N(-/(\Delta^{\mathrm{op}})) \cong N((\Delta/-)^{\mathrm{op}}) \cong N(\Delta/-)^{\mathrm{op}} \Rightarrow \Delta^\bullet$. Assuming for simplicity that the simplicial object $X_\bullet$ is pointwise cofibrant, by bifunctoriality of weighted colimits, it follows that there is a canonical map between the weighted colimits

$$\mathrm{hocolim}_{\Delta^{\mathrm{op}}} X \cong N(\Delta/-)^{\mathrm{op}} \otimes_{\Delta^{\mathrm{op}}} X \to \Delta^\bullet \otimes_{\Delta^{\mathrm{op}}} X \cong |X|.$$

This is called the **Bousfield–Kan map**. We see in Theorem 14.3.1, an extension of Theorem 5.2.3, that if X is Reedy cofibrant, then this map is a weak equivalence.

Another homotopy invariant construction on simplicial objects is described in Example 8.5.12.

Example 7.7.3 Dualizing, we can use the universal property as a weighted limit to better understand the totalization of a cosimplicial space $X^\bullet\colon \Delta \to \mathbf{Top}$. Writing $S\mathbf{Top}$ for the mapping spaces, as defined in Example 3.7.15, the defining universal property of the weighted limit is

$$S\underline{\mathbf{Top}}(m, \mathrm{Tot}(X^\bullet)) \cong \underline{\mathbf{sSet}}^\Delta(\Delta^\bullet, S\underline{\mathbf{Top}}(m, X^\bullet)).$$

Taking m to be a point, we obtain an isomorphism

$$S\mathrm{Tot}(X^\bullet) \cong \underline{\mathbf{sSet}}^\Delta(\Delta^\bullet, SX^\bullet) \cong S\underline{\mathbf{Top}}^\Delta(|\Delta^\bullet|, X^\bullet)$$

between the total singular complex of the space $\mathrm{Tot}(X^\bullet)$ and the simplicial set of natural transformations between the cosimplicial spaces $|\Delta^\bullet|$ and $X^\bullet$.[4]

Taking underlying sets, we see that a point in the totalization consists of a point in X^0, a path in X^1 connecting the two images of this point under the coface maps, a (topological) 2-simplex in X^2 whose boundary is given by the three images of this path, and so on, where this data is compatible with the codegeneracies in the following sense. The image of the path in X^1 under the map $X^1 \to X^0$ is the constant path at the chosen point in X^0. The images of the 2-simplex under $X^2 \rightrightarrows X^1$ are appropriately degenerate 2-simplices on the chosen path, and so on.

Example 7.7.4   In particular, let $F, G\colon \mathcal{D} \rightrightarrows \mathbf{sSet}$. Recall that $C(F, \mathcal{D}, G)$ is the totalization of the cosimplicial object $C^n(F, \mathcal{D}, G) = \prod_{\vec{d}\colon [n]\to\mathcal{D}} Gd_n^{Fd_0}$. So a point in $C(F, \mathcal{D}, G)$ consists of maps $Fd \to Gd$ for each d; a map $Fd \times \Delta^1 \to Gd'$ for each $d \to d'$ restricting to the previously chosen maps; a map $Fd \times \Delta^2 \to Gd''$ for each chain $d \to d' \to d''$ restricting to previously chosen homotopies; and so forth, subject to the condition that this higher-dimensional data are degenerate if they are indexed by appropriately degenerate simplices in $N\mathcal{D}$. We call this data a **homotopy coherent natural transformation** from F to G.

For any $F, G\colon \mathcal{D} \rightrightarrows \mathcal{M}$, there is a natural isomorphism

$$C(F, \mathcal{D}, G) \cong \underline{\mathcal{M}}^{\mathcal{D}}(B(\mathcal{D}, \mathcal{D}, F), G) \qquad\qquad (7.7.5)$$

obtained by manipulating the defining weighted limits. One interpretation of the role played by the left deformation $B(\mathcal{D}, \mathcal{D}, -)\colon \mathcal{M}^{\mathcal{D}} \to \mathcal{M}^{\mathcal{D}}$ is that it replaces a diagram F by a "fattened" diagram $B(\mathcal{D}, \mathcal{D}, F)$ so that ordinary natural transformations $B(\mathcal{D}, \mathcal{D}, F) \Rightarrow G$ correspond to homotopy coherent natural transformations from F to G. Applying this identity to a pair $*, X\colon G \to \mathbf{Top}$, we see that

$$\operatorname*{holim}_{G} X \cong C(*, G, X) \cong \underline{\mathbf{Top}}^{G}(B(G, G, *), X) = \underline{\mathbf{Top}}^{G}(EG, X)$$

is the space of **homotopy fixed points**. See Example 6.5.8.

---

[4] A reader who would prefer to ignore this simplicial structure and work with the analogous topological weighted limits is free to do so (see Section 8.2).

7.8 Balancing bar and cobar constructions

We have begun to see that our understanding of homotopy (co)limits as weighted (co)limits can help us understand the local universal property of these objects. Some of these explorations, such as the one based on equations such as (7.7.5), could have been given much sooner, but there is a reason we have neglected to tell this story until now. To explain, consider $F: \mathcal{D} \to \mathcal{M}$, $\mathcal{M}$ a simplicial model category, and permit us a quick calculation:

$$\underline{\mathcal{M}}(B(*, \mathcal{D}, F), m) \cong \underline{\mathcal{M}}\left(\int^{n \in \Delta} \Delta^n \otimes (\coprod_{\vec{d} \in N\mathcal{D}_n} F d_0), m\right)$$

$$\cong \int_{n \in \Delta} \underline{\mathcal{M}}(\Delta^n \otimes (\coprod_{\vec{d} \in N\mathcal{D}_n} F d_0), m)$$

$$\cong \int_{n \in \Delta} \underline{\mathcal{M}}(\coprod_{\vec{d} \in N\mathcal{D}_n} F d_0, m^{\Delta^n})$$

$$\cong \int_{n \in \Delta} \prod_{\vec{d} \in N\mathcal{D}_n} \underline{\mathcal{M}}(F d_0, m^{\Delta^n})$$

$$\cong \int_{n \in \Delta} \prod_{\vec{d} \in N\mathcal{D}_n} \underline{\mathcal{M}}(F d_0, m)^{\Delta^n}. \tag{7.8.1}$$

Here the first isomorphism is the definition of the bar construction; the second and fourth use the fact that representables preserve weighted colimits; and the third and fifth use the tensor–cotensor–hom adjunction, the latter requiring that this adjunction is simplicially enriched. The final line looks similar to the cobar construction applied to the functor $\underline{\mathcal{M}}(F, m)$ but for two issues. The first is that this functor is contravariant in $\mathcal{D}$, not covariant – but of course, the product over $N\mathcal{D}_n$ is equally the product over $N\mathcal{D}_n^{\text{op}}$. However, if we perform such a replacement, then the cosimplicial object is dualized: the outer coface maps are swapped, and so on. For some simplicial categories $\mathcal{M}$ such as **Top**, the totalization of a cosimplicial object is isomorphic to the totalization of its opposite, but in others, such as **sSet**, this is not the case.[5] So the last line is not quite $C(*, \mathcal{D}^{\text{op}}, \underline{\mathcal{M}}(F, m))$ but rather something quite close to it.

---

[5] For a general small category $\mathcal{D}$, the simplicial sets $N\mathcal{D}$ and $N\mathcal{D}^{\text{op}}$ are not isomorphic, only weak homotopy equivalent. One way to prove this is to observe that the twisted arrow category (introduced in Example 7.2.9) comes equipped with canonical source and target projections $\mathcal{D}^{\text{op}} \xleftarrow{s} \text{tw}\mathcal{D} \xrightarrow{t} \mathcal{D}$. The nerve $N(\text{tw}\mathcal{D})$ can be described as the **edgewise subdivision** of $N\mathcal{D}$, defined to be the simplicial set obtained by precomposing with the functor $\Delta \to \Delta$ that takes the linearly ordered set $[n]$ to the join $[n]^{\text{op}} \star [n] \cong [2n + 1]$. Any simplicial set is weak homotopy equivalent to its edgewise subdivision via the maps described in [77, Appendix 1].

With 6.6.1, there is a much simpler proof of the isomorphism (7.8.1). Immediately from the defining universal property of a weighted colimit,

$$\mathcal{M}(\text{colim}^{N(-/\mathcal{D})} F, m) \cong \lim^{N(-/\mathcal{D})} \mathcal{M}(F, m). \qquad (7.8.2)$$

If $\mathcal{M} = \mathbf{Top}$, its simplicial enrichment was defined by applying the total singular complex functor S to its mapping spaces. All spaces are fibrant, so this mapping space is a fibrant simplicial set. But, as previously, the right-hand side is not quite our definition of the homotopy limit, which was

$$\underset{\mathcal{D}^{\text{op}}}{\text{holim}} \, \mathcal{M}(F, m) \cong \lim^{N(\mathcal{D}^{\text{op}}/-)} \mathcal{M}(F, m) \cong \lim^{N(-/\mathcal{D})^{\text{op}}} \mathcal{M}(F, m).$$

We regret that our conventions do not allow the slogans expressing the local universal property of the homotopy colimit – the space of maps out of the homotopy colimit is the homotopy limit of the mapping spaces – to hold on the nose. Instead, as a consequence of 11.5.14, this is only true up to weak homotopy equivalence.

Remark 7.8.3 This might be a good time to remark about how our conventions compare with others in the literature.[6] The construction of homotopy limits in [36] agrees with ours, but homotopy colimits are constructed as a colimit of a pointwise cofibrant replacement of the diagram weighted by $N(-/\mathcal{D})^{\text{op}}$, not $N(-/\mathcal{D})$, as is our convention. Neither construction of [10] agrees with ours: homotopy colimits are as in [36], and homotopy limits of pointwise fibrant diagrams are limits weighted by $N(\mathcal{D}/-)^{\text{op}}$. These "op"s appear because the definition of the nerve of a category in [10] is the opposite of the modern convention.

Our conventions were chosen so as to have as few "op"s as possible, and also so that the constructions as weighted (co)limits are isomorphic to the constructions via the standard definitions of the two-sided (co)bar constructions. The isomorphism (7.8.2) motivates Philip Hirschhorn's conventions (cf. Remark [36, 18.1.11]).

---

[6] Of course, morally, a homotopy colimit or homotopy limit is only defined up to weak equivalence, so all conventions agree.

8

Categorical tools for homotopy (co)limit computations

In this chapter, we collect a few miscellaneous results aimed at simplifying computations of homotopy limits and colimits. Several of these techniques come directly from enriched category theory, providing further justification for our lengthy detour through the theory of weighted limits and colimits. The relative simplicity of the proofs in this chapter illustrates how easy it is to obtain computationally useful results with the theory we have developed. For instance, the reduction theorem, which provides a formula for the homotopy colimit of a restricted diagram, is an immediate corollary of a general result about colimits weighted by left Kan extensions.

After discussing a few simple applications of the theory of weighted limits and colimits to homotopy theory, we turn our attention directly to the base for enrichment. We observe that homotopy limits and colimits in a topological model category can be defined directly in that context without pulling the enrichment back to simplicial sets. Furthermore, our preferred models for the homotopy (co)limit functors are *isomorphic* in both cases, not merely weakly equivalent. Our arguments are formal and thus generalize to other enriched contexts.

Our final topic is the theory of homotopy initial and final functors, extending analogous results from ordinary and enriched category theory. In exploring this material, we take care to separate the homotopical results from the categorical (up-to-isomorphism) ones because we find such distinctions to be conceptually clarifying.

8.1 Preservation of weighted limits and colimits

The first result is no surprise.

Proposition 8.1.1   *Right $\mathcal{V}$-adjoints preserve $\mathcal{V}$-limits.*

Proof   Consider a $\mathcal{V}$-adjunction $L: \underline{\mathcal{M}} \rightleftarrows \underline{\mathcal{N}}: R$, a diagram $F: \underline{\mathcal{D}} \to \underline{\mathcal{M}}$, and a weight $W: \underline{\mathcal{D}} \to \underline{\mathcal{V}}$. The chain of isomorphisms

$$\underline{\mathcal{M}}(m, R\lim^W F) \cong \underline{\mathcal{N}}(Lm, \lim^W F) \cong \underline{\mathcal{V}}^{\underline{\mathcal{D}}}(W, \underline{\mathcal{N}}(Lm, F))$$

$$\cong \underline{\mathcal{V}}^{\underline{\mathcal{D}}}(W, \underline{\mathcal{M}}(m, RF)) \cong \underline{\mathcal{M}}(m, \lim^W RF)$$

proves the claim by the Yoneda lemma 3.5.12.   □

We can use Proposition 8.1.1 to give simple proofs of 6.5.5 and 7.6.5, which had previously been left as exercises for the reader.

Corollary 8.1.2   *Suppose $F: \mathcal{D} \to \mathbf{Top}_*$ is any diagram. Then the underlying space of its homotopy limit is the homotopy limit of the diagram $UF: \mathcal{D} \to \mathbf{Top}$ of underlying spaces.*

Proof   We prove that our preferred models for these spaces are isomorphic. It follows that the underlying spaces of any other models for the homotopy limits are weakly equivalent because the forgetful functor U is homotopical.

By Theorems 6.6.1 and 7.6.3, the homotopy limit of F is the limit of F weighted by $N(\mathcal{D}/-): \mathcal{D} \to \mathbf{sSet}$. The underlying space–disjoint basepoint adjunction constructed in 3.3.14 is simplicial by Theorem 3.7.11 because the left adjoint is strong monoidal. Now the result follows from Proposition 8.1.1.   □

Corollary 8.1.3   *Weighted limits commute with pointwise cotensors in a tensored $\mathcal{V}$-category.*

Proof   In a tensored $\mathcal{V}$-category, cotensors are right $\mathcal{V}$-adjoints.   □

Suppose now that $\mathcal{M}$ is $\mathcal{V}$-bicomplete and consider a pair of $\mathcal{V}$-functors $F: \underline{\mathcal{C}} \to \underline{\mathcal{M}}$ and $K: \underline{\mathcal{C}} \to \underline{\mathcal{D}}$, with $\underline{\mathcal{C}}$ small. Recall that the $\mathcal{V}$-enriched left and right Kan extensions of a $\mathcal{V}$-functor $F: \underline{\mathcal{C}} \to \underline{\mathcal{M}}$ along $K: \underline{\mathcal{C}} \to \underline{\mathcal{D}}$ are defined, in the case where $\underline{\mathcal{C}}$ is small, by the weighted colimit and weighted limit

$$\mathrm{Lan}_K F(d) := \mathrm{colim}^{\underline{\mathcal{D}}(K,d)} F \cong \underline{\mathcal{D}}(K, d) \otimes_{\underline{\mathcal{C}}} F$$

$$\mathrm{Ran}_K F(d) := \lim^{\underline{\mathcal{D}}(d,K)} F \cong \{\underline{\mathcal{D}}(d, K), F\}^{\underline{\mathcal{C}}}.$$

Lemma 4.4.3, which says geometric realizations of n-skeletal simplicial objects can be computed by the functor tensor product of the n-truncated simplicial object with the restricted Yoneda embedding, is a special case of a general result:

Lemma 8.1.4   *Suppose $\mathcal{M}$ is a tensored and cotensored $\mathcal{V}$-category and let $K: \underline{\mathcal{C}} \to \underline{\mathcal{D}}, F: \underline{\mathcal{C}} \to \underline{\mathcal{M}}$, and $W: \underline{\mathcal{D}}^{\mathrm{op}} \to \underline{\mathcal{V}}$ be $\mathcal{V}$-functors. Then*

$$\mathrm{colim}^{WK} F \cong \mathrm{colim}^W \mathrm{Lan}_K F,$$

assuming these weighted colimits exist.

Proof By the co-Yoneda lemma and Fubini's theorem, both weighted colimits are isomorphic to the enriched coend

$$\int^{\mathcal{C}} \int^{\mathcal{D}} (Wd \times \underline{\mathcal{D}}(Kc, d)) \otimes Fc. \qquad \square$$

In the special case where $\underline{\mathcal{C}}$ and $\underline{\mathcal{D}}$ are both free $\mathcal{V}$-categories and W is the constant functor at the monoidal unit, WK is again constant at this object. Lemma 8.1.4 shows that the conical colimits $\operatorname{colim}^* \operatorname{Lan}_K F \cong \operatorname{colim}^* F$ are isomorphic, generalizing an observation made in Section 4.4.

We make use of several "dual" versions of Lemma 8.1.4, which the reader should pause for a moment to discover.

Exercise 8.1.5 State and prove another version of this result, for instance, when the weight is a left or right Kan extension.

One dual version provides the conceptual underpinning for the proof of Lemma 3.8.3.

Corollary 8.1.6 *Let $\mathcal{M}$ be a tensored and cotensored simplicial category. Then geometric realization $| - |: \mathcal{M}^{\Delta^{op}} \to \mathcal{M}$ preserves the simplicial tensors of 3.7.18.*

Proof We apply a dual version of Lemma 8.1.4 in the case $\mathcal{V} = \mathbf{Set}$. By direct computation, the left Kan extension of the Yoneda embedding $\Delta^\bullet: \Delta \hookrightarrow \mathbf{sSet}$ along the diagonal functor $\Delta: \Delta \to \Delta \times \Delta$ is the functor $([n], [m]) \mapsto \Delta^n \times \Delta^m$. It follows that the colimit of a bisimplicial object weighted by this left Kan extension is isomorphic to the colimit of its diagonal weighted by the Yoneda embedding.

Let K be a simplicial set and X be a simplicial object in a tensored (so that geometric realization as defined in 4.1.6 makes sense) and cotensored (so that the simplicial tensor preserves colimits) category. It follows that

$$K \otimes |X| \cong (K_\bullet \otimes_\Delta \Delta^\bullet) \otimes (\Delta^\bullet \otimes_{\Delta^{op}} X_\bullet) \cong (\Delta^\bullet \times \Delta^\bullet) \otimes_{\Delta^{op} \times \Delta^{op}} (\bigsqcup_{K_\bullet} X_\bullet)$$

$$\cong \Delta^\bullet \otimes_{\Delta^{op}} (\bigsqcup_{K_\bullet} X_\bullet) \cong |K \otimes X|. \qquad \square$$

Another corollary is the following a priori mysterious result from Hirschhorn:

Corollary 8.1.7 ([36, 18.9.1]) *Let X be a simplicial set and let $Y: (\mathbf{el} X)^{op} \to \mathcal{M}$ be a pointwise cofibrant contravariant functor from its category of simplices to a simplicial model category $\mathcal{M}$. Define a simplicial object $Z: \Delta^{op} \to \mathcal{M}$ with $Z_n = \bigsqcup_{\sigma \in X_n} Y(\sigma)$. Then the homotopy colimits of Y and Z are isomorphic.*

Again, what we mean to say is that our preferred models of these homotopy colimits are isomorphic.

Proof   Both diagrams are pointwise cofibrant, so the homotopy colimits may be computed as functor tensor products and hence as weighted colimits. Using the observations made in the discussion motivating the simplicial bar construction in Section 4.4, it is easy to see that Z is the left Kan extension of Y along the discrete left fibration $\Sigma \colon (\mathbf{el}X)^{\mathrm{op}} \to \Delta^{\mathrm{op}}$. Using the fact that this functor is a discrete left fibration, it follows that the slice categories $\sigma/(\mathbf{el}X)^{\mathrm{op}}$ and $\Sigma\sigma/\Delta^{\mathrm{op}}$ are isomorphic for all $\sigma \in \mathbf{el}X$. Hence

$$\operatorname{hocolim} Z \cong \operatorname{colim}^{N(-/\Delta^{\mathrm{op}})} \operatorname{Lan}_\Sigma Y \cong \operatorname{colim}^{N(\Sigma-/\Delta^{\mathrm{op}})} Y$$

$$\cong \operatorname{colim}^{N(-/(\mathbf{el}X)^{\mathrm{op}})} Y \cong \operatorname{hocolim} Y. \qquad \square$$

Finally, the reduction theorem of [23] is a special case of one dual version of Lemma 8.1.4.

Theorem 8.1.8 (reduction theorem)   *Let $F \colon \mathcal{D} \to \mathcal{M}$ be a pointwise cofibrant diagram in a simplicial model category and let $K \colon \mathcal{C} \to \mathcal{D}$. Then*

$$\operatorname{hocolim}_\mathcal{C} FK \cong N(-/K) \otimes_\mathcal{D} F.$$

Proof   This result is an immediate corollary of a dual version of Lemma 8.1.4 because $N(-/K) \cong \operatorname{Lan}_K N(-/\mathcal{C})$. Just for fun, we also give an alternate proof following [37]. By direct inspection, for any $d \in \mathcal{D}$, $N(d/K) \cong B(*, \mathcal{C}, \mathcal{D}(d, K-))$, where we regard the functor $\mathcal{D}(d, K-) \colon \mathcal{C} \to \mathbf{sSet}$ as taking values in discrete simplicial sets. Hence

$$N(-/K) \otimes_\mathcal{D} F \cong B(*, \mathcal{C}, \mathcal{D}(-, K-)) \otimes_\mathcal{D} F$$

$$\cong B(*, \mathcal{C}, \mathcal{D}(-, K-) \otimes_\mathcal{D} F) \cong B(*, \mathcal{C}, FK),$$

the last isomorphism by 4.1.4. $\qquad \square$

8.2 Change of base for homotopy limits and colimits

In the proof of Lemma 7.5.1 in the last chapter, we used the known constructions of the ordinary and conical limit of a $\mathcal{V}$-valued diagram, the former as an equalizer and the latter as an enriched end, to show that the same object satisfied the two distinct defining universal properties. In this section, we repeat this trick with particular weighted colimits of a diagram $F \colon \mathcal{D} \to \mathbf{Top}$ to show, among other things, that two reasonable definitions of homotopy colimits produce the same object, even though the defining universal properties are a priori distinct. Here we focus on the convenient category of spaces for concreteness, but our proofs are quite general. At the conclusion of this section, we state a general form of this "change of base" result, whose proof is the same.

To illustrate the need for a result of this form, let us dualize the discussion of totalization from Example 7.7.3. Combining Definition 4.1.6 and Theorem 7.6.3, the geometric realization of a simplicial object $X \colon \Delta^{\mathrm{op}} \to \mathbf{Top}$ is the colimit weighted by the Yoneda embedding $\Delta^\bullet \colon \Delta \to \mathbf{sSet}$. Because our weight is valued in simplicial sets, here again we are implicitly using the simplicial structure on $\mathbf{Top}$ defined in 3.7.15. The defining universal property of this weighted colimit is therefore an isomorphism of simplicial sets

$$S\underline{\mathbf{Top}}(|X|, Z) \cong \underline{\mathbf{sSet}}^\Delta(\Delta^\bullet, S\underline{\mathbf{Top}}(X_\bullet, Z)) \cong S\underline{\mathbf{Top}}^\Delta(|\Delta^\bullet|, \underline{\mathbf{Top}}(X_\bullet, Z))$$

for any $Z \in \mathbf{Top}$. Here we write $\underline{\mathbf{Top}}$ for its internal hom as a closed monoidal category, which means that $S\underline{\mathbf{Top}}$ is the hom-object in its simplicial enrichment. The second isomorphism uses the fact that the adjunction $| - | \dashv S$ is simplicially enriched by Corollary 3.7.12 and Lemma 6.1.6.

But by construction,

$$|X| := \int^{n \in \Delta^{\mathrm{op}}} \Delta^n \otimes X_n := \int^{n \in \Delta^{\mathrm{op}}} |\Delta^n| \times X_n,$$

which, by Theorem 7.6.3, is the $\mathbf{Top}$-enriched colimit of X weighted by the composite functor $\Delta \xrightarrow{\Delta^\bullet} \mathbf{sSet} \xrightarrow{|-|} \mathbf{Top}$. (The hypotheses of Theorem 7.6.3 apply because $\mathbf{Top}$, as a closed monoidal category, is tensored and cotensored.) Thus $|X|$ also is characterized by the isomorphism of spaces

$$\underline{\mathbf{Top}}(|X|, Z) \cong \underline{\mathbf{Top}}^\Delta(|\Delta^\bullet|, \underline{\mathbf{Top}}(X_\bullet, Z)),$$

which feels much more natural.

The same argument proves the following theorem:

Theorem 8.2.1 *Let $F \colon \mathcal{D} \to \mathbf{Top}$. Then the homotopy colimit of F is computed by both the $\mathbf{sSet}$-enriched colimit weighted by $N(-/\mathcal{D})$ or the $\mathbf{Top}$-enriched colimit weighted by $|N(-/\mathcal{D})|$. Dually, the homotopy limit of F is both the $\mathbf{sSet}$-enriched limit weighted by $N(\mathcal{D}/-)$ and the $\mathbf{Top}$-enriched limit weighted by $|N(\mathcal{D}/-)|$.*

Proof By Theorem 7.6.3, the formulae for the pair of weighted colimits or the pair of weighted limits are identical, even though the defining universal properties differ. $\square$

The argument here is categorical, not homotopical. What we mean to assert is that our preferred homotopy colimit functors are isomorphic, not merely weakly equivalent. The argument just given also proves the following general "change of base" result:

Theorem 8.2.2 *Suppose we have an adjunction $F \colon \mathcal{V} \rightleftarrows \mathcal{U} \colon G$ between closed symmetric monoidal categories such that the left adjoint F is strong monoidal, and let $\mathcal{D}$ be any small unenriched category. Consider a diagram*

$H: \mathcal{D} \to \mathcal{U}$ *and a* $\mathcal{V}$-*valued weight* W, *either covariant or contravariant in* $\mathcal{D}$, *as appropriate to the setting. Writing subscripts to indicate whether the following are meant to be interpreted with respect to the canonical* $\mathcal{U}$-*enrichments or* $\mathcal{V}$-*enrichments, we have*

$$\lim_{\mathcal{V}}^{W} H \cong \lim_{\mathcal{U}}^{FW} H \quad and \quad \operatorname{colim}_{\mathcal{V}}^{W} H \cong \operatorname{colim}_{\mathcal{U}}^{FW} H.$$

Proof   The hypotheses of Theorem 7.6.3 apply to both the canonical $\mathcal{U}$-enrichment and to the $\mathcal{V}$-enrichment of $\mathcal{U}$ defined by Theorem 3.7.11. Using these formulae, the weighted limits and colimits in the two enriched contexts are defined by exactly the same functor cotensor or tensor products.   □

Under additional hypotheses designed to guarantee that certain weighted limits and colimits are homotopically meaningful, Theorem 8.2.2 gives a "change of base" result for computing homotopy limits and colimits. If $\mathcal{V}$ is cocomplete, a cosimplicial object $\Delta^{\bullet} : \Delta \to \mathcal{V}$ produces an adjunction $\mathbf{sSet} \rightleftarrows \mathcal{V}$, as described in 1.5.1. If this left adjoint is strong monoidal, then Theorem 3.7.11 implies that $\mathcal{U}$-enrichments, tensors, and cotensors can be pulled back first to $\mathcal{V}$ and then to simplicial sets. If $\mathcal{V}$ is assumed to be a $\mathcal{V}$-model category and $\mathcal{U}$ to be a $\mathcal{U}$-model category, and if both adjunctions are Quillen, then, as remarked at the end of Chapter 3, these structures guarantee that any $\mathcal{U}$-model category or $\mathcal{V}$-model category becomes a simplicial model category in a canonical way. We leave it to the reader to observe that the natural definitions of homotopy limits and colimits in any of these enriched contexts coincide.

8.3 Final functors in unenriched category theory

In some cases, it is possible to reduce a colimit over a particular diagram to a colimit over a simpler diagram. For instance:

Lemma 8.3.1   *If* $\mathcal{D}$ *has a terminal object* t *and* $F: \mathcal{D} \to \mathcal{M}$, *then* $\operatorname{colim}_{\mathcal{D}} F \cong Ft$.

Proof 1   Guess and check: it is straightforward to show that the obvious cone to Ft has the desired universal property.   □

Proof 2   The parallel pair

$$\coprod_{f: d \to d'} d \underset{\iota_{d'} \circ f}{\overset{\iota_d}{\rightrightarrows}} \coprod_{d} d$$

extends to a split coequalizer diagram

$$\coprod_{f:\, d\to d'} d \underset{\iota_{d'}\circ f}{\overset{\iota_d}{\rightrightarrows}} \coprod_{d} d \underset{!}{\overset{\iota_t}{\leftrightarrows}} t$$

where the ιs denote coproduct inclusions. Hence t is the colimit of the parallel pair, and furthermore, this colimit is preserved by any functor.[1] □

Proof 3 The functor $t \colon \mathbb{1} \to \mathcal{D}$ is a **final functor**. □

Definition 8.3.2 A functor $K \colon \mathcal{C} \to \mathcal{D}$ is **final** if, for any functor $F \colon \mathcal{D} \to \mathcal{M}$, the canonical map

$$\operatorname*{colim}_{\mathcal{C}} F K \overset{\cong}{\to} \operatorname*{colim}_{\mathcal{D}} F$$

is an isomorphism, both sides existing if either does. Dually, K is **initial** if, for any $F \colon \mathcal{D} \to \mathcal{M}$, the canonical map

$$\lim_{\mathcal{D}} F \overset{\cong}{\to} \lim_{\mathcal{C}} F K$$

is an isomorphism.

Remark 8.3.3 Final functors were originally called "cofinal," motivated by the notion of a cofinal subsequence (see Example 8.5.2). The terminology used here is the modern categorical consensus, adopted because the directionality of "cofinal" is confusing and the correct dual terminology is even more so.

As it stands, the third proof of Lemma 8.3.1 just rephrases the original claim. The substance is on account of the following simple characterization of final functors.

Lemma 8.3.4 *A functor $K \colon \mathcal{C} \to \mathcal{D}$ is final if and only if, for each $d \in \mathcal{D}$, the slice category d/K is non-empty and connected.*

Remark 8.3.5 A category is **connected** just when any pair of objects can be joined by a finite zigzag of arrows. Let $\pi_0 \colon \mathbf{Cat} \to \mathbf{Set}$ be the "path components" functor that sends a category to its collection of objects up to such zigzags. This functor is left adjoint to the inclusion $\mathbf{Set} \to \mathbf{Cat}$, whose right adjoint is the functor that takes a category to its underlying set of objects. A category $\mathcal{C}$ is non-empty and connected if and only if $\pi_0 \mathcal{C}$ is the singleton set.

[1] Such colimits are called **absolute**.

Proof of Lemma 8.3.4   Given $F : \mathcal{D} \to \mathcal{M}$, a cone under F immediately gives rise to a cone under FK. Conversely, a cone under FK induces a unique cone under F in the case where each d/K is non-empty and connected. Given a cone $\lambda_c : FKc \to m$, define the leg of the cone indexed by d by choosing any arrow $d \to Kc$, which is possible since d/K is non-empty, and composing its image under F with λ_c. Connectedness of d/K shows that any two choices can be connected by a zigzag of commutative triangles, the bases of which commute with the maps to m because λ is a cone over FK. Because the sets of cones under FK and F are isomorphic, so are the colimits.

For the converse, first note that for any $X : \mathcal{C} \to \mathbf{Set}$, there is an isomorphism $\pi_0(\mathbf{el}X) \cong \mathrm{colim}_{\mathcal{C}} X$ because each arrow connecting two objects in $\mathbf{el}X$ corresponds to a condition demanding that these elements are identified in any cone under X. Recall from Example 7.1.11 that $\mathbf{el}\mathcal{D}(d, K-) \cong d/K$. Now suppose K is final. Then $\pi_0(d/K) \cong \mathrm{colim}_{\mathcal{C}} \mathcal{D}(d, K-) \cong \mathrm{colim}_{\mathcal{D}} \mathcal{D}(d, -) \cong *$, by inspection or the co-Yoneda lemma. Hence d/K is non-empty and connected.   □

Proof 3, revised   When $t : \mathbb{1} \to \mathcal{D}$ is a terminal object, each d/t is the category $\mathbb{1}$.   □

Final functors can reduce large colimits to small colimits or, in special cases, to small limits:

Corollary 8.3.6   *For any category $\mathcal{D}$, the colimit of the identity functor $1_{\mathcal{D}} : \mathcal{D} \to \mathcal{D}$, if it exists, is a terminal object of $\mathcal{D}$, and conversely, any terminal object defines a colimit of $1_{\mathcal{D}} : \mathcal{D} \to \mathcal{D}$.*

Proof   If $\mathcal{D}$ has a terminal object t, then $t : \mathbb{1} \to \mathcal{D}$ is final, and hence $\mathrm{colim}\, 1_{\mathcal{D}} \cong \mathrm{colim}\, t \cong t$. Conversely, if t is a colimit of the identity functor, the component of the colimit cone at t is easily seen to be an idempotent e. But then e and 1_t define the same cone under the identity diagram, hence $e = 1_t$, and it follows easily that t is terminal.   □

Example 8.3.7   If $\mathcal{D}$ is a disjoint union of categories $\mathcal{D}_\alpha$, each containing a terminal object, then the natural inclusion of the discrete category on these objects is final. In particular, the colimit of a diagram of shape $\mathcal{D}$ is isomorphic to the coproduct of the images of the terminal objects in each component. We made use of this observation in our explorations of the simplicial bar construction in Section 4.4.

Example 8.3.8   Write 2_2 for the category $\bullet \rightrightarrows \bullet$ of a parallel pair of arrows. The functor $2_2 \to \Delta^{\mathrm{op}}$ with image $[1] \rightrightarrows [0]$ is final. Hence the colimit of a simplicial object X is just the coequalizer of the face maps $X_1 \rightrightarrows X_0$, which produces the set of path components of the geometric realization of X.

Example 8.3.9 The twisted arrow category introduced in Example 7.2.9 admits a final functor $K : C^\S \to \mathrm{tw}C$, where $C^\S$ has an object for each object or morphism of C and arrows so that $C^\S$ has the shape of the wedges displayed in (1.2.3). The functor K is surjective on objects, but its image contains very few morphisms. Nonetheless, it is final, and hence coends can be computed as (ordinary) colimits over the twisted arrow category or as colimits over $C^\S$, as described in [50, §IX.5].

8.4 Final functors in enriched category theory

The notions of initial and final functors can be generalized to enriched category theory. To simplify the discussion, we consider only *cartesian* closed symmetric monoidal categories $\mathcal{V}$. For such $\mathcal{V}$, the monoidal unit is terminal, and hence there exists a constant $\mathcal{V}$-functor $* : \mathcal{D} \to \mathcal{V}$ for any small $\mathcal{V}$-category $\mathcal{D}$.

For sake of contrast, let us dualize. We say a $\mathcal{V}$-functor $K : \mathcal{C} \to \mathcal{D}$ is **initial** if it satisfies the equivalent conditions of the following theorem.

Theorem 8.4.1 ([46, 4.67]) *Suppose $\mathcal{V}$ is cartesian closed and $K : \mathcal{C} \to \mathcal{D}$ is a $\mathcal{V}$-functor. The following are equivalent:*

(i) *For any $F : \mathcal{D} \to \mathcal{M}$, the conical limits $\lim^* F \cong \lim^* FK$ are isomorphic, either side existing if the other does.*

(ii) *For each $d \in \mathcal{D}$, $\mathrm{colim}^* \mathcal{D}(K-, d) = *$.*

(iii) *The constant map $* : \mathcal{D} \to \mathcal{V}$ is the left Kan extension of the constant map $* : \mathcal{C} \to \mathcal{V}$ along K.*

Proof The conical colimit of $\mathcal{D}(K-, d)$ is the functor tensor product

$$* \otimes_{\mathcal{C}^{op}} \mathcal{D}(K-, d) \cong \mathcal{D}(K-, d) \otimes_{\mathcal{C}} *,$$

which also computes the left Kan extension of the constant functor $* : \mathcal{C} \to \mathcal{V}$ along K. Hence the second and third conditions are equivalent. The third condition implies the first by one of the duals of Lemma 8.1.4.

To prove the second condition given the first, we use the defining universal property of the conical colimit and the fact that $\mathrm{colim}^* \mathcal{D}(-, d) = *$, as a consequence of the co-Yoneda lemma. Then

$$\mathcal{V}(\mathrm{colim}^* \mathcal{D}(K-, d), v) \cong \lim^* \mathcal{V}(\mathcal{D}(K-, d), v) \cong \lim^* \mathcal{V}(\mathcal{D}(-, d), v)$$

$$\cong \mathcal{V}(\mathrm{colim}^* \mathcal{D}(-, d), v) \cong \mathcal{V}(*, v),$$

whence, by Lemma 3.5.12, the result. $\qquad\square$

Example 8.4.2   If $K : \mathcal{C} \to \mathcal{D}$ has a right $\mathcal{V}$-adjoint R, then precomposition with K is *right* $\mathcal{V}$-adjoint to precomposition with R. The following sequence of isomorphisms and Lemma 3.5.12 prove that K is initial:

$$\underline{\mathcal{M}}(m, \lim{}^* FK) \cong \lim{}^* \underline{\mathcal{M}}(m, FK) \cong \lim{}^* \underline{\mathcal{M}}(mR, F)$$

$$\cong \lim{}^* \underline{\mathcal{M}}(m, F) \cong \underline{\mathcal{M}}(m, \lim{}^* F).$$

8.5 Homotopy final functors

Motivated by the definitions in Sections 8.3 and 8.4, we introduce the homotopical version of initial and final functors. For this we need a homotopical notion: a simplicial set is **contractible** if the unique map to the terminal object is a weak homotopy equivalence.

Definition 8.5.1   A functor $K : \mathcal{C} \to \mathcal{D}$ is **homotopy final** if the simplicial set $N(d/K)$ is contractible for all $d \in \mathcal{D}$ and **homotopy initial** if each $N(K/d)$ is contractible.

A cofinal sequence defines a homotopy final subcategory of the associated ordinal category.

Example 8.5.2   Let ω be the ordinal category $0 \to 1 \to 2 \to \cdots$ and let $K : \omega \to \omega$ be any final functor, equivalently, any functor whose object function is cofinal in the classical sense. We claim ω is homotopy final. For each $n \in \omega$, there is some m such that $n < K(m)$, and hence the slice category n/K is non-empty. For each such m, either $K(m) = K(m + 1)$, in which case m and $m + 1$ index isomorphic objects of n/K, or $K(m) < K(m + 1)$, in which case there is a unique arrow from the former object to the latter in n/K. The category n/K is equivalent to its skeleton, which is isomorphic to ω by the observation just made. The nerve functor $N : \mathbf{Cat} \to \mathbf{sSet}$ takes equivalences of categories to homotopy equivalences of simplicial sets because it takes natural transformations to homotopies (see the proof of Lemma 8.5.3). Hence $N(n/K) \simeq N(\omega)$. But ω has an initial object, so there is a homotopy from the constant endofunctor at the initial object to the identity endofunctor, implying that $N(\omega)$ and $N(\mathbb{1})$ are homotopy equivalent, and hence that $N(n/K) \simeq N(\omega) \simeq N(\mathbb{1}) = \Delta^0$. Thus a cofinal sequence in ω gives rise to a homotopy final functor.

The following lemma records one component of the argument we have just used.

Lemma 8.5.3   *If $\mathcal{D}$ is a category with an initial object, then $N\mathcal{D}$ is contractible.*

Proof If $\mathcal{D}$ has an initial object i, there is a natural transformation from the constant functor at the initial object to the identity functor. A natural transformation $\alpha: F \Rightarrow G$ between functors $F, G: \mathcal{C} \rightrightarrows \mathcal{D}$ can be encoded by a functor $\mathcal{C} \times 2 \to \mathcal{D}$ restricting to F and G on the subcategories corresponding to the two objects of the walking arrow category 2. Right adjoints, of course, preserve products; hence, taking nerves, natural transformations become homotopies. In particular, the natural transformation from the constant functor to the identity defines a simplicial homotopy equivalence between the geometric realizations of the maps $i: \mathbb{1} \to \mathcal{D}$ and $!: \mathcal{D} \to \mathbb{1}$. $\qquad\square$

Example 8.5.4 Lemma 8.5.3 gives another proof that the space EG defined in Example 4.5.5 is contractible. EG is the geometric realization of the nerve of the **translation groupoid** of the discrete group G. The translation groupoid is the category with elements of G as objects and a unique morphism in each hom-set; the intuition is that the map $g' \to g''$ represents the unique $g \in G$ so that $gg' = g''$. This category is equivalent to the terminal category and, in particular, has an initial object.

A category $\mathcal{C}$ is **filtered** if any finite diagram in $\mathcal{C}$ has a cone under it. Equivalently, $\mathcal{C}$ is filtered if

(i) for each pair of objects c, c' there is some c'' together with maps $c \to c''$ and $c' \to c''$, and if
(ii) for each pair of morphisms $c \rightrightarrows c'$ there is some $c' \to c''$ such that the two composites coincide.

Lemma 8.5.5 *If $K: \mathcal{C} \to \mathcal{D}$ is homotopy final, then K is final. Conversely, if $\mathcal{C}$ is filtered and $K: \mathcal{C} \to \mathcal{D}$ is a final functor, then K is homotopy final.*

Proof For any category $\mathcal{C}$, $\pi_0 \mathcal{C} \cong \pi_0 N\mathcal{C}$. If K is homotopy final, then each $N(d/K)$ is contractible, and $\pi_0(d/K) \cong \pi_0 N(d/K) = *$ implies that d/K is non-empty and connected. Now suppose that $\mathcal{C} \to \mathcal{D}$ is final and that $\mathcal{C}$ is filtered. Combining these facts, it is easy to show that d/K is also filtered: condition (ii) is immediate and condition (i) follows because a zigzag connecting a pair of objects $d \to Kc$ and $d \to Kc'$ defines a finite diagram in $\mathcal{C}$. The claim now follows from the classical result that nerves of filtered categories are contractible: the main idea of the proof is that a class in the nth homotopy group of a simplicial set X can be represented by a map $S \to X$, where the geometric realization of S is homotopy equivalent to S^n and has only finitely many non-degenerate simplices. $\qquad\square$

These definitions are justified by the following theorem:

Theorem 8.5.6 (homotopy finality)   *Let $F \colon \mathcal{D} \to \mathcal{M}$ be any diagram in a simplicial model category. If $K \colon \mathcal{C} \to \mathcal{D}$ is homotopy final, then $\operatorname{hocolim}_{\mathcal{C}} FK \to \operatorname{hocolim}_{\mathcal{D}} F$ is a weak equivalence.*

Proof   We use the homotopical aspects of the bar construction detailed in 5.2.5. When K is homotopy final, the natural map $N(d/K) \to N(d/\mathcal{D})$ is a weak equivalence because both spaces are contractible. Using the isomorphisms $N(-/K) \cong B(*, \mathcal{C}, \mathcal{D}(-, K-))$ and $N(-/\mathcal{D}) \cong B(*, \mathcal{D}, \mathcal{D})$ of 4.2.6 and the fact that $B(-, \mathcal{D}, QF)$ preserves weak equivalences, the top map in the following diagram is a weak equivalence:

$$
\begin{array}{ccc}
B(B(*, \mathcal{C}, \mathcal{D}(-, K-)), \mathcal{D}, QF) & \xrightarrow{\ \sim\ } & B(B(*, \mathcal{D}, \mathcal{D}), \mathcal{D}, QF) \\
\cong \downarrow & & \downarrow \cong \\
B(*, \mathcal{C}, B(\mathcal{D}(-, K-), \mathcal{D}, QF)) & \longrightarrow & B(*, \mathcal{D}, B(\mathcal{D}, \mathcal{D}, QF)) \\
\downarrow & & \downarrow \\
B(*, \mathcal{C}, QFK) & \longrightarrow & B(*, \mathcal{D}, QF) \\
\cong \downarrow & & \downarrow \cong \\
\operatorname{hocolim}_{\mathcal{C}} FK & \longrightarrow & \operatorname{hocolim}_{\mathcal{D}} F
\end{array}
$$

The associativity of the bar construction gives the vertical isomorphisms and hence the middle weak equivalence. The left vertical map is defined using the natural augmentation from the simplicial object $B_{\bullet}(\mathcal{D}(-, Kc), \mathcal{D}, QF)$ to $QFKc$; the right vertical map is a special case of this. Furthermore, both augmented simplicial objects have forward contracting homotopies, as detailed in Example 4.5.7. By Corollary 4.5.2, these maps define weak equivalences $B(\mathcal{D}(-, K-), \mathcal{D}, QF) \to QFK$ and $B(\mathcal{D}, \mathcal{D}, QF) \to QF$ between pointwise cofibrant diagrams, which are preserved by $B(*, \mathcal{C}, -)$ and $B(*, \mathcal{D}, -)$ by 5.2.5. Hence the vertical maps are weak equivalences, and therefore so is the bottom map by the 2-of-3 property. $\square$

An alternate proof of this theorem is given at the end of Section 11.5.

Exercise 8.5.7   Using Theorem 8.1.8, show that if $K \colon \mathcal{C} \to \mathcal{D}$ is any functor such that for any diagram $F \colon \mathcal{D} \to \mathbf{sSet}$, the natural map $\operatorname{hocolim}_{\mathcal{C}} FK \to \operatorname{hocolim}_{\mathcal{D}} F$ is a weak equivalence, then K is homotopy final.

Corollary 8.5.8 (Quillen's Theorem A)   *If $K \colon \mathcal{C} \to \mathcal{D}$ is homotopy final, then $N\mathcal{C} \to N\mathcal{D}$ is a weak equivalence.*

Proof   A map of simplicial sets is a weak equivalence just when the induced map of geometric realizations is a weak equivalence. But $|N\mathcal{C}| = B(*, \mathcal{C}, *) \cong \operatorname{hocolim}_{\mathcal{C}} *$. So the map in question, $|N\mathcal{C}| \to |N\mathcal{D}|$, is just the map between the homotopy colimit of the restricted diagram $C \xrightarrow{K} \mathcal{D} \xrightarrow{*} \mathbf{Top}$ and the homotopy colimit of $* \colon \mathcal{D} \to \mathbf{Top}$. $\square$

Example 8.5.9 If $\mathcal{D}$ has a terminal object t, the inclusion $t \colon \mathbb{1} \to \mathcal{D}$ is homotopy final, and hence the homotopy colimit of any diagram of shape $\mathcal{D}$ is weakly equivalent to the image of t. For instance, the mapping cylinder, a model for the homotopy colimit of an arrow $f \colon X \to Y$ in $\mathbf{Top}$, deformation retracts onto Y (cf. 6.4.4).

The author learned the following example from Omar Antolín Camarena:

Example 8.5.10 If M is a commutative monoid, its group completion G can be constructed as a quotient of $M \times M$ by the relation $(m, n) \sim (m', n')$ if there is $k \in M$, so that $m + n' + k = m' + n + k$. The map $m \mapsto (m, 0) \colon M \to G$ satisfies the expected universal property.

Regarding M and G as one-object categories, the functor $M \to G$ is homotopy final because the relevant slice category is filtered. Hence Corollary 8.5.8 tells us that the natural map $BM \to BG$ is a weak equivalence. For arbitrary (not necessarily commutative) monoids M, the fundamental group of BM is the group completion, as can be directly verified, but having a weak equivalence $BM \to BG$ is quite special; in fact, while BG is always a 1-type, Dusa McDuff proved that any connected homotopy type can be obtained as the classifying space of a monoid [61].

Exercise 8.5.11 Give an example to show that the functor of Example 8.3.8 is not homotopy final.

Example 8.5.12 Write $m \colon \overrightarrow{\Delta} \hookrightarrow \Delta$ for the inclusion subcategory containing all objects but only the monomorphisms (face maps). We claim this inclusion is homotopy initial. To show that each slice category $m/[n]$ is contractible, first define an endofunctor $S \colon m/[n] \to m/[n]$ that sends $\alpha \colon [k] \to [n]$ to the map $S\alpha \colon [k + 1] \to [n]$ with $(S\alpha)(0) = 0$ and $(S\alpha)(i) = \alpha(i - 1)$ for $i > 0$. Write $E \colon m/[n] \to m/[n]$ for the functor constant at the map $[0] \xrightarrow{0} [n]$. The maps d^0 define the components of a natural transformation $\mathrm{id} \Rightarrow S$, and the maps 0 define a natural transformation $E \Rightarrow S$. Taking nerves, these natural transformations define a zigzag of homotopies from the identity to E, exhibiting contractibility of $N(m/[n])$. Hence $\overrightarrow{\Delta} \hookrightarrow \Delta$ is homotopy initial. The dual inclusion is thus homotopy final, and Theorem 8.5.6 implies that the homotopy colimit of a simplicial object in a simplicial model category is weakly equivalent to the homotopy colimit of the subfunctor that "forgets the degeneracy maps."

Our interest in this observation stems from the fact that, for pointwise cofibrant simplicial objects, the homotopy colimit of the restricted diagram is weakly equivalent to its **fat geometric realization**, the functor tensor product with the restricted Yoneda embedding $\overrightarrow{\Delta} \to \mathbf{Set}^{\Delta^{op}}$ (see Example 11.5.6). The fat geometric realization is constructed like ordinary geometric realization, except that it forgets to collapse the degenerate simplices. It follows from the 2-of-3 property that the homotopy colimit of a pointwise cofibrant simplicial object in a simplicial model category is weakly equivalent to its fat geometric realization, which is generally easier to compute.

Example 8.5.13   The diagonal map $\Delta: \Delta \to \Delta \times \Delta$ is homotopy initial. To prove this, we must show $\Delta/([p], [q])$ is contractible. Extend the notation of 8.5.12 to define an analogous functor $S: \Delta/([p], [q]) \to \Delta/([p], [q])$ by

$$(\alpha: [k] \to [p], \beta: [k] \to [q]) \mapsto (S\alpha: [k+1] \to [p], S\alpha: [k+1] \to [q]).$$

Define $E: \Delta/[p], [q] \to \Delta/[p], [q]$ to be the functor constant at $([0] \xrightarrow{0} [p], [0] \xrightarrow{0} [q])$. As before, there exist natural transformations $\mathrm{id} \Rightarrow S \Leftarrow E$ that exhibit the homotopy contractibility of $\Delta/([p], [q])$. The upshot is that the homotopy colimit of a bisimplicial object in a simplicial model category is weakly equivalent to the homotopy colimit of its diagonal.

Exercise 8.5.14   Let Δ_∞ and $\Delta_{-\infty}$ denote the wide subcategories of Δ containing only those maps that preserve the top and bottom elements, respectively, in each ordinal. Freely adjoining a top or bottom element defines natural inclusions $\Delta \hookrightarrow \Delta_\infty$ and $\Delta \hookrightarrow \Delta_{-\infty}$. Extending a cosimplicial object along one of these inclusions produces an augmentation and either a forward or backward contracting homotopy.

Show that these inclusions are homotopy initial, proving the result promised in Remark 4.5.4: the homotopy colimit of a simplicial object admitting an augmentation and contracting homotopy is weakly equivalent to its augmentation.

To motivate the topic of the next chapter, we note that an alternate proof of the Homotopy Finality Theorem 8.5.6 should be available. In the Reduction Theorem 8.1.8, we showed that the homotopy colimit of a diagram of the form $\mathcal{C} \xrightarrow{K} \mathcal{D} \xrightarrow{F} \mathcal{M}$ is isomorphic to the homotopy colimit of the diagram F weighted by $N(-/K): \mathcal{D}^{op} \to \mathbf{sSet}$. If K is homotopy final, then the natural map $N(-/K) \to N(-/\mathcal{D})$ between the weights is a weak equivalence. If we knew that weakly equivalent weights induced weakly equivalent weighted colimits of pointwise cofibrant diagrams, we would be done.

It turns out this is not quite true; the weights must also be "fat" enough, or stated precisely, the weights must be cofibrant in a suitable sense. Our hoped-for simple proof will appear as an immediate corollary to a theorem

(Theorem 11.5.1) in Chapter 11, once we have the appropriate language to state and prove it. But in the next chapter, we take an important first step and prove this result for weighted *homotopy* limits and colimits, generalizing Theorem 5.1.1 and Corollary 5.1.3 to enriched category theory. In particular, we show that for any diagram in a simplicial model category, any weighted homotopy colimits formed with weakly equivalent weights are weakly equivalent.

9

Weighted homotopy limits and colimits

In Chapter 5, we proved that the bar construction gives a uniform way to construct homotopy limits and colimits of diagrams of any shape, defined to be derived functors of the appropriate limit or colimit functor in a simplicial model category. But, on account of Corollary 7.6.4, such categories admit a richer class of limits and colimits – the weighted limits and colimits for any simplicially enriched weight, computed as enriched functor cotensor or tensor products.

In this chapter we will prove that the enriched version of the two-sided bar construction can be used to construct a derived functor of the enriched functor tensor product. This provides a notion of weighted homotopy colimit that enjoys the same formal properties of our homotopy colimit functor. The construction and proofs closely parallel those of Chapter 5, though more sophisticated hypotheses will be needed to guarantee that the two-sided bar construction is homotopically well behaved. These results, due to Shulman [79], partially motivated our earlier presentation.

The notion of derived functor used here is precisely the one introduced in Chapter 2. This is to say, a derived functor of a $\mathcal{V}$-functor between $\mathcal{V}$-categories whose underlying categories are homotopical is defined to be a derived functor of the underlying unenriched functor. A priori, and indeed in many cases of interest, the point-set-derived functor defining the weighted homotopy colimit will not be a $\mathcal{V}$-functor. However, under reasonable hypotheses, its *total* derived functor inherits a natural enrichment – just not over $\mathcal{V}$. This will be the subject of the next chapter.

9.1 The enriched bar and cobar construction

Recall that the functor tensor product generalizes seamlessly to the enriched context. The coend used to define the functor tensor product is replaced by the enriched coend, the difference being that the coproduct indexed by sets of arrows is replaced by a coproduct indexed by ordered pairs of objects and tensors with the appropriate hom-object. A similar modification defines the two-sided enriched bar construction associated to a pair of $\mathcal{V}$-functors $G \colon \mathcal{D}^{op} \to \mathcal{V}$, $F \colon \mathcal{D} \to \mathcal{M}$ and a tensored and cotensored cocomplete $\mathcal{V}$-category $\mathcal{M}$.

The simplicial two-sided bar construction extends the diagram inside the coequalizer of (7.6.1) to a simplicial object in $\mathcal{M}$. Note that because Δ is unenriched, a simplicial object in a $\mathcal{V}$-category is the same thing as a simplicial object in the underlying unenriched category.

Definition 9.1.1 Given a small $\mathcal{V}$-category $\mathcal{D}$, a tensored $\mathcal{V}$-category $\mathcal{M}$, and $\mathcal{V}$-functors $F \colon \mathcal{D} \to \mathcal{M}$, $G \colon \mathcal{D}^{op} \to \mathcal{V}$, the **enriched simplicial bar construction** is a simplicial object $B_\bullet(G, \mathcal{D}, F)$ in $\mathcal{M}$ defined by

$$B_n(G, \mathcal{D}, F) = \coprod_{d_0,\dots,d_n \in \mathcal{D}} \left(Gd_n \times \underline{\mathcal{D}}(d_{n-1}, d_n) \times \cdots \times \underline{\mathcal{D}}(d_0, d_1) \right) \otimes Fd_0.$$

For $0 < i < n$, the ith face map $B_n(G, \mathcal{D}, F) \to B_{n-1}(G, \mathcal{D}, F)$ is induced by the composition map

$$\circ \colon \underline{\mathcal{D}}(d_i, d_{i+1}) \times \underline{\mathcal{D}}(d_{i-1}, d_i) \to \underline{\mathcal{D}}(d_{i-1}, d_{i+1}),$$

an arrow in $\mathcal{V}$. For $0 \le i \le n$, the ith degeneracy $B_n(G, \mathcal{D}, F) \to B_{n+1}(G, \mathcal{D}, F)$ is induced by the identity arrow $* \to \underline{\mathcal{D}}(d_i, d_i)$. The outer face maps have similar definitions: the 0th face map is induced by

$$
\begin{array}{ccc}
(Gd_n \times \underline{\mathcal{D}}(d_{n-1}, d_n) \times \cdots \times \underline{\mathcal{D}}(d_0, d_1)) \otimes Fd_0 & \hookrightarrow & B_n(G, \mathcal{D}, F) \\
\cong \downarrow & & \downarrow \\
(Gd_n \times \underline{\mathcal{D}}(d_{n-1}, d_n) \times \cdots \times \underline{\mathcal{D}}(d_1, d_2)) \otimes (\underline{\mathcal{D}}(d_0, d_1) \otimes Fd_0) & & \downarrow \\
1 \otimes ev \downarrow & & \downarrow \\
(Gd_n \times \underline{\mathcal{D}}(d_{n-1}, d_n) \times \cdots \times \underline{\mathcal{D}}(d_1, d_2)) \otimes Fd_1 & \hookrightarrow & B_{n-1}(G, \mathcal{D}, F)
\end{array}
$$

where $ev \colon \underline{\mathcal{D}}(d_0, d_1) \otimes Fd_0 \to Fd_1$ is adjunct, in the adjunction defining the tensor, to the arrow $\underline{\mathcal{D}}(d_0, d_1) \to \underline{\mathcal{M}}(Fd_0, Fd_1)$ specified as part of the data that makes F a $\mathcal{V}$-functor. The contravariant $\mathcal{V}$-functor G plays the role of F in the definition of the nth face map.

It is straightforward to check that the face and degeneracy maps satisfy the desired relations to make $B_\bullet(G, \mathcal{D}, F)$ a simplicial object in $\mathcal{M}$. Dually:

Definition 9.1.2 Given $\mathcal{V}$-functors $G \colon \mathcal{D} \to \mathcal{V}$ and $F \colon \mathcal{D} \to \mathcal{M}$, the **enriched cosimplicial cobar construction** $C^\bullet(G, \mathcal{D}, F)$ is a cosimplicial object in $\mathcal{M}$. Using exponential notation for the cotensor, the object of n-simplices is

$$C^n(G, \underline{\mathcal{D}}, F) = \prod_{d_0, \ldots, d_n} F d_n^{\underline{\mathcal{D}}(d_{n-1}, d_n) \times \cdots \times \underline{\mathcal{D}}(d_0, d_1) \times G d_0}$$

Remark 9.1.3 Before defining the maps that make $C^\bullet(G, \underline{\mathcal{D}}, F)$ a cosimplicial object in $\mathcal{M}$, recall that an arrow $V \to W$ in $\mathcal{V}$ induces an arrow $m^W \to m^V$ in $\mathcal{M}$ by the defining universal property. The intuition is that cotensors are similar to homs; indeed, if $\mathcal{V} = \mathcal{M} = \mathbf{Set}$, then $m^V = \prod_V m = \mathbf{Set}(V, m)$. This is the essential reason why $C^\bullet(G, \underline{\mathcal{D}}, F)$ is cosimplicial rather than simplicial.

As for the enriched simplicial bar construction, the degeneracy and inner face maps are defined using the $\mathcal{V}$-category structure on $\underline{\mathcal{D}}$, for example, composition $\underline{\mathcal{D}}(d_i, d_{i+1}) \times \underline{\mathcal{D}}(d_{i-1}, d_i) \to \underline{\mathcal{D}}(d_{i-1}, d_{i+1})$ induces the ith face map $C^{n-1}(G, \underline{\mathcal{D}}, F) \to C^n(G, \underline{\mathcal{D}}, F)$. The nth face map $C^{n-1}(G, \underline{\mathcal{D}}, F) \to C^n(G, \underline{\mathcal{D}}, F)$ is induced by the map

where coev: $F d_{n-1} \to F d_n^{\underline{\mathcal{D}}(d_{n-1}, d_n)}$ is adjunct, in the adjunction defining the cotensor, to the arrow $\underline{\mathcal{D}}(d_{n-1}, d_n) \to \underline{\mathcal{M}}(F d_{n-1}, F d_n)$ specified as part of the data that makes F a $\mathcal{V}$-functor. The definition of the 0th face map is similar.

Remark 9.1.4 If $\mathcal{D}$ is unenriched, this definition agrees with the ordinary simplicial bar construction and cosimplicial cobar construction introduced in 4.2.1 and 4.3.2.

Just as in the unenriched case, given any cosimplicial object $\Delta^\bullet \colon \Delta^{\mathrm{op}} \to \mathcal{V}$, we can define geometric realization of simplicial objects and totalization of cosimplicial objects in any $\mathcal{V}$-category.

Definition 9.1.5 Fixing a cosimplicial object $\Delta^\bullet \colon \Delta^{\mathrm{op}} \to \mathcal{V}$, the **enriched bar construction** and **enriched cobar construction** are defined to be the functor tensor product and functor cotensor product, as appropriate:

$$B(G, \underline{\mathcal{D}}, F) := \Delta^\bullet \otimes_{\Delta^{\mathrm{op}}} B_\bullet(G, \underline{\mathcal{D}}, F) \quad C(G, \underline{\mathcal{D}}, F) := \{\Delta^\bullet, C^\bullet(G, \underline{\mathcal{D}}, F)\}^\Delta$$

Note that the left-hand and right-hand $\mathcal{V}$-functors G have conflicting variance.

Example 9.1.6 The enriched version of the two-sided bar construction can be used to extend the constructions in Examples 4.2.4 and 4.5.5 of spaces $BG = B(*, G, *)$ and $EG = B(G, G, *)$ associated to a discrete group G to **topological groups**, represented as a one-object **Top**-categories $\underline{G}$ equipped with an involution that is used to define inverses.

A continuous (topologically enriched) functor $X \colon \underline{G} \to \mathbf{Top}$ whose domain is a topological group encodes a $\underline{G}$-space X. Its conical colimit, which can be defined because **Top** is cartesian closed, is the orbit space of X. The results in the next section will tell us how to construct instead the conical homotopy colimit, a model for the homotopy orbit space.

9.2 Weighted homotopy limits and colimits

Let $\mathcal{M}$ be a complete and cocomplete category that is tensored, cotensored, and enriched over a complete and cocomplete closed symmetric monoidal category $(\mathcal{V}, \times, *)$. We use the notation $(\mathcal{M}^{\mathcal{D}})_0$ to distinguish the (unenriched) category of $\mathcal{V}$-functors $\underline{\mathcal{D}} \to \underline{\mathcal{M}}$ and $\mathcal{V}$-natural transformations from the category $\mathcal{M}^{\mathcal{D}}$ of functors and natural transformations between the underlying categories of $\mathcal{D}$ and $\underline{\mathcal{M}}$.

By Theorem 7.6.3, weighted colimits and weighted limits are computed by the functors

$$- \otimes_{\underline{\mathcal{D}}} - \colon (\mathcal{V}^{\mathcal{D}^{\mathrm{op}}})_0 \times (\mathcal{M}^{\mathcal{D}})_0 \to \mathcal{M} \quad \text{and} \quad \{-, -\}^{\underline{\mathcal{D}}} \colon (\mathcal{V}^{\mathcal{D}})_0^{\mathrm{op}} \times (\mathcal{M}^{\mathcal{D}})_0 \to \mathcal{M}.$$

To discuss derived functors, we must have some notion of weak equivalence in $\mathcal{V}$ and $\mathcal{M}$. When the underlying categories $\mathcal{V}$ and $\mathcal{M}$ have a homotopical structure, these functors may or may not preserve pointwise weak equivalences. The main theorem of this section describes deformations that exist under suitable conditions that may be used to construct a left derived functor of the weighted colimit bifunctor and a right derived functor of the weighted limit bifunctor. We call this left derived functor the **weighted homotopy colimit** and this right derived functor the **weighted homotopy limit**.

On the basis of previous experience in Chapter 5, we might guess part of the answer. The colimit of $F \colon \underline{\mathcal{D}} \to \underline{\mathcal{M}}$ weighted by $G \colon \underline{\mathcal{D}}^{\mathrm{op}} \to \underline{\mathcal{V}}$ is computed by the enriched functor tensor product

$$\mathrm{colim}^G F \cong G \otimes_{\underline{\mathcal{D}}} F$$

$$:= \mathrm{coeq}\left(\coprod_{d, d' \in \underline{\mathcal{D}}} (Gd' \times \underline{\mathcal{D}}(d, d')) \otimes Fd \rightrightarrows \coprod_{d \in \underline{\mathcal{D}}} Gd \otimes Fd \right).$$

Observe that the diagram contained inside the coequalizer is a truncation of the enriched simplicial bar construction. On account of the co-Yoneda isomorphisms

$$G \otimes_{\underline{D}} B(\underline{D}, \underline{D}, F) \cong B(G, \underline{D}, F) \cong B(G, \underline{D}, \underline{D}) \otimes_{\underline{D}} F, \qquad (9.2.1)$$

we think of the enriched bar construction as a fattened version of the enriched functor tensor product. These isomorphisms say $B(G, \underline{D}, F)$ is the colimit of F weighted by a fattened weight $B(G, \underline{D}, \underline{D})$ or the colimit of a fattened diagram $B(\underline{D}, \underline{D}, F)$ weighted by G. The remainder of our discussion is aimed at supplying conditions so that this construction is homotopical, in which case it will follow from the isomorphisms (9.2.1) that the enriched two-sided bar construction supplies a deformation for the weighted colimit functor, and hence computes the appropriate left derived functor.

The main task is to understand the homotopical properties of the two-sided enriched bar construction. As a first step, observe that the cosimplicial object $\Delta^\bullet \colon \Delta \to \mathcal{V}$ used to define geometric realization and totalization extends to an adjunction

$$|-| \colon \mathbf{sSet} \; \underset{\perp}{\overset{}{\rightleftarrows}} \; \mathcal{V} \colon \mathcal{V}(\Delta^\bullet, -) \qquad (9.2.2)$$

because $\mathcal{V}$ was supposed to be cocomplete (see 1.5.1). If we further suppose that this left adjoint is strong monoidal, then by Theorem 3.7.11, the $\mathcal{V}$-enriched categories $\mathcal{V}$ and $\mathcal{M}$ become simplicially enriched, tensored, and cotensored. It follows from Corollary 3.8.4 that geometric realization preserves simplicial homotopy equivalences of the form described in 3.8.2.

Furthermore, if $\mathcal{V}$ is a monoidal model category and the adjunction (9.2.2) is Quillen as well as strong monoidal, then any $\mathcal{V}$-model category $\mathcal{M}$ becomes a simplicial model category [38, §4.2]. The definitions of **monoidal model category** and $\mathcal{V}$-**model category** will be given in 11.4.6 and 11.4.7. For now, we axiomatize a few consequences of those definitions in analogy to our axioms for a simplicial model category in 3.8.6:

Lemma 9.2.3 *A monoidal model category $\mathcal{V}$, or a $\mathcal{V}$-model category $\mathcal{M}$ in the case where $\mathcal{V}$ is a monoidal model category*

(i) *is complete and cocomplete*
(ii) *is tensored, cotensored, and enriched over $\mathcal{V}$*
(iii) *has subcategories of **cofibrant** and **fibrant** objects preserved, respectively, by tensoring or cotensoring with any cofibrant object of $\mathcal{V}$*
(iv) *admits a left deformation Q into the cofibrant objects and a right deformation R into the fibrant objects*
(v) *is a saturated homotopical category*

(vi) has the property that the internal hom preserves weak equivalences between cofibrant objects in its first variable, provided the second variable is fibrant, and preserves weak equivalences between fibrant objects in its second variable, provided its first variable is cofibrant

Examples of such $\mathcal{V}$ include **Top**, **sSet**$_*$, and **Top**$_*$. This is how we concluded that the convenient categories of spaces described in Section 6.1 were simplicial model categories. If the homotopical structures on $\mathcal{V}$ and $\mathcal{M}$ come from these model structures, then simplicial homotopy equivalences are necessarily weak equivalences under these hypotheses.

The next step is to obtain an analog of Corollary 5.2.5, which gives conditions under which the unenriched two-sided bar construction preserves weak equivalences in each variable. In the enriched setting, the necessary hypotheses are somewhat more technical.

Lemma 9.2.4 ([79, 23.12]) *Suppose $\mathcal{V}$ is a monoidal model category with a strong monoidal Quillen adjunction (9.2.2); $\mathcal{M}$ is a $\mathcal{V}$-model category; the hom-objects $\underline{\mathcal{D}}(d, d') \in \mathcal{V}$ are cofibrant; and the unit maps $* \to \underline{\mathcal{D}}(d, d)$ are cofibrations in $\mathcal{V}$. Then*

- $B(-, \underline{\mathcal{D}}, -) : (\mathcal{V}^{\underline{\mathcal{D}}^{op}})_0 \times (\mathcal{M}^{\underline{\mathcal{D}}})_0 \to \mathcal{M}$ *preserves weak equivalences between pointwise cofibrant diagrams*
- $B(-, \underline{\mathcal{D}}, \underline{\mathcal{D}}) : (\mathcal{V}^{\underline{\mathcal{D}}^{op}})_0 \to (\mathcal{V}^{\underline{\mathcal{D}}^{op}})_0$ *and* $B(\underline{\mathcal{D}}, \underline{\mathcal{D}}, -) : (\mathcal{M}^{\underline{\mathcal{D}}})_0 \to (\mathcal{M}^{\underline{\mathcal{D}}})_0$ *preserve pointwise cofibrant diagrams*
- $C(-, \underline{\mathcal{D}}, -) : (\mathcal{V}^{\underline{\mathcal{D}}})_0^{op} \times (\mathcal{M}^{\underline{\mathcal{D}}})_0 \to \mathcal{M}$ *preserves weak equivalences between pointwise cofibrant diagrams in the first variable and pointwise fibrant diagrams in the second*
- $C(\underline{\mathcal{D}}, \underline{\mathcal{D}}, -) : (\mathcal{M}^{\underline{\mathcal{D}}})_0 \to (\mathcal{M}^{\underline{\mathcal{D}}})_0$ *preserves pointwise fibrant diagrams*

Note that when $\mathcal{V} = \mathbf{sSet}$ and $\mathcal{M}$ is a simplicial model category, the hypotheses on $\underline{\mathcal{D}}$ are automatically satisfied. When $\mathcal{V} = \mathbf{Top}$ and $\mathcal{M}$ is a topological model category, the hypotheses on $\underline{\mathcal{D}}$ are satisfied when the hom-spaces are cell complexes and the unit map is a relative cell complex. Replacing the usual model structure on **Top** by the mixed model structure of Exercise 11.3.8, also a monoidal model category with a strong monoidal Quillen adjunction (9.2.2), it suffices that the hom-spaces of $\underline{\mathcal{D}}$ are homotopy equivalent to cell complexes.[1]

Proof sketch The statement in [79, §23] is more general, presenting a precise axiomatization of the "goodness" conditions needed for $B(-, \underline{\mathcal{D}}, -)$ to be homotopical. The general strategy is the following: First use these axioms to show that, for pointwise cofibrant diagrams, the enriched simplicial bar

---

[1] Thanks to Angela Klamt for pointing this out.

construction is "Reedy cofibrant" in the sense that the latching maps are "cofibrations" in $\mathcal{M}$. Second, note that the axioms imply that pointwise equivalences between pointwise cofibrant diagrams give rise to pointwise equivalences between simplicial objects. Finally, show that the axioms imply that pointwise weak equivalences between "Reedy cofibrant" simplicial objects realize to weak equivalences. □

One issue remains in using the enriched bar construction to define a left derived functor of the weighted colimit: our diagram $F \colon \mathcal{D} \to \mathcal{M}$ and weight $G \colon \mathcal{D}^{\mathrm{op}} \to \mathcal{V}$ might not be pointwise cofibrant. We have assumed that $\mathcal{M}$ and $\mathcal{V}$ are $\mathcal{V}$-model categories, which means that there exist left deformations into the subcategories of cofibrant objects. However, these deformations are not necessarily $\mathcal{V}$-functors and hence cannot be used to construct weakly equivalent pointwise cofibrant $\mathcal{V}$-functors.[2] It might be the case that some deformation $Q \colon \mathcal{M} \to \mathcal{M}$ into the subcategory of cofibrant objects just happens to be a $\mathcal{V}$-functor, in which case we obtain the desired deformation on $\mathcal{M}^{\mathcal{D}}$ just by postcomposition. Alternatively, this issue can be avoided in the following settings:

(i) When all objects of $\mathcal{M}$ and $\mathcal{V}$ are cofibrant, of course, there is no problem.

(ii) When $\mathcal{D}$ is unenriched, any unenriched functor $\mathcal{D} \to \mathcal{M}$ is equally a $\mathcal{V}$-functor $\mathcal{D} \to \mathcal{M}$ whose domain is implicitly the free $\mathcal{V}$-category on $\mathcal{D}$. So any deformation $Q \colon \mathcal{M} \to \mathcal{M}$, even if unenriched, will suffice.

(iii) When all objects of $\mathcal{V}$ are cofibrant and $\mathcal{M}$ is a cofibrantly generated $\mathcal{V}$-model category, Theorem 13.2.1 will show that there exist $\mathcal{V}$-functorial left and right deformations in $\mathcal{M}$.

To unify these conditions, we formally suppose that there exists a functor $Q \colon (\mathcal{M}^{\mathcal{D}})_0 \to (\mathcal{M}^{\mathcal{D}})_0$ equipped with a natural weak equivalence $Q \Rightarrow 1$ such that the $\mathcal{V}$-functors in its image are pointwise cofibrant. We require similar hypotheses on $\mathcal{V}^{\mathcal{D}^{\mathrm{op}}}$. For the reader's convenience, let us summarize the assumptions we have made.

Assumptions 9.2.5 The standing hypotheses are as follows:

(i) $(\mathcal{V}, \times, *)$ is a closed monoidal model category with a strong monoidal Quillen adjunction $|-| \colon \mathbf{sSet} \rightleftarrows \mathcal{V} \colon \mathcal{V}(\Delta^\bullet, -)$.

(ii) $\mathcal{M}$ is a $\mathcal{V}$-model category; in particular, $\mathcal{M}$ is enriched, tensored, and cotensored over $\mathcal{V}$ and over $\mathbf{sSet}$, and simplicial homotopy equivalences are weak equivalences in $\mathcal{M}$.

---

[2] Note that, we do not require a "pointwise cofibrant replacement" $\mathcal{V}$-functor $\mathcal{M}^{\mathcal{D}} \to \mathcal{M}^{\mathcal{D}}$, but we do need a functorial way to replace a $\mathcal{V}$-functor $F \in \mathcal{M}^{\mathcal{D}}$ by a weakly equivalent $\mathcal{V}$-functor that is pointwise cofibrant.

(iii) There exist deformations for $\underline{\mathcal{M}}^{\mathcal{D}}$ and $\underline{\mathcal{V}}^{\mathcal{D}^{\mathrm{op}}}$ into pointwise cofibrant or fibrant $\mathcal{V}$-functors, as appropriate.

(iv) The hom-objects of $\mathcal{D}$ are cofibrant, and the unit maps are cofibrations in $\mathcal{V}$.

Theorem 9.2.6 ([79, 13.7,13.14]) *Under the standing hypotheses, we have a left deformation*

$$Q \times B(\mathcal{D}, \mathcal{D}, Q) \colon (\underline{\mathcal{V}}^{\mathcal{D}^{\mathrm{op}}})_0 \times (\underline{\mathcal{M}}^{\mathcal{D}})_0 \to (\underline{\mathcal{V}}^{\mathcal{D}^{\mathrm{op}}})_0 \times (\underline{\mathcal{M}}^{\mathcal{D}})_0$$

for the enriched functor tensor product $- \otimes_{\mathcal{D}} -$ *and a right deformation*

$$Q \times C(\mathcal{D}, \mathcal{D}, R) \colon (\underline{\mathcal{V}}^{\mathcal{D}})_0^{\mathrm{op}} \times (\underline{\mathcal{M}}^{\mathcal{D}})_0 \to (\underline{\mathcal{V}}^{\mathcal{D}})_0^{\mathrm{op}} \times (\underline{\mathcal{M}}^{\mathcal{D}})_0$$

for the enriched functor cotensor product $\{-, -\}^{\mathcal{D}}$.

Proof Exactly like the proof of Theorem 5.1.1. Note that the proof of Lemma 5.2.6 works equally with a pointwise cofibrant functor G in place of $*$. □

The following corollary is then immediate from Theorem 2.2.8:

Corollary 9.2.7 ([79, 13.12,13.17]) *Under the hypotheses of 9.2.5, $B(Q, \mathcal{D}, Q)$ is a left derived functor of the weighted colimit bifunctor and $C(Q, \mathcal{D}, R)$ is a right derived functor of the weighted limit bifunctor computing the weighted homotopy colimit and the weighted homotopy limit. That is, given weights $G \in \underline{\mathcal{V}}^{\mathcal{D}^{\mathrm{op}}}$ and $H \in \underline{\mathcal{V}}^{\mathcal{D}}$ and a diagram $F \in \underline{\mathcal{M}}^{\mathcal{D}}$, we have*

$$\mathrm{hocolim}^G F := \mathbb{L}\,\mathrm{colim}^G F \cong B(QG, \mathcal{D}, QF)$$

$$\mathrm{holim}^H F := \mathbb{R}\,\mathrm{lim}^H F \cong C(QH, \mathcal{D}, RF).$$

Remark 9.2.8 A point made in Section 6.3 is even more salient here. By 9.2.6 and 9.2.7, the existence of any deformations into pointwise cofibrant and pointwise fibrant diagrams, no matter how complicated, means that if the given weights and diagrams happen to be pointwise cofibrant or fibrant already, then the enriched two-sided bar or cobar construction already has the correct homotopy type to compute the weighted homotopy colimit or limit without making use of the deformations Q and R.

Furthermore, our proof, using the homotopical properties of the two-sided bar and cobar construction as detailed in Lemma 9.2.4, shows that a pointwise weak equivalence between pointwise cofibrant weights induces a weak equivalence between the associated weighted homotopy limits or colimits. When $\mathcal{V}$ is the usual simplicial model category of simplicial sets, all weights are pointwise cofibrant, so we have now proven the claim made at the end of Chapter 8: weighted homotopy limits and weighted homotopy colimits formed with pointwise weakly equivalent weights are weakly equivalent.

Warning 9.2.9 The simplification for the case of topological spaces described in Remark 6.3.4 and proven in Section 14.5, which allowed us to compute homotopy colimits without first taking a pointwise cofibrant replacement of the diagram, does not appear to extend to weighted homotopy colimits. As far as we know, it really is necessary to replace a generic diagram by a pointwise cofibrant one. See Remark 14.5.10.

Example 9.2.10 The homotopy colimit of a continuous functor $X \colon \underline{G} \to$ **Top**, whose domain is a topological group and whose target space X is a cell complex, is the space $B(*, \underline{G}, X) \cong \mathrm{colim}^*(B(\underline{G}, \underline{G}, X))$ of **homotopy orbits**. The homotopy limit, using the enriched analog of the isomorphism (7.7.5), is the space $C(*, \underline{G}, X) \cong \lim^* C(\underline{G}, \underline{G}, X) \cong \underline{\mathbf{Top}}^{\underline{G}}(B(*, \underline{G}, \underline{G}), X)$ of **homotopy fixed points**.

Example 9.2.11 Using the weighted colimit and limit formulae (7.6.7) for left and right enriched Kan extension, we can define **homotopy Kan extensions** in enriched settings satisfying the hypotheses of 9.2.5. Suppose we have a $\mathcal{V}$-functor $K \colon \underline{C} \to \underline{D}$ between small $\mathcal{V}$-categories so that the hom-objects of both $\underline{C}$ and $\underline{D}$ are cofibrant. Then the weights for left and right Kan extension are pointwise cofibrant, and we define, for $F \in \underline{\mathcal{M}}^{\underline{C}}$,

$$\mathrm{hoLan}_K F := \mathbb{L}\mathrm{Lan}_K F \cong \mathbb{L}\,\mathrm{colim}^{\underline{D}(K-,-)} F \cong B(\underline{D}(K-,-), \underline{C}, QF)$$

$$\mathrm{hoRan}_K F := \mathbb{R}\mathrm{Ran}_K F \cong \mathbb{R}\lim^{\underline{D}(-,K-)} F \cong C(\underline{D}(-,K-), \underline{C}, RF).$$

Note that the weighted homotopy colimit and limit functors defined by Corollary 9.2.7 require composition with the deformations Q and R. Hence these point-set derived functors are not necessarily $\mathcal{V}$-functors. This rather unsatisfying state of affairs is addressed in the next chapter.

10

Derived enrichment

Suppose $\mathcal{M}$ is a $\mathcal{V}$-model category with cofibrant replacement $Q\colon \mathcal{M} \to \mathcal{M}$. As noted earlier, on account of the free $\mathcal{V}$-category–underlying category adjunction, even if Q is not a $\mathcal{V}$-functor, postcomposition transforms a $\mathcal{V}$-functor $F\colon \mathcal{D} \to \underline{\mathcal{M}}$ into a pointwise cofibrant $\mathcal{V}$-functor, $QF\colon \mathcal{D} \to \underline{\mathcal{M}}$, provided that the domain $\mathcal{D}$ is unenriched. In this way, postcomposition defines a functor $Q\colon (\underline{\mathcal{M}}^{\mathcal{D}})_0 \to (\underline{\mathcal{M}}^{\mathcal{D}})_0$ that replaces any diagram by a pointwise weakly equivalent objectwise cofibrant $\mathcal{V}$-functor. Although this Q is sufficient for the purposes of constructing derived functors of the weighted limit and colimit functors, it is likely not a $\mathcal{V}$-functor, and hence the derived functors constructed in this way will not be enriched. For example, the standard procedure of replacing a space by a CW complex is not continuous [35, §4.1].[1]

Indeed, a reasonable case can be made that $\mathcal{V}$-enrichments are too much to ask for: after all, derived functors are merely point-set level lifts of functors between homotopy categories, which we would not expect to be enriched. But on the contrary, in good situations, homotopy categories of enriched categories and total derived functors between them admit canonical enrichments – just not over the closed monoidal category $\mathcal{V}$ with which we started.

Our presentation largely follows [79, §16]. As a starting point for this discussion, we state a theorem that we generalize (and prove) subsequently.

Theorem 10.0.1 *If $\mathcal{M}$ is a simplicial model category, then $\mathrm{Ho}\,\mathcal{M}$ is enriched, tensored, and cotensored over the homotopy category of spaces.*

By a result of Quillen [65], the total derived functors of $|-| \dashv S$ form an adjoint equivalence of homotopy categories. We write

$$\mathcal{H} := \mathrm{Ho}\,\mathbf{sSet} \cong \mathrm{Ho}\,\mathbf{Top}$$

---

[1] Indeed, it is not a functor at all, but this is not the point here.

for the **homotopy category of spaces**, equivalent to the category of CW complexes and homotopy classes of maps. Taking $\mathcal{M} = \textbf{sSet}$, this result asserts that $\mathcal{H}$ admits the structure of a closed monoidal category. In this case, the monoidal structure on $\mathcal{H}$ is given by the cartesian product (which is also the homotopy product; cf. Remark 6.3.1) with the singleton serving as the monoidal unit. Its internal hom is somewhat more subtle. In this chapter, we see how such monoidal structures arise in general.

We close this chapter with a two-section epilogue exploring a notion of weak equivalence that arises formally from enrichment over the family of homotopical categories introduced in Section 10.2. Under mild hypotheses, this notion of equivalence is quite well behaved and provides a useful stepping-stone to the broader class of weak equivalences previously studied, mirroring classical results in homotopy theory that describe the relationship between homotopy equivalences and weak homotopy equivalences of topological spaces.

10.1 Enrichments encoded as module structures

We have already defined the homotopy categories and total derived functors we now assert are enriched. To prove this, we use a technique, previewed in Section 3.7, for recognizing when a priori unenriched functors admit enrichments. The general strategy is as follows: Point-set level enrichments can be encoded as module structures with respect to a tensor or cotensor. These module structures consist of unenriched functors and natural transformations satisfying suitable coherence conditions. In good situations, axiomatized in Definition 10.2.9, these functors can be derived, in which case we can use them to recognize when other derived functors – including the defining tensors and cotensors themselves – admit enrichments.

We leave aside the derived functors until the necessary enriched category theory is in place. Recall that an (unenriched) **two-variable adjunction** consists of bifunctors

$$F: \mathcal{M} \times \mathcal{N} \to \mathcal{P} \qquad G: \mathcal{M}^{\mathrm{op}} \times \mathcal{P} \to \mathcal{N} \qquad H: \mathcal{N}^{\mathrm{op}} \times \mathcal{P} \to \mathcal{M}$$

that are pointwise adjoints in a compatible way, that is, so that there exist hom-set isomorphisms

$$\mathcal{P}(F(m, n), p) \cong \mathcal{N}(n, G(m, p)) \cong \mathcal{M}(m, H(n, p)) \tag{10.1.1}$$

natural in each variable. Our main examples will be a closed monoidal structure (in which case G and H coincide) or a tensor–cotensor–hom two-variable adjunction.

Remark 10.1.2 In both of these examples, the categories involved are all $\mathcal{V}$-categories, and the usual definition demands something stronger: that the functors are $\mathcal{V}$-bifunctors and that the isomorphisms (10.1.1) lie in the enriching category $\mathcal{V}$. This will be an important consequence of the hypotheses we assume later – the derived tensor–cotensor–hom two-variable adjunctions will consist of enriched bifunctors and enriched natural transformations – but one of the main points is that we will not need to assume this hypothesis. This enrichment will follow from coherence conditions with respect to module structures on $\mathcal{M}$, $\mathcal{N}$, and $\mathcal{P}$.

As always, let $(\mathcal{V}, \times, *)$ be a closed symmetric monoidal category.

Definition 10.1.3 A **closed** $\mathcal{V}$-**module** $\underline{\mathcal{M}}$ consists of

- a two-variable adjunction $(\otimes, \{\}, \underline{\mathrm{hom}}) \colon \mathcal{V} \times \mathcal{M} \to \mathcal{M}$
- a natural isomorphism $v \otimes (w \otimes m) \cong (v \times w) \otimes m$ for all $v, w \in \mathcal{V}$ and $m \in \mathcal{M}$
- a natural isomorphism $* \otimes m \cong m$ for all $m \in \mathcal{M}$.

As the notation suggests, the two-variable adjunction should be thought of as a candidate tensor–cotensor–hom adjunction, but it is a priori unenriched. We saw in Lemma 3.7.7 that in a tensored and cotensored $\mathcal{V}$-category, the tensor is associative and unital. In other words, a tensored and cotensored $\mathcal{V}$-category is a closed $\mathcal{V}$-module. Conversely:

Proposition 10.1.4 *A closed $\mathcal{V}$-module $\underline{\mathcal{M}}$ is uniquely a tensored, cotensored, and enriched $\mathcal{V}$-category in such a way that the given two-variable adjunction underlies the tensor–cotensor–hom two-variable adjunction.*

Proof The proof, relying principally on Lemma 3.5.12, is similar to many of the proofs in Section 3.7 and is left as an exercise for the reader. □

Note that we have elected to use the same notation for an enriched category and a closed $\mathcal{V}$-module. Dually, the enrichment can be encoded by a "cotensor $\mathcal{V}$-module" structure. We will have more to say about this dual encoding in Remark 10.1.6.

Proposition 10.1.5 *Let $F \colon \mathcal{M} \to \mathcal{N}$ be a functor between closed $\mathcal{V}$-modules. The following are equivalent:*

- *a $\mathcal{V}$-functor $F \colon \underline{\mathcal{M}} \to \underline{\mathcal{N}}$ with underlying functor F*
- *natural transformations $\alpha \colon v \otimes Fm \to F(v \otimes m)$, associative and unital with respect to the closed $\mathcal{V}$-module structures on $\underline{\mathcal{M}}$ and $\underline{\mathcal{N}}$ in the*

sense that the diagrams

$$
1 \otimes Fm \xrightarrow{\ \alpha\ } F(1 \otimes m) \qquad v \otimes (w \otimes Fm) \xrightarrow{\ 1 \otimes \alpha\ } v \otimes F(w \otimes m) \xrightarrow{\ \alpha\ } F(v \otimes (w \otimes m))
$$

$$
Fm \qquad (v \times w) \otimes Fm \xrightarrow{\hspace{5cm}\alpha\hspace{5cm}} F((v \times w) \otimes m)
$$

with $\cong$ maps on the left, $\cong$ vertical maps, and $\cong$ on the right.

commute with the specified isomorphisms.

Proof If F is a $\mathcal{V}$-functor, the map α is defined to be

$$
v \xrightarrow{\ \eta_v\ } \underline{\mathcal{M}}(m, v \otimes m) \xrightarrow{\ F_{m,v \otimes m}\ } \underline{\mathcal{N}}(Fm, F(v \otimes m))
$$

where the first arrow is a component of the unit of the adjunction $- \otimes m \dashv \underline{\mathcal{M}}(m, -)$. The coherence conditions follow from naturality of this adjunction and a diagram chase.

Conversely, given such α, we must define morphisms $\underline{\mathcal{M}}(m, m') \to \underline{\mathcal{N}}(Fm, Fm')$ in $\mathcal{V}$ for each pair $m, m' \in \underline{\mathcal{M}}$ satisfying the two diagrams that encode the functoriality axioms. The requisite morphisms are adjunct to the composite

$$
\underline{\mathcal{M}}(m, m') \otimes Fm \xrightarrow{\ \alpha\ } F(\underline{\mathcal{M}}(m, m') \otimes m) \xrightarrow{\ F\text{ev}\ } Fm'
$$

where the map "ev" is adjunct to the identity arrow in $\mathcal{V}$ at the hom-object $\underline{\mathcal{M}}(m, m')$. The triangle for α implies the unit condition and the pentagon the associativity condition so that these maps assemble into a $\mathcal{V}$-functor.

Finally, we argue that the underlying functor of this $\mathcal{V}$-functor is the original functor F. We evaluate at an arrow $f : m \to m'$ in the underlying category by precomposing $\underline{\mathcal{M}}(m, m') \to \underline{\mathcal{N}}(Fm, Fm')$ with the representing morphism $f : * \to \underline{\mathcal{M}}(m, m')$. The resulting arrow $* \to \underline{\mathcal{N}}(Fm, Fm')$ defines the image of f with respect to the underlying functor. Its adjunct is the composite around the top right, which by naturality agrees with the composite around the left bottom:

$$
\begin{array}{ccc}
* \otimes Fm & \xrightarrow{\ f \otimes \mathrm{id}\ } & \underline{\mathcal{M}}(m, m') \otimes Fm \\[0.3em]
{\scriptstyle \alpha}\downarrow & & \downarrow{\scriptstyle \alpha} \\[0.3em]
F(* \otimes m) & \xrightarrow{\ F(f \otimes \mathrm{id})\ } & F(\underline{\mathcal{M}}(m, m') \otimes m) \\[0.3em]
{\scriptstyle F\text{ev}}\downarrow & & \downarrow{\scriptstyle F\text{ev}} \\[0.3em]
Fm & \xrightarrow{\ Ff\ } & Fm'
\end{array}
$$

The left-hand evaluation map is exactly the unit isomorphism, so the composite down the left-hand side is similarly the unit isomorphism. Hence the adjunct of this composite is precisely $Ff: * \to \underline{\mathcal{N}}(Fm, Fm')$, as claimed. □

Remark 10.1.6 Dually, a $\mathcal{V}$-functor can be uniquely encoded as a lax functor between closed "cotensor $\mathcal{V}$-modules." This is particularly useful for right adjoints. Combining Propositions 10.1.5 and 3.7.10, if a left adjoint preserves tensors up to coherent natural isomorphism, that is, if the α of Proposition 10.1.5 is an isomorphism, then any right adjoint inherits a canonical enrichment so that the resulting adjunction is a $\mathcal{V}$-adjunction. The dual theorem says that if a functor preserves cotensors up to coherent isomorphism, then any left adjoint admits a canonical enrichment.

Similarly, a two-variable $\mathcal{V}$-adjunction can be encoded as a $\mathcal{V}$-**bilinear functor**. We leave the definition as an exploratory exercise for the intrepid; or see [79, 14.8–12].

10.2 Derived structures for enrichment

Our aim is to use the definitions just presented to axiomatize the conditions necessary to obtain derived enrichments. Before beginning, we need one final preliminary: an extension of Theorem 2.2.11, which told us that the total derived functors of a deformable adjunction are themselves adjoint, to two-variable adjunctions. Using the notation of (10.1.1), suppose $\mathcal{M}$, $\mathcal{N}$, and $\mathcal{P}$ are homotopical categories such that $\mathcal{M}$ and $\mathcal{N}$ have left deformations and $\mathcal{P}$ has a right deformation.

Definition 10.2.1 The two-variable adjunction (10.1.1) is **deformable** just when F is left deformable and G and H are right deformable.[2]

Lemma 10.2.2 ([79, 15.2]) *If $(F, G, H): \mathcal{M} \times \mathcal{N} \to \mathcal{P}$ is a deformable two-variable adjunction, then the total derived functors constructed from the deformations form a two-variable adjunction*

$$(\mathbf{L}F, \mathbf{R}G, \mathbf{R}H): \mathrm{Ho}\mathcal{M} \times \mathrm{Ho}\mathcal{N} \to \mathrm{Ho}\mathcal{P}$$

between the associated homotopy categories.

Proof Writing Q for the left deformations on both $\mathcal{M}$ and $\mathcal{N}$ and R for the right deformation on $\mathcal{P}$, it follows from Theorem 2.2.8 that $F(Q-, Q-)$, $G(Q-, R-)$, $H(Q-, R-)$ are derived functors. Fixing $m \in \mathcal{M}, n \in \mathcal{N}$,

---

[2] The left deformations on $\mathcal{M}$ and $\mathcal{N}$ become right deformations on $\mathcal{M}^{\mathrm{op}}$ and $\mathcal{N}^{\mathrm{op}}$.

$p \in \mathcal{P}$, Theorem 2.2.11 gives us adjunctions

$$\mathbf{L}F(m, -) \dashv \mathbf{R}G(m, -), \quad \mathbf{L}F(-, n) \dashv \mathbf{R}H(n, -), \quad \mathbf{R}G(-, p) \dashv \mathbf{R}H(-, p),$$

where, for example, $\mathbf{L}F(m, -)$ is the total derived functor whose point-set level lift is $F(Qm, Q-)$. The uniqueness statement in Theorem 2.2.11 says that the hom-set isomorphisms exhibiting these adjunctions are the unique ones compatible with the upstairs adjunctions. Thus the upstairs compatibility implies that these total derived adjunctions assemble into a two-variable adjunction between the homotopy categories.   □

Example 10.2.3   The axioms for a simplicial model category, as listed in Lemma 3.8.6, or more generally for a monoidal model category or a $\mathcal{V}$-model category, as described in Lemma 9.2.3, demand that Q and R form deformations for the two-variable adjunction $(\otimes, \{\}, \underline{\mathrm{hom}})$.

Now that we know how to derive two-variable adjunctions, let us try to prove that $\mathcal{H}$ is a closed symmetric monoidal category. By Lemma 10.2.2, the two-variable adjunction $(\times, \underline{\mathrm{hom}}, \underline{\mathrm{hom}})$ defining the closed monoidal structure on simplicial sets induces a total derived two-variable adjunction $(\overset{\mathbf{L}}{\times}, \mathbf{R}\underline{\mathrm{hom}}, \mathbf{R}\underline{\mathrm{hom}})$ on $\mathcal{H}$ where the homotopy product and internal hom are defined by

$$X \overset{\mathbf{L}}{\times} Y := QX \times QY \qquad \mathbf{R}\underline{\mathrm{hom}}(X, Y) := \underline{\mathrm{hom}}(QX, RY).$$

In this particular case, the left deformations are not necessary, but we leave them in place so that our proof extends to the general case. We require an associativity isomorphism $- \overset{\mathbf{L}}{\times} (- \overset{\mathbf{L}}{\times} -) \cong (- \overset{\mathbf{L}}{\times} -) \overset{\mathbf{L}}{\times} -$. If the product of two cofibrant objects is again cofibrant, then the obvious candidate map

$$X \overset{\mathbf{L}}{\times} (Y \overset{\mathbf{L}}{\times} Z) = QX \times Q(QY \times QZ) \overset{1 \times q}{\longrightarrow} QX \times (QY \times QZ)$$

$$\cong (QX \times QY) \times QZ \overset{(q \times 1)^{-1}}{\longrightarrow} Q(QX \times QY) \times QZ$$

$$= (X \overset{\mathbf{L}}{\times} Y) \overset{\mathbf{L}}{\times} Z$$

is an isomorphism in $\mathcal{H}$. We also require a unit isomorphism $* \overset{\mathbf{L}}{\times} - \cong -$. If the unit object $*$ is cofibrant, or if tensoring with cofibrant objects preserves the weak equivalence $Q* \to *$, then the map

$$X \overset{\mathbf{L}}{\times} * = QX \times Q* \overset{1 \times q}{\longrightarrow} QX \times * \cong QX \overset{q}{\longrightarrow} X$$

is an isomorphism in $\mathcal{H}$.

This situation is axiomatized in the following definition.

Definition 10.2.4 A **closed symmetric monoidal homotopical category** is a closed symmetric monoidal category $(\mathcal{V}, \times, *)$ that is also homotopical and is equipped with a specified left deformation $(Q: \mathcal{V} \to \mathcal{V}_Q, q: Q \Rightarrow 1)$ and right deformation $(R: \mathcal{V} \to \mathcal{V}_R, r: 1 \Rightarrow R)$ such that

(i) these functors define a deformation for the tensor-hom two-variable adjunction
(ii) the tensor preserves cofibrant objects, i.e., the tensor restricts to a functor

$$- \times -: \mathcal{V}_Q \times \mathcal{V}_Q \to \mathcal{V}_Q$$

(iii) if v is cofibrant, the natural map $Q* \times v \to * \times v \cong v$ is a weak equivalence
(iv) the internal hom from a cofibrant object to a fibrant object is fibrant
(v) if v is fibrant, the natural map $v \cong \underline{\text{hom}}(*, v) \to \underline{\text{hom}}(Q*, v)$ is a weak equivalence

The symmetry isomorphism tells us that $v \times Q* \to v$ is also a weak equivalence. Conditions (iii) and (v) follow from (i) if the monoidal unit $*$ is a cofibrant object in $\mathcal{V}$.

Example 10.2.5 A monoidal model category, axiomatized in 9.2.3 and defined in 11.4.6, is a closed monoidal homotopical category. In particular, **sSet, sSet$_*$, Top, Top$_*$** are all examples, as is the category **Ch$_\bullet$(R)** of unbounded chain complexes over any commutative ring R.

Theorem 10.2.6 ([79, 15.4]) *The homotopy category of a closed symmetric monoidal homotopical category is a closed monoidal category, with structures given by the total derived functors of the upstairs structure functors.*

Proof The discussion commenced earlier proves that the total left derived functor of the monoidal product defines a monoidal product for the homotopy category with $Q*$ serving as the monoidal unit. In the notation of Definition 10.2.4, define the internal hom to be $\underline{\text{hom}}(Q-, R-)$, the total right derived functor of the upstairs internal hom. By Lemma 10.2.2, these total derived functors define a closed monoidal structure on the homotopy category. See also the proof of [38, 4.3.2]. □

Remark 10.2.7 The proof just given made no use of the axioms (iv) and (v), but we have good reason for their inclusion, explained in Remark 10.2.15.

Remark 10.2.8 As a quick reality check, note that the hom-sets in the underlying category of the enriched category Ho$\mathcal{V}$ coincide with the original hom-sets:

$$(\underline{\text{Ho}\mathcal{V}})_0(v, w) := \text{Ho}\mathcal{V}(*, \mathbf{R}\underline{\mathcal{V}}(v, w)) \cong \text{Ho}\mathcal{V}(* \overset{L}{\otimes} v, w) \cong \text{Ho}\mathcal{V}(v, w).$$

Of course this is necessarily the case for any closed monoidal category (cf. Lemma 3.4.9).

It is essential in what follows that the deformations of a closed symmetric monoidal homotopical category be taken as part of the structure of that category $\mathcal{V}$. In this context:

Definition 10.2.9 A $\mathcal{V}$-**homotopical category** $\underline{\mathcal{M}}$ is a tensored and cotensored $\mathcal{V}$-category whose underlying category $\mathcal{M}$ is homotopical and equipped with deformations $\mathcal{M}_Q, \mathcal{M}_R$ such that, together with the specified deformations $\mathcal{V}_Q, \mathcal{V}_R$ for $\mathcal{V}$,

(i) the specified functors form a deformation for the unenriched two-variable adjunction $(\otimes, \{\}, \underline{\text{hom}}) \colon \mathcal{V} \times \mathcal{M} \to \mathcal{M}$
(ii) the tensor preserves cofibrant objects, that is, the tensor restricts to a functor

$$- \otimes - \colon \mathcal{V}_Q \times \mathcal{M}_Q \to \mathcal{M}_Q$$

(iii) the natural map $Q * \otimes m \to * \otimes m \cong m$ is a weak equivalence for all $m \in \mathcal{M}_Q$
(iv) the cotensor from a cofibrant object to a fibrant object is fibrant
(v) the natural map $m \cong \{*, m\} \to \{Q*, m\}$ is a weak equivalence for all $m \in \mathcal{M}_R$

Example 10.2.10 A closed symmetric monoidal homotopical category $\mathcal{V}$ is a $\mathcal{V}$-homotopical category.

Example 10.2.11 A $\mathcal{V}$-model category, axiomatized in 9.2.3 and defined in 11.4.7, is a $\mathcal{V}$-homotopical category. In particular, a simplicial model category is a **sSet**-homotopical category.

The following result, whose proof extends the proof of Theorem 10.2.6, has Theorem 10.0.1 as a special case.

Theorem 10.2.12 ([79, 16.2]) *If $\mathcal{M}$ is a $\mathcal{V}$-homotopical category, then $\text{Ho}\mathcal{M}$ is the underlying category of a $\text{Ho}\mathcal{V}$-enriched, tensored, and cotensored category with enrichment given by the total derived two-variable adjunction*

$$(\overset{L}{\otimes}, \mathbf{R}\{\}, \mathbf{R}\underline{\text{hom}}) \colon \text{Ho}\mathcal{V} \times \text{Ho}\mathcal{M} \to \text{Ho}\mathcal{M}.$$

Proof It follows from Definition 10.2.9 and Lemma 10.2.2 that the derived functors of $(\otimes, \{\}, \underline{\text{hom}})$ define a two-variable adjunction $(\overset{L}{\otimes}, \mathbf{R}\{\}, \mathbf{R}\underline{\text{hom}}) \colon \text{Ho}\mathcal{V} \times \text{Ho}\mathcal{M} \to \text{Ho}\mathcal{M}$. By Theorem 10.2.6, we know $(\text{Ho}\mathcal{V}, \overset{L}{\times}, *)$ is a closed symmetric monoidal category. It remains to show that this derived two-variable adjunction gives Ho$\mathcal{M}$ the structure of a tensored and cotensored Ho$\mathcal{V}$-category. We prove this by exhibiting Ho$\mathcal{M}$ as a closed Ho$\mathcal{V}$-module and appealing to Proposition 10.1.4.

For this, we must define natural isomorphisms

$$v \overset{L}{\otimes} (w \overset{L}{\otimes} m) \cong (v \overset{L}{\times} w) \overset{L}{\otimes} m \quad \text{and} \quad * \overset{L}{\otimes} m \cong m \quad \forall v, w \in \text{Ho}\mathcal{V}, m \in \text{Ho}\mathcal{M}.$$

For the latter, by 10.2.9.(iii), we have a natural weak equivalence

$$* \overset{L}{\otimes} m = Q * \otimes Qm \overset{\sim}{\to} * \otimes Qm \cong Qm \overset{\sim}{\to} m$$

in $\mathcal{M}$, which descends to the desired isomorphism. The former makes use of 10.2.9.(ii) and the fact that cofibrant replacement in $\mathcal{V}$ defines a common left deformation for the monoidal product and the tensor. The two non-trivial maps in the zigzag

$$Qv \otimes Q(Qw \otimes Qm) \overset{\sim}{\to} Qv \otimes (Qw \otimes Qm) \cong (Qv \times Qw) \otimes Qm$$

$$\overset{\sim}{\leftarrow} Q(Qv \times Qw) \otimes Qm$$

are weak equivalences because the bifunctors $\times$ and $\otimes$ preserve cofibrant objects. In this way, the associativity isomorphism exhibits $\mathcal{M}$ and as a closed $\mathcal{V}$-module descends to make Ho$\mathcal{M}$ a closed Ho$\mathcal{V}$-module. $\qquad \square$

With this result, we now understand why the homotopy categories of any simplicial model category are enriched, tensored, and cotensored over $\mathcal{H}$. Now we would like to see under what conditions total derived functors become $\mathcal{H}$ enriched.

Warning 10.2.13 The deformations attached to a $\mathcal{V}$-homotopical category should be regarded as a fixed part of its structure, much like enrichments are specified, rather than chosen anew each time. The point is that the process of deriving the coherence isomorphisms used to show that particular total derived functors admit enrichments is somewhat delicate. The axioms of Definitions 10.2.4 and 10.2.9 are chosen so that there exist common deformations for all of the relevant structures. See [79, §16] for more on this point.

Proposition 10.2.14 ([79, 16.4]) *If $F: \mathcal{M} \to \mathcal{N}$ is a $\mathcal{V}$-functor between $\mathcal{V}$-homotopical categories and is left deformable and preserves cofibrant objects, then its total left derived functor $\mathbf{L}F: \text{Ho}\mathcal{M} \to \text{Ho}\mathcal{N}$ is canonically $\text{Ho}\mathcal{V}$ enriched.*

We give two proofs, the first a direct and the second a more formal argument, which provide different perspectives illustrating the same point.

Proof 1 Using Proposition 10.1.5, we can obtain a Ho$\mathcal{V}$-enrichment on $\mathbf{L}F$ by means of a natural transformation

$$\hat{\alpha}: v \overset{\mathbf{L}}{\otimes} \mathbf{L}Fm \Rightarrow \mathbf{L}F(v \overset{\mathbf{L}}{\otimes} m).$$

The obvious candidate is

$$Qv \otimes QFQm \overset{1 \times q}{\to} Qv \otimes FQm \overset{\alpha_{Q,Q}}{\to} F(Qv \otimes Qm) \overset{(Fq)^{-1}}{\to} FQ(Qv \otimes Qm).$$

The right map is a weak equivalence in $\mathcal{N}$ because we have assumed that the tensor on $\mathcal{M}$ preserves cofibrant objects. If F preserves cofibrant objects, the left map is also a weak equivalence in $\mathcal{N}$. Hence, in Ho$\mathcal{N}$, $\hat{\alpha}$ is isomorphic to $\alpha_{Q,Q}$ and therefore satisfies the appropriate coherence conditions to define an enrichment of $\mathbf{L}F$. □

Proof 2 By hypothesis, the bifunctors $- \otimes F-$ and $F(- \otimes -)$ are both homotopical on $\mathcal{V}_Q \times \mathcal{M}_Q$. It follows that these composite functors admit total left derived functors

$$\mathbf{L}(- \otimes F-) = Q - \otimes FQ - \qquad \text{and} \qquad \mathbf{L}F(- \otimes -) = F(Q - \otimes Q-).$$

As remarked in the proof of Theorem 2.2.9, it is not automatically the case that the total left derived functor of a composite functor agrees with the composite of the total left derived functors of the components, even if all these derived functors exist. This does hold if the first functor in the composable pair preserves cofibrant objects. For the right-hand side, this was part of the hypotheses in the $\mathcal{V}$-homotopical category $\mathcal{M}$. For the left-hand side, we include this condition as a specific hypothesis on F. In general, a natural transformation, such as α, admits a total left derived natural transformation whenever its domain and codomain admit a common deformation, so the rest of the argument is formal. □

Remark 10.2.15 Because we have included the axioms (iv) and (v) involving cotensors in our definitions of closed symmetric monoidal homotopical category and $\mathcal{V}$-homotopical category, a dual version of this result is available: if F is a $\mathcal{V}$-functor between $\mathcal{V}$-homotopical categories that is right deformable and preserves fibrant objects, then its total right derived functor is canonically Ho$\mathcal{V}$ enriched.

Proposition 10.2.16 ([79, 16.8]) *If* $F: \mathcal{M} \rightleftarrows \mathcal{N}: G$ *is a $\mathcal{V}$-adjunction between $\mathcal{V}$-homotopical categories and a deformable adjunction, then if either F preserves cofibrant objects or if G preserves fibrant objects, then there is a total derived $\mathcal{V}$-adjunction*

$$\mathbf{L}F: \underline{\text{Ho}\mathcal{M}} \rightleftarrows \underline{\text{Ho}\mathcal{N}}: \mathbf{R}G.$$

When the hypotheses of Proposition 10.2.16 are satisfied, we say the adjunction $F \dashv G$ is $\mathcal{V}$-**deformable**. The right-hand statement is only true in the presence of the cotensor axioms (iv) and (v) of Definitions 10.2.4 and 10.2.9.

Exercise 10.2.17 Write out a (sketched) proof of this proposition in a way that highlights the main points of the argument.

We record a final result for the sake of completeness.

Proposition 10.2.18 ([79, 16.13]) *Suppose we have a two-variable $\mathcal{V}$-adjunction*

$$ F : \underline{\mathcal{M}} \otimes \underline{\mathcal{N}} \to \underline{\mathcal{P}} \qquad G : \underline{\mathcal{M}}^{\mathrm{op}} \otimes \underline{\mathcal{P}} \to \underline{\mathcal{N}} \qquad \underline{\mathcal{N}}^{\mathrm{op}} \otimes \underline{\mathcal{P}} \to \underline{\mathcal{M}} $$

between $\mathcal{V}$-homotopical categories whose underlying unenriched two-variable adjunction is deformable and such that one of the three bifunctors preserves (co)fibrant objects, as appropriate. Then the total derived functors form a two-variable $\mathrm{Ho}\mathcal{V}$-adjunction.

10.3 Weighted homotopy limits and colimits, revisited

We have seen that if $\mathcal{V}$ is a closed symmetric monoidal homotopical category, then $\mathrm{Ho}\mathcal{V}$ is a closed monoidal category and hence a suitable base for enrichment. Also, if $\mathcal{M}$ is a $\mathcal{V}$-homotopical category, then its homotopy category is $\mathrm{Ho}\mathcal{V}$-enriched, tensored, and cotensored. Furthermore, there exist elementary conditions under which $\mathrm{Ho}\mathcal{V}$-enrichments extend to total derived functors of $\mathcal{V}$-functors between $\mathcal{V}$-homotopical categories: the total left derived functor of a left deformable $\mathcal{V}$-functor that preserves cofibrant objects is $\mathrm{Ho}\mathcal{V}$-enriched. Similarly, given a deformable $\mathcal{V}$-adjunction, if either the left adjoint preserves cofibrant objects or the right adjoint preserves fibrant objects, then the total derived functors define a $\mathrm{Ho}\mathcal{V}$-adjunction; this is what it means to say the adjoint pair is $\mathcal{V}$ deformable.

Our goal is now to apply these results to the total derived functors of the weighted homotopy colimit and weighted homotopy limit bifunctors constructed in Chapter 10. To do so, we must combine the hypotheses from these chapters. The assumptions of 9.2.5, in which we supposed that $\mathcal{V}$ is a monoidal model category and that $\mathcal{M}$ is a $\mathcal{V}$-model category, imply that $\mathcal{V}$ is a closed symmetric monoidal homotopical category and that $\mathcal{M}$ is $\mathcal{V}$-homotopical. It remains only to provide a $\mathcal{V}$-homotopical structure on the diagram category $(\underline{\mathcal{M}}^{\mathcal{D}})_0$ for any small $\mathcal{V}$-category $\underline{\mathcal{D}}$, which we do not assume supports any relevant model structure. It turns out no additional hypotheses are needed, though this claim will require a bit of justification.

Recall that if $\mathcal{M}$ is a homotopical category, then $\mathcal{M}^{\mathcal{D}}$ is a homotopical category with its weak equivalences defined pointwise. Furthermore:

Lemma 10.3.1 *If $\mathcal{M}$ is a saturated homotopical category, then $\mathcal{M}^{\mathcal{D}}$ is saturated.*

Proof Because the weak equivalences in $\mathcal{M}^{\mathcal{D}}$ were defined pointwise, for each $d \in \mathcal{D}$, the functor $\mathrm{ev}_d \colon \mathcal{M}^{\mathcal{D}} \to \mathcal{M}$ is homotopical. Hence we have a commutative diagram

$$
\begin{array}{ccc}
\mathcal{M}^{\mathcal{D}} & \overset{\mathrm{ev}_d}{\longrightarrow} & \mathcal{M} \\
\downarrow & & \downarrow \\
\mathrm{Ho}(\mathcal{M}^{\mathcal{D}}) & \overset{\mathrm{ev}_d}{\longrightarrow} & \mathrm{Ho}\mathcal{M}
\end{array}
$$

where the vertical arrows are the localizations defined after 2.1.6. The component at d of any map which becomes an isomorphism in $\mathrm{Ho}(\mathcal{M}^{\mathcal{D}})$ is therefore an isomorphism in $\mathrm{Ho}\mathcal{M}$ and hence a weak equivalence if $\mathcal{M}$ is saturated. So $\mathcal{M}^{\mathcal{D}}$ is also saturated under these hypotheses. □

The same argument works for the underlying category of an enriched functor category $\underline{\mathcal{M}}^{\underline{\mathcal{D}}}$. Recall that simplicial homotopies in $(\underline{\mathcal{M}}^{\underline{\mathcal{D}}})_0$ were defined using the $\mathcal{V}$-tensor structure on diagrams, which is defined pointwise (cf. Remark 3.8.2). In particular, simplicial homotopy equivalences in $(\underline{\mathcal{M}}^{\underline{\mathcal{D}}})_0$ are weak equivalences if this is true for $\mathcal{M}$.

It remains to define the deformations that make $\underline{\mathcal{M}}^{\underline{\mathcal{D}}}$ a $\mathcal{V}$-homotopical category. The following is a direct consequence of the homotopical properties of the two-sided enriched bar construction described in Lemma 9.2.4.

Lemma 10.3.2 ([79, 20.8]) *If $\underline{\mathcal{M}}$ and $\underline{\mathcal{D}}$ are $\mathcal{V}$-categories satisfying the assumptions of 9.2.5, then $\underline{\mathcal{M}}^{\underline{\mathcal{D}}}$ is a $\mathcal{V}$-homotopical category when equipped with either*

(i) the left deformation $B(\underline{\mathcal{D}}, \underline{\mathcal{D}}, Q-)$ and the right deformation R or
(ii) the right deformation $C(\underline{\mathcal{D}}, \underline{\mathcal{D}}, R-)$ and the left deformation Q

We call (i) and (ii) the **bar $\mathcal{V}$-homotopical structure** and the **cobar $\mathcal{V}$-homotopical structure**. Note that this result does not rely on the existence of any model structure on the diagram category $(\underline{\mathcal{M}}^{\underline{\mathcal{D}}})_0$. This is an advantage of axiomatizing the conditions we need to derive enrichment in the definition of a $\mathcal{V}$-homotopical category. The following results, which require the same hypotheses, are immediate consequences.

Corollary 10.3.3 ([79, 20.9]) *The homotopy category of the category* $(\underline{\mathcal{M}}^{\mathcal{D}})_0$ *of* $\mathcal{V}$-functors and $\mathcal{V}$-natural transformations is tensored, cotensored, and enriched over* Ho$\mathcal{V}$.

Corollary 10.3.4 ([79, 21.1–2]) *If* W *is a pointwise cofibrant weight, then the total left derived functor of* colim^W *and the total right derived functor of* lim^W *are* Ho$\mathcal{V}$-enriched.

Now suppose $\Phi \colon \underline{\mathcal{M}} \rightleftarrows \underline{\mathcal{N}} \colon \Psi$ is a $\mathcal{V}$-deformable adjunction. Because left $\mathcal{V}$-adjoints commute with the enriched bar construction and right $\mathcal{V}$-adjoints commute with the enriched cobar construction, the induced $\mathcal{V}$-adjunction $\Phi^{\mathcal{D}} \colon \underline{\mathcal{M}}^{\mathcal{D}} \rightleftarrows \underline{\mathcal{N}}^{\mathcal{D}} \colon \Psi^{\mathcal{D}}$ is also $\mathcal{V}$-deformable. In particular:

Theorem 10.3.5 ([79, 22.2-3]) *Let* W *be a pointwise cofibrant weight. Under the standing hypotheses, there is a* Ho$\mathcal{V}$-natural isomorphism

$$\mathbf{L} \operatorname{colim}^W (\mathbf{L}\Phi^{\mathcal{D}}(F)) \cong \mathbf{L}\Phi(\mathbf{L} \operatorname{colim}^W F).$$

In words, total left derived functors of left $\mathcal{V}$-adjoints preserve weighted homotopy colimits.*

A dual Ho$\mathcal{V}$-natural isomorphism shows that total right derived functors of right $\mathcal{V}$-adjoints preserve weighted homotopy limits.

Proof The only thing to check is that the derived functors can be composed. This is straightforward (or see [79, §21]). □

We conclude by mentioning a final result concerning enriched Kan extensions. The proof makes use of the flexible handedness of Proposition 10.2.16: the homotopical behavior of precomposition, sometimes a right adjoint and sometimes a left adjoint, is much easier to understand than the homotopical behavior of either Kan extension.

Theorem 10.3.6 ([79, 22.2-22.3]) *If* $K \colon \underline{\mathcal{C}} \to \underline{\mathcal{D}}$ *is a* $\mathcal{V}$-functor between small $\mathcal{V}$-categories and $\underline{\mathcal{V}}$ and $\underline{\mathcal{M}}$ satisfy 9.2.5, then the $\mathcal{V}$-adjunctions

$$\operatorname{Lan}_K \colon \underline{\mathcal{M}}^{\underline{\mathcal{C}}} \rightleftarrows \underline{\mathcal{M}}^{\mathcal{D}} \colon K^* \qquad \text{and} \qquad K^* \colon \underline{\mathcal{M}}^{\mathcal{D}} \rightleftarrows \underline{\mathcal{M}}^{\underline{\mathcal{C}}} \colon \operatorname{Ran}_K$$

are $\mathcal{V}$-deformable.

More precisely, $\operatorname{Lan}_K \dashv K^*$ is $\mathcal{V}$-deformable with respect to the bar $\mathcal{V}$-homotopical structures, and $K^* \dashv \operatorname{Ran}_K$ is $\mathcal{V}$-deformable with respect to the cobar $\mathcal{V}$-homotopical structures. In the former case, this means that the point-set level left derived functor of Lan_K, applied to $F \in \underline{\mathcal{M}}^{\underline{\mathcal{C}}}$, is

$$\mathbb{L}\mathrm{Lan}_K F \cong \mathrm{Lan}_K B(\underline{\mathcal{C}}, \underline{\mathcal{C}}, QF) \cong \underline{\mathcal{D}}(K-, -) \otimes_{\underline{\mathcal{C}}} B(\underline{\mathcal{C}}, \underline{\mathcal{C}}, QF)$$
$$\cong B(\underline{\mathcal{D}}(K-, -) \otimes_{\underline{\mathcal{C}}} \underline{\mathcal{C}}, \underline{\mathcal{C}}, QF)$$
$$\cong B(\underline{\mathcal{D}}(K-, -), \underline{\mathcal{C}}, QF),$$

which was the definition of the homotopy left Kan extension functor given in Example 9.2.11. Dually, using the deformations of the cobar $\mathcal{V}$-homotopical structure, the point-set level right derived functor of Ran_K is

$$\mathbb{R}\mathrm{Ran}_K F \cong \mathrm{Ran}_K C(\underline{\mathcal{C}}, \underline{\mathcal{C}}, RF) \cong \{\underline{\mathcal{D}}(-, K-), C(\underline{\mathcal{C}}, \underline{\mathcal{C}}, RF)\}^{\underline{\mathcal{C}}}$$
$$\cong C(\underline{\mathcal{D}}(-, K-) \otimes_{\underline{\mathcal{C}}^{\mathrm{op}}} \underline{\mathcal{C}}, \underline{\mathcal{C}}, RF)$$
$$\cong C(\underline{\mathcal{D}}(-, K-), \underline{\mathcal{C}}, RF),$$

which recovers our formula for homotopy right Kan extension.

Because restriction along K is homotopical, the total left and right derived functors of K^* are both isomorphic to K^*. A priori, the derived *enrichments* produced by the two $\mathcal{V}$-deformable adjunctions might differ; however, as [79, §22] explains, this turns out not to be the case. Hence:

Corollary 10.3.7 ([79, 22.5])   *There is a derived* $\mathrm{Ho}\mathcal{V}$*-adjunction*

$$\mathbb{L}\mathrm{Lan}_K \dashv \mathbb{R}K^* \cong K^* \cong \mathbb{L}K^* \dashv \mathbb{R}\mathrm{Ran}_K.$$

10.4 Homotopical structure via enrichment

We conclude this chapter with two short sections that explore homotopical structures on the underlying category of an enriched category that arise from the enrichment in a closed symmetric monoidal homotopical category. The highlight will be a generalization of the classical comparison between the homotopy equivalences (the homotopical structure determined by the topological enrichment) and weak homotopy equivalences of topological spaces.

As a starting point, though not fully in the spirit of what is to follow, we note the following somewhat curious result.

Lemma 10.4.1   *If* $\mathrm{Ho}\mathcal{V}$ *is locally small, the homotopy category of any* $\mathcal{V}$*-homotopical category is also locally small.*

Proof   Let $\underline{\mathcal{M}}$ be a $\mathcal{V}$-homotopical category. The claim holds because the underlying category of the $\mathrm{Ho}\mathcal{V}$-enriched category $\mathrm{Ho}\underline{\mathcal{M}}$ is the homotopy category of the underlying category of $\underline{\mathcal{M}}$. Tautologically, both categories have the same objects. By definition, the set of arrows from m to m' in the underlying category of $\underline{\text{Ho}\mathcal{M}}$ is

$$\text{Ho}\mathcal{V}(*, \underline{\text{Ho}\mathcal{M}}(m, m')) = \text{Ho}\mathcal{V}(*, \mathbf{R}\underline{\text{hom}}(m, m'))$$

$$\cong \underline{\text{Ho}\mathcal{M}}(* \overset{L}{\otimes} m, m') \cong \text{Ho}\mathcal{M}(m, m'). \qquad \square$$

In some sense, we seem to have gotten something for nothing. To quote [79], "The presence of enrichment often simplifies, rather than complicates, the study of homotopy theory, since enrichment over a suitable category automatically provides well-behaved notions of homotopy and homotopy equivalence."

To explore what this means, let us first simplify and suppose $\mathcal{M}$ is a tensored and cotensored $\mathcal{V}$-category, where $\mathcal{V}$ is a closed symmetric monoidal homotopical category, but where $\mathcal{M}$ is not known to be homotopical in its own right. By Theorem 10.2.6, $\text{Ho}\mathcal{V}$ is a closed symmetric monoidal category with monoidal product $\overset{L}{\times}$ represented on the point-set level by the bifunctor $- \overset{\mathbb{L}}{\times} - := Q - \times Q-$. The natural map

$$X \overset{\mathbb{L}}{\times} Y = QX \times QY \overset{q \times q}{\longrightarrow} X \times Y$$

descends along the localization functor $\gamma \colon \mathcal{V} \to \text{Ho}\mathcal{V}$ to define a natural arrow

$$\gamma X \overset{L}{\times} \gamma Y \to \gamma(X \times Y) \qquad (10.4.2)$$

that makes γ a lax monoidal functor. It follows from Lemma 3.4.3 that any $\mathcal{V}$-enriched category is canonically $\text{Ho}\mathcal{V}$-enriched.

Let $\underline{h\mathcal{M}}$ be the $\text{Ho}\mathcal{V}$-category arising from a $\mathcal{V}$-category $\mathcal{M}$, with hom-objects defined by applying the localization functor γ and composition given by the map

$$\gamma\underline{\mathcal{M}}(y, z) \overset{L}{\times} \gamma\underline{\mathcal{M}}(x, y) \to \gamma(\underline{\mathcal{M}}(y, z) \times \underline{\mathcal{M}}(x, y)) \to \gamma\underline{\mathcal{M}}(x, z)$$

in $\text{Ho}\mathcal{V}$. Note that the functor γ is the identity on objects. Thus the hom-objects of $\underline{h\mathcal{M}}$ are represented by the same hom-objects of $\mathcal{M}$, but the structure of the former enriched category is given by maps in $\text{Ho}\mathcal{V}$, not in $\mathcal{V}$. Employing the category theorists' philosophy that an object in a category is fundamentally a stand-in for its isomorphism class, the objects in $\text{Ho}\mathcal{V}$ can be thought of as weak equivalence classes of objects of $\mathcal{V}$. For $\mathcal{V} = \mathbf{sSet}$ or $\mathbf{Top}$, the hom-objects of $\mathcal{M}$ are spaces, whereas the hom-objects of $\underline{h\mathcal{M}}$ are typically thought of as homotopy types.

We are not interested in $\underline{h\mathcal{M}}$ directly; rather, we care about its underlying category $h\mathcal{M}$ with hom-sets

$$h\mathcal{M}(x, y) := \text{Ho}\mathcal{V}(*, \underline{h\mathcal{M}}(x, y)) = \text{Ho}\mathcal{V}(*, \underline{\mathcal{M}}(x, y)).$$

Note that this notation is consistent with the introduction to Chapter 3.

Definition 10.4.3 Call a morphism in $\mathcal{M}$ a $\mathcal{V}$-**equivalence** if its image in $h\mathcal{M}$ is an isomorphism.

Because isomorphisms satisfy the 2-of-6 property, the $\mathcal{V}$-equivalences give any category $\mathcal{M}$ enriched over a closed symmetric monoidal homotopical category $\mathcal{V}$ the structure of a homotopical category. There is a canonical functor $\mathcal{M} \to h\mathcal{M}$ that we call "localization." This terminology will be justified by Theorem 10.4.6.

Example 10.4.4 When $\mathcal{V} = \textbf{Top}, \textbf{sSet}$, or $\textbf{Ch}_{\bullet}(R)$, $h\mathcal{M}$ is the category whose morphisms are given by taking points in the hom-spaces up to, respectively, homotopy, simplicial homotopy, or chain homotopy. Hence the $\mathcal{V}$-equivalences are precisely the homotopy equivalences, simplicial homotopy equivalences, or chain homotopy equivalences.

Just as any topological functor preserves homotopy equivalences (though not necessarily weak homotopy equivalences), and any additive functor preserves chain homotopy equivalences (though not necessarily quasi-isomorphisms), we can prove in general that any $\mathcal{V}$-functor automatically preserves $\mathcal{V}$-equivalences.

Lemma 10.4.5 ([79, 17.4]) *Any $\mathcal{V}$-functor $F : \underline{\mathcal{M}} \to \underline{\mathcal{N}}$ between $\mathcal{V}$-categories preserves $\mathcal{V}$-equivalences.*

Proof We claim that the functor underlying F extends along the localizations to produce a commutative diagram

$$
\begin{array}{ccc}
\mathcal{M} & \xrightarrow{\ F\ } & \mathcal{N} \\
\downarrow & & \downarrow \\
h\mathcal{M} & \overset{hF}{\dashrightarrow} & h\mathcal{N}
\end{array}
$$

This assertion is obvious on objects. Given $x, y \in \mathcal{M}$, the map hF is defined on hom-sets by applying the functor $\text{Ho}\mathcal{V}(*, -)$ to $F : \underline{\mathcal{M}}(x, y) \to \underline{\mathcal{N}}(Fx, Fy)$, as indicated:

$$
h\mathcal{M}(x, y) = \text{Ho}\mathcal{V}(*, \underline{\mathcal{M}}(x, y)) \longrightarrow \text{Ho}\mathcal{V}(*, \underline{\mathcal{N}}(Fx, Fy)) = h\mathcal{N}(Fx, Fy).
$$

The rest of the argument is elementary: a $\mathcal{V}$-equivalence in $\mathcal{M}$ becomes an isomorphism in $h\mathcal{M}$, hence remains an isomorphism in $h\mathcal{N}$, and hence its image in $\mathcal{N}$ is a $\mathcal{V}$-equivalence. $\qquad\square$

The $\mathcal{V}$-equivalences make $\mathcal{M}$ a homotopical category, but is it a $\mathcal{V}$-homotopical category? And if so, is the associated $\text{Ho}\mathcal{V}$ enrichment associated to the homotopy category $\text{Ho}\mathcal{M}$ related to $\underline{h\mathcal{M}}$? We have no reason to suppose that $\mathcal{M}$ has well-behaved left or right deformations, so we will attempt to provide a $\mathcal{V}$-homotopical structure with all objects of $\mathcal{M}$ fibrant and cofibrant. This is not as unreasonable as it sounds: the tensor and cotensor with any $v \in \mathcal{V}$, being $\mathcal{V}$-functors, preserve all $\mathcal{V}$-equivalences and hence are homotopical. Similarly, by definition, the internal hom $h\underline{M}(m, m') = \gamma \underline{M}(m, m')$ takes $\mathcal{V}$-equivalences to isomorphisms in Ho$\mathcal{V}$, so if $\mathcal{V}$ is saturated, then the internal hom also preserves weak equivalences in both variables.

Supposing that the monoidal unit is cofibrant (or that the unit conditions 10.2.9 (iii) and (v) are satisfied), it remains only to check that tensoring with or cotensoring into any $m \in \mathcal{M}$ preserves weak equivalences between cofibrant objects in $\mathcal{V}$. It turns out this holds if the $\mathcal{V}$-category $\underline{M}$ is locally fibrant, that is, if the hom-objects $\underline{M}(m, m')$ are fibrant in $\mathcal{V}$.

Theorem 10.4.6 ([79, 17.5]) *If $\mathcal{M}$ is locally fibrant and if the unit conditions (iii) and (v) are satisfied, then the $\mathcal{V}$-equivalences make $\mathcal{M}$ a saturated $\mathcal{V}$-homotopical category in such a way that we have an isomorphism $h\underline{M} \cong$ Ho$\mathcal{M}$ of Ho$\mathcal{V}$-categories. Therefore $h\mathcal{M}$ is the localization of $\mathcal{M}$ at the $\mathcal{V}$-equivalences.*

Proof Suppose $f : v \to w$ is a weak equivalence between cofibrant objects. We use the Yoneda lemma to show that $f \otimes m : v \otimes m \to w \otimes m$ is an isomorphism in $h\mathcal{M}$, and hence a $\mathcal{V}$-equivalence. By naturality, the following diagram commutes in $\mathcal{V}$:

$$
\begin{array}{ccc}
\underline{M}(w \otimes m, n) & \overset{\cong}{\longrightarrow} & \underline{V}(w, \underline{M}(m, n)) \\
{\scriptstyle (f \otimes m)^*} \Big\downarrow & & \Big\downarrow {\scriptstyle f^*} \\
\underline{M}(v \otimes m, n) & \underset{\cong}{\longrightarrow} & \underline{V}(v, \underline{M}(m, n))
\end{array}
$$

The internal hom in $\mathcal{V}$ preserves weak equivalences between cofibrant objects in its first variable when its second variable is fibrant. Hence the right vertical map and thus also the left vertical map is a weak equivalence. Applying $\gamma : \mathcal{V} \to$ Ho$\mathcal{V}$, we obtain an isomorphism $h\underline{M}(w \otimes m, n) \cong h\underline{M}(v \otimes m, n)$ in Ho$\mathcal{V}$. Because the map $(f \otimes m)^*$ is $\mathcal{V}$-natural, this isomorphism is Ho$\mathcal{V}$-natural. Hence the Yoneda lemma implies that $f \otimes m : v \otimes m \to w \otimes m$ is an isomorphism in $h\mathcal{M}$, and hence $f \otimes m$ is a $\mathcal{V}$-equivalence, as desired.

For the last part of the statement, because the deformations on $\mathcal{M}$ are taken to be identities, the definitions of hom-objects in the Ho$\mathcal{V}$-categories $h\underline{M}$ and Ho$\underline{M}$ agree:

$$
h\underline{M}(m, m') = \gamma \underline{M}(m, m') = \underline{M}(m, m') = \text{Ho}\underline{M}(m, m'),
$$

and tautologously $\mathcal{M}$ is saturated. $\qquad\qquad\square$

This completes the justification of the discussion from the beginning of Chapter 3. If $\mathcal{M}$ is a topologically enriched, tensored, and cotensored category, its homotopy products, defined representably, are products in its homotopy category $h\mathcal{M} \cong \mathrm{Ho}\mathcal{M}$.

10.5 Homotopy equivalences versus weak equivalences

Now suppose $\mathcal{M}$ is known to be a $\mathcal{V}$-homotopical category. The weak equivalences specified by its homotopical structure are likely distinct from the $\mathcal{V}$-equivalences just defined. Nonetheless, we show that just as in classical homotopy theory, the $\mathcal{V}$-equivalences are a useful stepping-stone to the weak equivalences. The following remarkable theorem of Shulman is a generalization of classical results that describe the relationship between homotopy equivalences and weak homotopy equivalences of topological spaces (see also Digression 11.3.13).

Theorem 10.5.1 ([79, 18.1]) *Suppose $\underline{\mathcal{M}}$ is a $\mathcal{V}$-homotopical category. Then there is a* $\mathrm{Ho}\mathcal{V}$*-functor* $\phi\colon \underline{h\mathcal{M}} \to \underline{\mathrm{Ho}\mathcal{M}}$ *whose underlying unenriched functor commutes with both localizations*

$$(10.5.2)$$

Moreover:

(i) *The restriction of ϕ to the fibrant–cofibrant objects is fully faithful.*
(ii) *If either Q preserves fibrant objects or R preserves cofibrant objects, then the restriction of ϕ to the fibrant-cofibrant objects is an equivalence.*
(iii) *A weak equivalence between fibrant–cofibrant objects is a $\mathcal{V}$-equivalence.*
(iv) *If $\mathcal{M}$ is saturated, then all $\mathcal{V}$-equivalences are weak equivalences.*

Note all three functors in (10.5.2) are identity-on-objects; the difference is in the morphisms. The first localization "takes homotopy classes of maps." The second localization formally inverts the weak equivalences. In classical terminology, (i) says that maps in the homotopy category between fibrant–cofibrant objects are homotopy classes of maps, and (ii) says that the homotopy category is equivalent to the category of homotopy classes of maps between fibrant–cofibrant objects, supposing that fibrant–cofibrant replacements exist.

Proof of Theorem 10.5.1   The functor ϕ is defined on hom-objects by the natural map

$$\underline{h\mathcal{M}}(m, m') = \underline{\mathcal{M}}(m, m') \to \underline{\mathcal{M}}(Qm, Rm') = \underline{\mathrm{Ho}\mathcal{M}}(m, m') \qquad (10.5.3)$$

given by precomposing with $Qm \to m$ and postcomposing with $m' \to Rm'$. We leave it as an exercise to check that this defines a Ho$\mathcal{V}$-functor.

For (i), if m is cofibrant and m' is fibrant, then (10.5.3) is a weak equivalence and hence an isomorphism in Ho$\mathcal{V}$; hence ϕ is fully faithful. For (ii), if either Q preserves fibrant objects or R preserves cofibrant objects, then every object in $\mathcal{M}$ is weakly equivalent to a fibrant–cofibrant one, and hence every object is isomorphic in Ho$\mathcal{M}$ to one in the image of $h\mathcal{M}_{QR}$. The restriction of ϕ is therefore an equivalence of Ho$\mathcal{V}$-categories.

For (iii), a weak equivalence between fibrant–cofibrant objects is an isomorphism in Ho$\mathcal{M}$ and hence also in $h\mathcal{M}$ because ϕ is fully faithful on this subcategory. Hence such a map is a $\mathcal{V}$-equivalence. Finally, for (iv), any $\mathcal{V}$-equivalence in $\mathcal{M}$ is an isomorphism in $h\mathcal{M}$ and hence an isomorphism in Ho$\mathcal{M}$, and hence a weak equivalence, if $\mathcal{M}$ is saturated. □

As a corollary, we note that in certain conditions, $\mathcal{V}$-functors are extraordinarily easy to derive.

Corollary 10.5.4 ([79, 18.3])   *Suppose $\mathcal{M}$ and $\mathcal{N}$ are $\mathcal{V}$-homotopical categories such that $\mathcal{N}$ is saturated. If every object of $\mathcal{M}$ is fibrant, then any $\mathcal{V}$-functor $F: \underline{\mathcal{M}} \to \underline{\mathcal{N}}$ has a left derived functor. If every object of $\mathcal{M}$ is cofibrant, then any such $\mathcal{V}$-functor has a right derived functor.*

Proof   Suppose all objects are fibrant. By Theorem 10.5.1, all weak equivalences between fibrant–cofibrant objects are $\mathcal{V}$-equivalences, which any $\mathcal{V}$-functor preserves. Because $\mathcal{N}$ is saturated, all $\mathcal{V}$-equivalences in $\mathcal{N}$ are weak equivalences. It follows that F is homotopical on the cofibrant objects, and hence is left deformable, and hence has a left derived functor. □

In fact, any $\mathcal{V}$-functor between $\mathcal{V}$-homotopical categories whose target is saturated has a **middle derived functor**. See [79, §4] for the definition.

Part III

Model categories and weak factorization systems

11

Weak factorization systems in model categories

Having exhausted the theory of derived functors, we shift our focus to model categories, the context in which they are most commonly constructed. This contextualization appears in Section 11.3, where the definition of a model category finally appears, but first we focus on the parts of a model structure invisible to the underlying homotopical category. We think this perspective nicely complements standard presentations of model category theory, for example, [24, 36, 38]. (The newer [58] bears a familial resemblance to our presentation.) Those sources allow our treatment here to be quite brief. Where we have nothing of substance to contribute, rather than retrace well-trodden ground, we leave the standard parts of the theory to existing literature.

The reward for our work comes in the last section, in which we prove a theorem with a number of important consequences. Specifically, we describe the homotopical properties of the weighted limit and weighted colimit bifunctors in model category language. Using this, we give a simple proof of the homotopy finality theorem 8.5.6 and finally show that the different "op" conventions for the homotopy limit and homotopy colimit functors discussed in Remark 7.8.3 give weakly equivalent results. Finally, we prove that several familiar constructions for particular homotopy colimits have the appropriate universal properties.

11.1 Lifting problems and lifting properties

Classically, **cofibrations** and **fibrations**, technical terms in the context of any model structure, refer to classes of continuous functions of topological spaces characterized by certain lifting properties. Our story begins by explaining the common features of any class of maps defined in this way. We give an "algebraic" characterization of such classes in Chapter 12.

Let i and f be arrows in a fixed category $\mathcal{M}$. A **lifting problem** between i and f is simply a commutative square:

A **lift** or **solution** is a dotted arrow, as indicated, making both triangles commute. If any lifting problem between i and f has a solution, we say that i has the **left lifting property** with respect to f and, equivalently, that f has the **right lifting property** with respect to i. We use the suggestive symbolic notation $i \boxtimes f$ to encode these equivalent assertions.

Example 11.1.1   A map of sets has the right lifting property against the unique map $\emptyset \to *$ if and only if the map is an epimorphism. A map of sets has the right lifting property against the unique map $* \sqcup * \to *$ if and only if the map is a monomorphism.

Example 11.1.2   A discrete right fibration, defined in 7.1.9, is a functor that has the right lifting property with respect to the codomain inclusion of the terminal category into the walking arrow $\mathbb{1} \to \mathbb{2}$.

Suppose $\mathcal{L}$ is a class of maps in $\mathcal{M}$. We write $\mathcal{L}^{\boxtimes}$ for the class of arrows that have the right lifting property against each element of $\mathcal{L}$. Dually, we write $^{\boxtimes}\mathcal{R}$ for the class of arrows that have the left lifting property against a given class $\mathcal{R}$.

Example 11.1.3   Writing i_0 and p_0 for the obvious maps induced by the inclusion of the 0th endpoint of the standard unit interval I, the **Hurewicz fibrations** are defined to be $\{i_0 : A \to A \times I\}^{\boxtimes}$ and the **Hurewicz cofibrations** are $^{\boxtimes}\{p_0 : A^I \to A\}$, where the classes defining these lifting properties are indexed by all topological spaces A. Restricting to the subset of cylinder inclusions on disks, $\{i_0 : D^n \to D^n \times I\}^{\boxtimes}$ is the class of **Serre fibrations**.

Any class of maps that is defined by a left lifting property is **weakly saturated**, meaning it is closed under the following constructions:

Lemma 11.1.4   *Any class of arrows of the form $^{\boxtimes}\mathcal{R}$ is closed under coproducts, pushouts, transfinite composition, retracts, and contains the isomorphisms.*

A diagram whose domain is the ordinal α is called an α-**composite** if, for each limit ordinal $\beta < \alpha$, the subdiagram indexed by β is a colimit cone. An ω-composite might also be called a countable composite. To say that $^{\boxtimes}\mathcal{R}$

is closed under transfinite composition means that if each arrow in an α-composite between the images of some ordinal and its successor is in $^\boxtimes\mathcal{R}$, then the "composite" arrow from the image of zero to the image of α is also in $^\boxtimes\mathcal{R}$.

Proof   All of the arguments are similar. For instance, suppose j is a retract of $i \in {}^\boxtimes\mathcal{R}$. By precomposing a lifting problem with the retract diagram, displayed on the left, we obtain a solution as indicated:

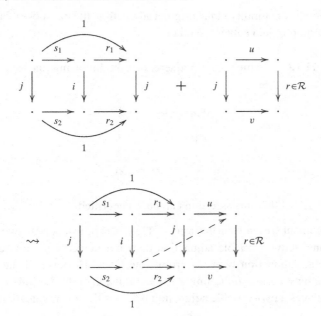

So j lifts against $\mathcal{R}$ and is therefore in $^\boxtimes\mathcal{R}$.   □

Lemma 11.1.4, of course, has a dual version describing the closure properties of classes of arrows characterized by a right lifting property, but we prefer not to state it because we do not know what to call the dual notion of transfinite composition.

If $\mathcal{L}$ and $\mathcal{R}$ are two classes of maps, we write $\mathcal{L} \boxtimes \mathcal{R}$ to mean that $\mathcal{L} \subset {}^\boxtimes\mathcal{R}$ and, equivalently, that $\mathcal{R} \subset \mathcal{L}^\boxtimes$. Observe that the operators $(-)^\boxtimes$ and $^\boxtimes(-)$ form a Galois connection with respect to inclusion.

It will be useful to note that lifting properties interact nicely with adjunctions.

Lemma 11.1.5   *Suppose $F\colon \mathcal{M} \rightleftarrows \mathcal{N} : U$ is an adjunction and let $\mathcal{A}$ be a class of arrows in $\mathcal{M}$ and $\mathcal{B}$ be a class of arrows in $\mathcal{N}$. Then $F\mathcal{A} \boxtimes \mathcal{B}$ if and only if $\mathcal{A} \boxtimes U\mathcal{B}$.*

Proof   An adjunction $F \dashv U$ induces an adjunction $F\colon \mathcal{M}^2 \rightleftarrows \mathcal{N}^2 : U$ between the respective arrow categories; the adjoints are defined pointwise by postcomposition. In particular, any lifting problem $Fi \Rightarrow f$ in $\mathcal{N}$ has an adjunct lifting problem $i \Rightarrow Uf$ in $\mathcal{M}$:

Furthermore, by naturality of the original adjunction, the transpose of a solution to one lifting problem solves the other. $\qquad\square$

Remark 11.1.6 Lemma 11.1.5 asserts that the following diagram is a pullback in **Set**:

$$
\begin{array}{ccc}
(F\mathcal{A})^{\boxtimes} & \xrightarrow{\ U\ } & \mathcal{A}^{\boxtimes} \\
\downarrow & \lrcorner & \downarrow \\
\mathcal{N}^2 & \xrightarrow[\ U\ ]{} & \mathcal{M}^2
\end{array}
$$

We see in 12.6.4 that this assertion can be categorified.

The argument presented in the proof of 11.1.5 can be extended to two-variable adjunctions. A two-variable adjunction does not induce a pointwise-defined two-variable adjunction between arrow categories. However, if the ambient categories have certain finite limits and colimits, then the **Leibniz construction**[1] produces a two-variable adjunction between the arrow categories.

Construction 11.1.7 (Leibniz construction) Consider a two-variable adjunction

$$- \otimes -: \mathcal{M} \times \mathcal{N} \to \mathcal{P} \quad \{-,-\}: \mathcal{M}^{\mathrm{op}} \times \mathcal{P} \to \mathcal{N}$$

$$\underline{\hom}(-,-): \mathcal{N}^{\mathrm{op}} \times \mathcal{P} \to \mathcal{M}$$

$$\mathcal{P}(m \otimes n, p) \cong \mathcal{N}(n, \{m, p\}) \cong \mathcal{M}(m, \underline{\hom}(n, p)).$$

If $\mathcal{P}$ has pushouts and $\mathcal{M}$ and $\mathcal{N}$ have pullbacks, there is an induced two-variable adjunction

$$- \hat{\otimes} -: \mathcal{M}^2 \times \mathcal{N}^2 \to \mathcal{P}^2 \quad \{-,\hat{-}\}: (\mathcal{M}^2)^{\mathrm{op}} \times \mathcal{P}^2 \to \mathcal{N}^2$$

$$\hat{\underline{\hom}}(-,-): (\mathcal{N}^2)^{\mathrm{op}} \times \mathcal{P}^2 \to \mathcal{M}^2$$

---

[1] The name refers to Leibniz's formula for the boundary of the product of two polygons $\partial(A \times B) = (\partial A \times B) \cup_{(\partial A \times \partial B)} (A \times \partial B)$ (cf. Exercise 15.0.2).

between the arrow categories. The left adjoint $\hat{\otimes}$ is the **pushout-product** of $i: m \to m' \in \mathcal{M}^2$ and $j: n \to n' \in \mathcal{N}^2$ as defined by the diagram

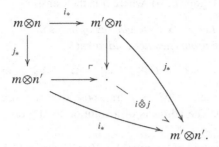

The right adjoints $\{\hat{,}\}$ and $\underline{\hat{\mathrm{hom}}}$, called **pullback-cotensors** and **pullback-homs**, are defined dually for $f: p \to p'$ by the pullbacks in $\mathcal{N}$ and $\mathcal{M}$:

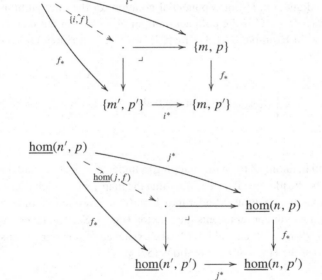

Example 11.1.8 (degenerate cases) The pushout-product of a map $j: n \to n'$ with the unique map $\emptyset \to m$ is the map $m \otimes j: m \otimes n \to m \otimes n'$. The pullback-hom of the map j with the unique map $p \to *$ is the map $j^*: \underline{\mathrm{hom}}(n', p) \to \underline{\mathrm{hom}}(n, p)$. The pullback-cotensor of the map $\emptyset \to m$ with $f: p \to p'$ is the map $f_*: \{m, p\} \to \{m, p'\}$.

Exercise 11.1.9 Prove that $\hat{\otimes}$, $\{\hat{,}\}$, and $\underline{\hat{\mathrm{hom}}}$ define a two-variable adjunction between the arrow categories by writing down explicit hom-set bijections

$$\mathcal{P}^2(i \hat{\otimes} j, f) \cong \mathcal{N}^2(j, \{i, \hat{f}\}) \cong \mathcal{M}^2(i, \underline{\hat{\mathrm{hom}}}(j, f)).$$

A warm-up exercise might be in order: Consider $m \otimes n' \xrightarrow{h} p$ and let $n \xrightarrow{j} n'$. Show that the transpose of $m \otimes n \xrightarrow{m \otimes j} m \otimes n' \xrightarrow{h} p$ is the composite $m \xrightarrow{\bar{h}} \underline{\hom}(n', p) \xrightarrow{j^*} \underline{\hom}(n, p)$, where $\bar{h}$ is the transpose of h.

Lemma 11.1.10 *Let $\mathcal{A}, \mathcal{B}, \mathcal{C}$ be classes of maps in $\mathcal{M}, \mathcal{N}, \mathcal{P}$, respectively. The following lifting properties are equivalent*

$$\mathcal{A} \hat{\otimes} \mathcal{B} \boxtimes \mathcal{C} \quad \Leftrightarrow \quad \mathcal{B} \boxtimes \{\hat{\mathcal{A}, \mathcal{C}}\} \quad \Leftrightarrow \quad \mathcal{A} \boxtimes \underline{\hom}(\mathcal{B}, \mathcal{C}).$$

Proof Lifting problems $i \hat{\otimes} j \Rightarrow f$ transpose to lifting problems $j \Rightarrow \{i, \hat{f}\}$ and $i \Rightarrow \underline{\hom}(j, f)$. The adjuncts of a solution to any one of these solve the other two. □

A powerful application of Lemma 11.1.10 is Theorem 11.5.1; an experienced reader might wish to skip to there directly. But simpler applications are also of interest.

Example 11.1.11 The closed monoidal structure on **Top** defines a two-variable adjunction. Using exponential notation for the internal hom, the projection $p_0 \colon Z^I \to Z$ is the pullback-hom of $Z \to *$ with the map $i_0 \colon * \to I$. As defined in Example 11.1.3, a map $j \colon A \to X$ is a Hurewicz cofibration if and only if

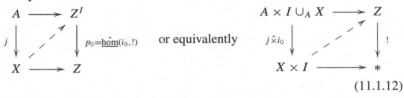

$$(11.1.12)$$

In particular, taking Z to be the mapping cylinder $A \times I \cup_A X$ and the top right arrow to be the identity, it is necessary that the map $j \hat{\times} i_0$, which is the canonical inclusion of the mapping cylinder into $X \times I$, has a retraction. Indeed, because the mapping cylinder represents the functor **Top** $\to$ **Set** that takes a space Z to the set of lifting problems (11.1.12), the existence of this retraction suffices to characterize the Hurewicz cofibrations [57, §6.4].

11.2 Weak factorization systems

In a model category, the lifting properties defining the cofibrations and fibrations are supplemented with a factorization axiom in the following manner:

Definition 11.2.1 A **weak factorization system** on a category is a pair $(\mathcal{L}, \mathcal{R})$ of classes of morphisms such that

(factorization) every arrow can be factored as an arrow of $\mathcal{L}$ followed by an arrow of $\mathcal{R}$

(lifting) $\mathcal{L} \boxtimes \mathcal{R}$

(closure) furthermore, $\mathcal{L} = {}^{\boxtimes}\mathcal{R}$ and $\mathcal{R} = \mathcal{L}^{\boxtimes}$

Remark 11.2.2 Of course, the third axiom subsumes the second. Indeed, the third axiom makes it clear that either class of a weak factorization system determines the other.

We list them separately to facilitate the comparison with an alternate definition:

Lemma 11.2.3 (retract argument) *In the presence of the first two axioms, the third can be replaced by*

(closure′) *the classes $\mathcal{L}$ and $\mathcal{R}$ are closed under retracts*

Proof Lemma 11.1.4 proves that (closure) $\Rightarrow$ (closure′). The converse is the so-called "retract argument," familiar from the model category literature: Suppose $k \in {}^{\boxtimes}\mathcal{R}$. In particular, it lifts against its right factor in the factorization guaranteed by the first axiom. Any solution w to this canonical lifting problem can be rearranged into a retract diagram:

$$\tag{11.2.4}$$

Because $\mathcal{L}$ is closed under retracts, it follows that $k \in \mathcal{L}$. $\qquad\square$

Example 11.2.5 (Thomas Goodwillie via MathOverflow) The category **Set** admits exactly six weak factorization systems. Let $\mathcal{A}, \mathcal{E}, \mathcal{I}, \mathcal{M}, \mathcal{N}$ denote the classes of all maps, epimorphisms, isomorphisms, monomorphisms, and maps with empty (null) domain and non-empty codomain. Then

$$(\mathcal{A}, \mathcal{I}), \quad (\mathcal{I}, \mathcal{A}), \quad (\mathcal{E}, \mathcal{M}), \quad (\mathcal{M}, \mathcal{E}), \quad (\mathcal{A}\backslash\mathcal{N}, \mathcal{I} \cup \mathcal{N}), \quad \text{and} \quad (\mathcal{M}\backslash\mathcal{N}, \mathcal{E} \cup \mathcal{N})$$

are weak factorization systems, the first three of which satisfy the stricter condition that the postulated factorizations and liftings are unique.[2]

Example 11.2.6 In practice, most weak factorization systems are generated by a set of arrows $\mathcal{J}$ in the following manner. The right class is $\mathcal{J}^{\boxtimes}$ and the left class is ${}^{\boxtimes}(\mathcal{J}^{\boxtimes})$. Tautologically, ${}^{\boxtimes}(\mathcal{J}^{\boxtimes}) \boxtimes \mathcal{J}^{\boxtimes}$ and $\mathcal{J}^{\boxtimes} = ({}^{\boxtimes}(\mathcal{J}^{\boxtimes}))^{\boxtimes}$. If the category satisfies certain set theoretical conditions, the **small object argument**, which is the subject of Chapter 12, produces appropriate factorizations, in which case the weak factorization system $({}^{\boxtimes}(\mathcal{J}^{\boxtimes}), \mathcal{J}^{\boxtimes})$ is called **cofibrantly generated**.

---

[2] The appellation "weak" is intended to distinguish from **orthogonal factorization systems**, sometimes called simply "factorization systems," which have unique factorizations and unique liftings. The prototypical example is $(\mathcal{E}, \mathcal{M})$ on **Set**.

11.3 Model categories and Quillen functors

Quillen's **closed model categories** of [65] are called **model categories** by the modern literature. Because a given category can admit multiple model category structures, we prefer to use the term **model structure** when referring to particular classes of maps that define a model category.

The following definition, perhaps first due to [43], a source of several useful facts about model categories, is equivalent to the usual one.

Definition 11.3.1   A **model structure** on a complete and cocomplete homotopical category $(\mathcal{M}, \mathcal{W})$ consists of two classes of morphisms $\mathcal{C}$ and $\mathcal{F}$ such that $(\mathcal{C} \cap \mathcal{W}, \mathcal{F})$ and $(\mathcal{C}, \mathcal{F} \cap \mathcal{W})$ are weak factorization systems.

The maps in $\mathcal{C}$ are called **cofibrations** and the maps in $\mathcal{F}$ are called **fibrations**. The maps in $\mathcal{C} \cap \mathcal{W}$ are called **trivial cofibrations** or **acyclic cofibrations**, and the maps in $\mathcal{F} \cap \mathcal{W}$ are called **trivial fibrations** or **acyclic fibrations**. The model structure is said to be **cofibrantly generated** if both of its weak factorization systems are.

Remark 11.3.2   The usual definition asks that $\mathcal{W}$ satisfies the weaker 2-of-3 property. By Remark 2.1.3, if $\mathcal{W}$ does not also satisfy the 2-of-6 property, then the pair $(\mathcal{M}, \mathcal{W})$ will not admit a model structure, so you would do well to stop trying to find one.

This issue resolved, the only remaining subtle point in the proof that this definition is equivalent to the one the reader might have in mind is the demonstration that our weak equivalences are closed under retracts. The following proof due to [43, 7.8] can also be found in [58, 14.2.5], the argument in the former source transmitted via this author to the latter.

Lemma 11.3.3   *If $(\mathcal{C} \cap \mathcal{W}, \mathcal{F})$ and $(\mathcal{C}, \mathcal{F} \cap \mathcal{W})$ are weak factorization systems on a complete and cocomplete category, and if $\mathcal{W}$ satisfies the 2-of-3 property, then $\mathcal{W}$ is closed under retracts.*

Proof   Let f be a retract of $w \in \mathcal{W}$, and suppose temporarily that $f \in \mathcal{F}$. Factor w as $w = vu$ using either weak factorization system; by the 2-of-3 property, $u \in \mathcal{C} \cap \mathcal{W}$ and $v \in \mathcal{F} \cap \mathcal{W}$. Define arrows s and t:

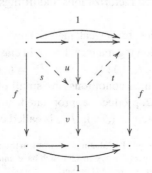

the former by composing and the latter as a solution to the lifting problem between $u \in \mathcal{C} \cap \mathcal{W}$ and $f \in \mathcal{F}$. By commutativity of the two triangles, $ts = 1$, which means that f is a retract of v. As $v \in \mathcal{F} \cap \mathcal{W}$, f is a well by the dual of Lemma 11.1.4.

Now we prove the general case, dropping the hypothesis that f is a fibration. Factor f as $f = hg$ with $g \in \mathcal{C} \cap \mathcal{W}$ and $h \in \mathcal{F}$ and construct the indicated pushout:

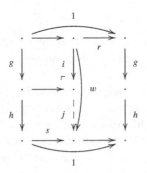

By Lemma 11.1.4, $i \in \mathcal{C} \cap \mathcal{W}$. The arrows w and sh form a cone over the pushout diagram, so there is a unique morphism j, as shown, such that $ji = w$. By the 2-of-3 property, $j \in \mathcal{W}$. Similarly, gr and the identity form a cone over the pushout diagram, so there is a unique morphism y,

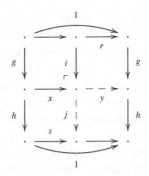

such that yx is the identity. The lower two squares now display h as a retract of j. As $j \in \mathcal{W}$ and $h \in \mathcal{F}$, the previous argument shows that $h \in \mathcal{W}$. But g is already in $\mathcal{W}$, so by the 2-of-3 property, $f = hg \in \mathcal{W}$, as desired. $\square$

When convenient, we use a tilde to decorate arrows that represent weak equivalences, a tail to decorate cofibrations, and an extra tip to decorate fibrations. For instance, the pair of factorizations provided by the two weak factorization systems of a model structure might be displayed as a commutative square

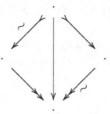

Exercise 11.3.4   Use Example 11.2.5 to show that **Set** admits only finitely many model structures. How many? What are they?

Example 11.3.5   Quillen's original manuscript [65] establishes a model structure on simplicial sets in which the cofibrations are the monomorphisms, the fibrations are the **Kan fibrations**, and the weak equivalences are the weak homotopy equivalences. This model structure is cofibrantly generated. The generating cofibrations are the set of inclusions of the boundary of an n-simplex into that simplex, for each n. The generating trivial cofibrations are the set of inclusions $\{\Lambda^n_k \to \Delta^n \mid n \geq 0, 0 \leq k \leq n\}$ of each horn into the appropriate simplex. The trivial cofibrations are called **anodyne** maps. The fibrant objects are called **Kan complexes**.

Another model structure with the same cofibrations but whose fibrant objects are the quasi-categories, due to André Joyal, is introduced in Part IV.

Example 11.3.6   There is a model structure on **Top**, also due to [65], whose cofibrations are retracts of relative cell complexes, fibrations are Serre fibrations, and weak equivalences are weak homotopy equivalences. The generating cofibrations are the inclusions of the spheres of each dimension into the disks they bound. The generating trivial cofibrations are described in Example 11.1.3.

Another model structure due to Arne Strøm is formed by the Hurewicz cofibrations, Hurewicz fibrations, and homotopy equivalences [82].

Example 11.3.7   The category $\mathbf{Ch}_\bullet(R)$ of unbounded chain complexes of modules over a ring R has a model structure, due in this context to Mark Hovey [38, §2.3], whose weak equivalences are quasi-isomorphisms and whose trivial fibrations and fibrations are defined by the lifting properties $\{S^{n-1} \to D^n \mid n \in \mathbb{Z}\}^\boxslash$ and $\{0 \to D^n \mid n \in \mathbb{Z}\}^\boxslash$. Here S^n is the chain complex with R in degree n and zeros elsewhere, and D^n has R in degrees n, $n-1$ with an identity differential.

In parallel with the topological setting, there is another model structure given by the Hurewicz cofibrations, Hurewicz fibrations, and chain homotopy equivalences. These notions are defined using an interval object. If the ring is non-commutative, this should be defined to be the chain complex of abelian groups with $\mathbb{Z}$ in degree one, $\mathbb{Z} \oplus \mathbb{Z}$ in degree zero, and $1 \mapsto (1, -1)$ as the only non-zero differential [58, §18].

Example 11.3.7 shows that classical homological algebra is subsumed by Quillen's "homotopical algebra." There are many other model structures relevant to homological algebra, for instance, on chain complexes that are bounded above or bounded below, or for unbounded chain complexes taking values in other abelian categories; see [12].

Exercise 11.3.8   Show that the categories **Top** and **Ch.**(R) each admits a third "mixed" model structure with weak homotopy equivalences or quasi-isomorphisms and Hurewicz fibrations. This observation was originally made by Michael Cole [13].

Example 11.3.9   There is a "folk model structure" on **Cat** whose weak equivalences are categorical equivalences. The cofibrations are functors injective on objects, and the fibrations are the **isofibrations**, that is, functors lifting on the right against the inclusion, into the free-standing isomorphism, of one of the two objects.

Exercise 11.3.10   There is a model structure on the category **sSet-Cat** of simplicially enriched categories and simplicially enriched functors due to Julia Bergner [6]. The weak equivalences are the DK-equivalences introduced in Section 3.5. The fibrant objects are simplicial categories that are **locally Kan**, that is, for which each hom-space is a Kan complex. See Theorem 16.1.2.

Morphisms between model categories come in two flavors: **left** and **right Quillen functors**. A left Quillen functor is a cocontinuous functor[3] that preserves cofibrations and trivial cofibrations. Dually, a right Quillen functor preserves limits, fibrations, and trivial fibrations. Frequently, such functors occur in an adjoint pair, in which case the pair is called a **Quillen adjunction**.

Lemma 11.3.11   *An adjunction $F : \mathcal{M} \rightleftarrows \mathcal{N} : U$ between model categories is a Quillen adjunction if any of the following equivalent conditions hold:*

- *F is left Quillen*
- *U is right Quillen*
- *F preserves cofibrations and U preserves fibrations*
- *F preserves trivial cofibrations and U preserves trivial fibrations*

---

[3] This hypothesis is perhaps not standard; however, we cannot think of any examples of interesting left Quillen functors that fail to preserve colimits.

Proof   This is an immediate consequence of Lemma 11.1.5.   □

Remark 11.3.12   We like emphasizing weak factorization systems in the context of model structures for a few reasons. The overdetermination of the model category axioms and the closure properties of the classes of cofibrations and fibrations are consequences of analogous characteristics of the constituent weak factorization systems described in Remark 11.2.2 and Lemma 11.1.4. The equivalence of various definitions of a Quillen adjunction has to do with the individual interactions between the adjunction and each weak factorization system, as shown in Lemma 11.1.5. Finally, the small object argument, mentioned in Example 11.2.6, constructs a functorial factorization for a cofibrantly generated weak factorization system; the model structure context is beside the point.

Digression 11.3.13 (the homotopy category of a model category)   If a homotopical category $(\mathcal{M}, \mathcal{W})$ admits a model structure, any model structure, then its homotopy category admits a simple description, in precise analogy to that presented in Theorem 10.5.1. Because we are not assuming that $\mathcal{M}$ is enriched, the proofs are somewhat more delicate. The appropriate notions of homotopy have to be conjured from mid air by factoring either the diagonal or the fold map. But this story is by now quite standard, and we happily defer to [24].

We should at least state the upshot. A model structure on $(\mathcal{M}, \mathcal{W})$ in particular gives a notion of **fibrant** and **cofibrant** objects – more about which in just a moment. An object is fibrant just when the map to the terminal object is a fibration and cofibrant just when the map from the initial object is a cofibration. In a model category, it is an elementary exercise to show that

- every object is weakly equivalent to one that is both fibrant and cofibrant

With more care, one can show that

- weak equivalences between fibrant-cofibrant objects are precisely "homotopy equivalences"

where "homotopy" is an equivalence relation defined on such hom-sets with respect to appropriate "cylinder" or "path" objects, as alluded to earlier.

In particular, $\text{Ho}\mathcal{M}$ is equivalent to the homotopy category of the subcategory of fibrant–cofibrant objects, and this latter homotopy category is obtained by just taking "homotopy classes of maps" as arrows. This was precisely the conclusion of Theorem 10.5.1 (under somewhat different hypotheses).

Now we will provide the promised contextualization of the theory of derived functors. Recall from 2.2.4 that a functor is left deformable if it is homotopical when restricted to a subcategory of "cofibrant" objects. The following lemma implies that any left Quillen functor is left deformable with respect to the cofibrant objects.

Lemma 11.3.14 (Ken Brown's lemma)   *Let* M *and* N *be model categories and suppose* $F\colon M \to N$ *sends trivial cofibrations between cofibrant objects to weak equivalences. Then* F *is homotopical on the full subcategory of cofibrant objects.*

Proof   Suppose $w\colon A \to B$ is a weak equivalence between cofibrant objects. Factor the map $(w, 1)\colon A \sqcup B \to B$ as a cofibration followed by a trivial fibration. The pushout square shows that coproduct inclusions of cofibrant objects are cofibrations:

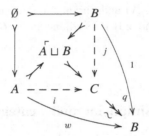

Hence the maps $i\colon A \to C$ and $j\colon B \to C$ are seen to be cofibrations, by commutativity of the inner triangles, and also weak equivalences, by the 2-of-3 property applied to the outer ones. In particular, we have constructed a diagram

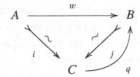

in which the maps i and j are trivial cofibrations and q is a retraction of j that factors w through i.

The rest of the argument is elementary. By hypothesis, the maps Fi and Fj are weak equivalences, as is the image of the identity at B. Hence Fq is a weak equivalence by the 2-of-3 property and, consequently, so is Fw by 2-of-3 again, as desired. $\square$

Applying the cofibration–trivial fibration factorization to maps of the form $\emptyset \to m$, we obtain cofibrant objects Qm together with a natural weak equivalence $q_m\colon Qm \to m$. If the factorization is functorial,[4] as is commonly supposed to be the case, this procedure defines a left deformation (Q, q). The

---

[4] This notion will be defined in 12.1.1.

upshot is that the left derived functor of any left Quillen functor F is defined by FQ. Note further that left Quillen functors preserve cofibrant objects, so composites of these left derived functors of left Quillen functors are again left derived functors (cf. Theorem 2.2.9).

Dual arguments show that right Quillen functors are right deformable with respect to the subcategories of fibrant objects, which they preserve. A right deformation is obtained by applying the trivial cofibration–fibration factorization to maps to the terminal object.

Exercise 11.3.15   A **Quillen equivalence** is a Quillen adjunction $F : \mathcal{M} \rightleftarrows \mathcal{N} : U$ so that either of the equivalent conditions are satisfied:

(i) The total derived functors $\mathbf{L}F \dashv \mathbf{R}U$ form an equivalence of categories.
(ii) For any cofibrant $m \in \mathcal{M}$ and fibrant $n \in \mathcal{N}$, a map $Fm \to n$ is a weak equivalence in $\mathcal{N}$ if and only if its adjunct $m \to Un$ is a weak equivalence in $\mathcal{M}$.

Prove this.

11.4 Simplicial model categories

A proper definition of a simplicial model category is surely overdue. First, we need one more ancillary notion, a left Quillen bifunctor, whose definition makes use of the Leibniz construction 11.1.7. Using the notation appropriate to this setting:

Definition 11.4.1   A bifunctor $-\otimes-$ between model categories is a **left Quillen bifunctor** if it preserves colimits in both variables and if the associated pushout-product bifunctor $-\hat{\otimes}-$ maps pairs of cofibrations to a cofibration that is acyclic if either of the domain cofibrations are.

At first glance, this definition seems a bit odd. An important point is the following lemma:

Lemma 11.4.2   *Left Quillen bifunctors are homotopical on the subcategories of cofibrant objects and, furthermore, preserve cofibrant objects.*

Proof   These claims follow from Example 11.1.8 and Ken Brown's Lemma 11.3.14. First note that the map $\emptyset \to m \otimes n$ is the pushout-product of the maps $\emptyset \to m$ and $\emptyset \to n$; hence, if the latter two are cofibrations, so is the former.

Now suppose $i : m \to m'$ and $j : n \to n'$ are weak equivalences between cofibrant objects. The map $i \otimes j : m \otimes n \to m' \otimes n'$ factors as $i \otimes n$ followed by $m' \otimes j$. So it suffices to prove that tensoring with a cofibrant object preserves weak equivalences between cofibrant objects. If i is a trivial cofibration, then

$i \otimes n$ is the pushout-product of i with the cofibration $\emptyset \to n$ and is hence a trivial cofibration and in particular a weak equivalence. The conclusion follows from Lemma 11.3.14. □

Right Quillen bifunctors are defined dually – the meaning of "dual" in this context perhaps merits some explanation. Here a bifunctor, which we'll call "hom," contravariant in its first variable and covariant in its second, is right Quillen if the associated "pullback-hom" sends a cofibration in the first variable and fibration in the second to a fibration that is acyclic if either of these maps is also a weak equivalence. The duality has to do with the fact that a model structure on $\mathcal{M}$ gives rise to a model structure on $\mathcal{M}^{op}$ with the cofibrations and fibrations swapped.

Lemma 11.4.3 *If $(\otimes, \{, \}, \underline{\text{hom}})$ is a two-variable adjunction, then $\otimes$ is a left Quillen bifunctor if and only if $\{, \}$ is a right Quillen bifunctor if and only if $\underline{\text{hom}}$ is a right Quillen bifunctor.*

In this case we say that $(\otimes, \{, \}, \underline{\text{hom}})$ is a **Quillen two-variable adjunction**.

Proof Lemma 11.1.10 implies this result and also more refined statements à la Lemma 11.3.11. □

Definition 11.4.4 A **simplicial model category** is a model category $\mathcal{M}$ that is tensored, cotensored, and simplicially enriched and such that $(\otimes, \{, \}, \underline{\text{hom}})$ is a Quillen two-variable adjunction.

The three statements encoded in the assertion that $(\otimes, \{, \}, \underline{\text{hom}})$ is a Quillen two-variable adjunction are frequently referred to as the **SM7 axiom**.

Exercise 11.4.5 Prove the assertions in Lemma 3.8.6, except (vi) that simplicial homotopy equivalences are weak equivalences. This follows from Theorem 10.5.1 or a direct argument that can be found, for instance, in [36, 9.5.15–16].

Quillen's definition has been generalized by Hovey [38, §4.2].

Definition 11.4.6 A **monoidal model category** is a closed (symmetric) monoidal category $(\mathcal{V}, \times, *)$ with a model structure so that the monoidal product and hom define a Quillen two-variable adjunction and, furthermore, so that the maps

$$Q * \times v \to * \times v \cong v \quad \text{and} \quad v \times Q* \to v \times * \cong v$$

are weak equivalences if v is cofibrant.

Definition 11.4.7   A $\mathcal{V}$-**model category** is a model category $\mathcal{M}$ that is tensored, cotensored, and $\mathcal{V}$-enriched in such a way that $(\otimes, \{, \}, \underline{\mathrm{hom}})$ is a Quillen two-variable adjunction and the maps

$$Q * \otimes m \rightarrow * \otimes m \cong m$$

are weak equivalences if m is cofibrant.

Exercise 11.4.8   Using Definition 11.4.6 or Definition 11.4.7, prove the appropriate case of Lemma 9.2.3.

The conditions on the cofibrant replacement of the monoidal unit (which are implied by the Quillen two-variable adjunction if the monoidal unit is cofibrant) are included so that a monoidal model category is a closed monoidal homotopical category and a $\mathcal{V}$-model category is a $\mathcal{V}$-homotopical category. Ultimately these conditions are necessary for the proofs of Theorems 10.2.6 and 10.2.12, which show that the homotopy categories are again closed monoidal and enriched, respectively. See the discussion in Section 10.2.

11.5 Weighted colimits as left Quillen bifunctors

To wrap up this chapter, we use our facility with weak factorization systems and two-variable adjunctions to prove some cool results about weighted (and thus homotopy) limits and colimits in simplicial model categories. For this we need one last preliminary. If $\mathcal{M}$ is a model category and $\mathcal{D}$ is a small category, the **projective model structure** on $\mathcal{M}^{\mathcal{D}}$ (which may or may not exist) has weak equivalences and fibrations defined pointwise and the **injective model structure** on $\mathcal{M}^{\mathcal{D}}$ (which may or may not exist) has weak equivalences and cofibrations defined pointwise. Projective model structures exist whenever $\mathcal{M}$ is cofibrantly generated (cf. Theorem 12.3.2). Injective model structures exist when $\mathcal{M}$ is additionally locally presentable, in which case we say that the model structure is **combinatorial** [4]. This is the case for Quillen's model structure on simplicial sets.

The following theorem of Nicola Gambino has an elementary proof. Nonetheless, it has a number of powerful consequences, which we devote the rest of this section to exploring.

Theorem 11.5.1 ([29])   *If $\mathcal{M}$ is a simplicial model category, and $\mathcal{D}$ is a small category, then the weighted colimit functor*

$$- \otimes_{\mathcal{D}} - : \mathbf{sSet}^{\mathcal{D}^{\mathrm{op}}} \times \mathcal{M}^{\mathcal{D}} \rightarrow \mathcal{M}$$

is left Quillen if the domain has the (injective, projective) or (projective, injective) model structure. Similarly, the weighted limit functor

$$\{-, -\}^{\mathcal{D}} : (\mathbf{sSet}^{\mathcal{D}})^{\mathrm{op}} \times \mathcal{M}^{\mathcal{D}} \rightarrow \mathcal{M}$$

is right Quillen if the domain has the (projective, projective) or (injective, injective) model structure.

Proof Note we have used Theorem 7.6.3 to express weighted colimits as functor tensor products and weighted limits as functor cotensor products. By Lemma 11.4.3, we can prove both statements in adjoint form. The functor tensor product $\otimes_{\mathcal{D}}$ has a right adjoint (used to express the defining universal property of the weighted colimit)

$$\underline{\mathcal{M}}(-,-) \colon (\mathcal{M}^{\mathcal{D}})^{\mathrm{op}} \times \mathcal{M} \to \mathbf{sSet}^{\mathcal{D}^{\mathrm{op}}},$$

which sends $F \in \mathcal{M}^{\mathcal{D}}$ and $m \in \mathcal{M}$ to $\underline{\mathcal{M}}(F-,m) \colon \mathcal{D}^{\mathrm{op}} \to \mathbf{sSet}$.

To prove the statement when $\mathbf{sSet}^{\mathcal{D}^{\mathrm{op}}}$ has the projective and $\mathcal{M}^{\mathcal{D}}$ has the injective model structure, we must show that this is a right Quillen bifunctor with respect to the pointwise (trivial) cofibrations in $\mathcal{M}^{\mathcal{D}}$, (trivial) fibrations in $\mathcal{M}$, and pointwise (trivial) fibrations in $\mathbf{sSet}^{\mathcal{D}^{\mathrm{op}}}$. Because the limits involved in the definition of right Quillen bifunctors are also formed pointwise, this follows immediately from the corresponding property of the simplicial hom bifunctor, which was part of the definition of a simplicial model category. The other cases are similar. $\qquad\square$

Remark 11.5.2 The previous argument works equally well if $\mathcal{D}$ is a small simplicial category and the diagram categories are the categories of simplicial functors and simplicial natural transformations. Because our applications will involve unenriched diagrams, we have opted for the simpler statement.

This result has immediate consequences for the theory of derived functors, including, in particular, homotopy (co)limits – ordinary (co)limits being functor tensor products with the terminal weight – giving alternate proofs of some of the key theorems from Part I and Part II of this book. A special case of Theorem 11.5.1 asserts that homotopy colimits can be computed by taking a projective cofibrant replacement of the weight and a pointwise cofibrant replacement of the diagram (our usual approach) or by taking a projective cofibrant replacement of the diagram. The latter approach typically involves replacing certain maps in the diagram by cofibrations, a well-known strategy for producing homotopy colimits in particular cases.

For ease of reference, let us record this new observation as a corollary.

Corollary 11.5.3 *Let $\mathcal{M}$ be a simplicial model category and let $\mathcal{D}$ be a small category. If $\mathcal{M}$ is cofibrantly generated, then projective cofibrant replacement defines a left deformation for* colim: $\mathcal{M}^{\mathcal{D}} \to \mathcal{M}$ *and hence the homotopy colimit of a diagram may be computed as the colimit of any projective cofibrant replacement. Dually, if $\mathcal{M}$ is combinatorial, then injective fibrant replacement defines a right deformation for* lim: $\mathcal{M}^{\mathcal{D}} \to \mathcal{M}$, *and hence the homotopy* *limit of a diagram may be computed as the limit of an injective fibrant replacement.*

To apply Theorem 11.5.1, we need a better understanding of the projective cofibrations.

Lemma 11.5.4   *Let $A \to B$ be any cofibration in a model category $\mathcal{M}$. Then the induced map from the copowers $\mathcal{D}(d, -) \cdot A \to \mathcal{D}(d, -) \cdot B$ is a projective cofibration in $\mathcal{M}^{\mathcal{D}}$.*

To explain the notation, the copower bifunctor $- \cdot -: \mathbf{Set} \times \mathcal{M} \to \mathcal{M}$ extends to a bifunctor $\mathbf{Set}^{\mathcal{D}} \times \mathcal{M} \to \mathcal{M}^{\mathcal{D}}$, where morphisms in $\mathcal{D}$ act by reindexing coproducts.

Proof   It suffices to show that this map lifts against all pointwise trivial fibrations. By the defining universal property of the copower and the Yoneda lemma, there is adjunction $\mathcal{D}(d, -) \cdot -: \mathcal{M} \rightleftarrows \mathcal{M}^{\mathcal{D}}: \mathrm{ev}_d$ whose right adjoint evaluates at an object $d \in \mathcal{D}$. By definition, $A \to B$ lifts against all trivial fibrations, and the conclusion follows from Lemma 11.1.5. $\square$

Remark 11.5.5   Combining Lemma 11.5.4 and Lemma 11.1.4 allows us to recognize a large class of projective cofibrations: any retract of a transfinite composite of pushouts of coproducts of the maps of Lemma 11.5.4 lifts against any pointwise trivial fibration and is hence a projective cofibration. Indeed, by Corollary 12.2.4, *all* projective cofibrations admit this description.

Example 11.5.6   Let $y \colon \vec{\Delta} \to \mathbf{sSet}$ denote the restriction of the usual cosimplicial object to the wide subcategory of monomorphisms. We claim that y is projective cofibrant with respect to the Quillen model structure on simplicial sets. Using Remark 11.5.5, we define y to be the colimit of a sequence of pushouts of projective cofibrations $\vec{\Delta}([n], -) \cdot (\partial\Delta^n \to \Delta^n)$.

The initial map in this sequence is $\emptyset \to \vec{\Delta}([0], -)$. The second is constructed via the pushout

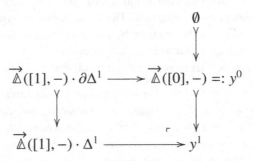

The top horizontal "attaching map" is the obvious one: by the Yoneda lemma, this natural transformation is determined by two points in the discrete simplicial

set $y_1^0 = \overrightarrow{\Delta}([0], [1])$, and we choose the elements corresponding to the faces of $\partial \Delta^1$.

Observe that the simplicial set y_2^1 has a non-degenerate 1-simplex corresponding to each monomorphism $[1] \to [2]$. These assemble into the boundary of a 2-simplex and in this way can be used to define the attaching map for $\overrightarrow{\Delta}([2], -) \cdot (\partial \Delta^2 \to \Delta^2)$; the pushout defines the functor y^2. By construction, the simplicial set y_{n+1}^n contains a non-degenerate $\partial \Delta^{n+1}$ whose consistent n-simplices correspond to the monomorphisms $[n] \to [n+1]$. These are used to freely attach an $(n + 1)$-simplex. By construction, $y \cong \mathrm{colim}_n \, y^n$ is thus projectively cofibrant. Theorem 11.5.1 now proves the assertions of Example 8.5.12.

Example 11.5.7   Let $\mathcal{D}$ be the category $b \xleftarrow{f} a \xrightarrow{g} c$ indexing pushout diagrams and let $F \colon \mathcal{D} \to \mathcal{M}$ take values in a cofibrantly generated simplicial model category. If Fa is cofibrant and Ff and Fg are cofibrations, then F is projectively cofibrant (and Fb and Fc are cofibrant objects). To see this, we factor $\emptyset \to F$ as a composite of three projective cofibrations:

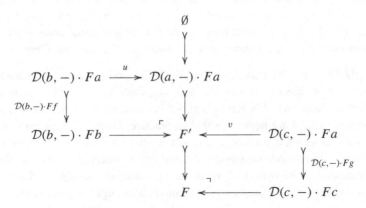

The vertical arrows are cofibrations by Lemma 11.5.4 and Lemma 11.1.4. The attaching maps u and v are adjunct, under different adjunctions, to the identity at Fa.

By this observation and Corollary 11.5.3, the ordinary pushout of a pair of cofibrations between cofibrant objects is a homotopy pushout. Given a generic pushout diagram $Y \xleftarrow{h} X \xrightarrow{k} Z$, its projective cofibrant replacement may be formed by taking a cofibrant replacement $q \colon X' \to X$ of X and then factoring the composites hq and kq as a cofibration followed by a trivial fibration:

$$
\begin{array}{ccccc}
Y' & \longleftarrow\!\!\!< & X' & >\!\!\!\longrightarrow & Z' \\
{\scriptstyle \wr}\downarrow & & {\scriptstyle \wr}\downarrow q & & \downarrow {\scriptstyle \wr} \\
Y & \xleftarrow{\ h\ } & X & \xrightarrow{\ k\ } & Z
\end{array}
\qquad (11.5.8)
$$

By Corollary 11.5.3, the pushout of the diagram $Y' \leftarrow X' \rightarrow Z'$ is the homotopy pushout of the original diagram.

Remark 11.5.9   If all that is desired of our notion of homotopy pushout is that it preserves pointwise weak equivalences between diagrams (and not that homotopy pushouts represent homotopy coherent cones), the hypothesis that $\mathcal{M}$ is a *simplicial* model category can be dropped: the ordinary pushout functor is homotopical when restricted to diagrams of cofibrations between cofibrant objects.

To prove this, note that the argument just given shows that such diagrams are projectively cofibrant. The pushout functor is easily seen to be left Quillen with respect to the projective model structure: its right adjoint, the constant diagram functor, is manifestly right Quillen. The conclusion follows by Ken Brown's Lemma 11.3.14.

With more care, we can also drop the hypothesis that the model structure is cofibrantly generated. Even if the projective model structure does not exist, the shape of the pushout diagram allows us to construct functorial "projective cofibrant replacements" nonetheless.[5] The proof in Corollary 14.3.2 and Remark 14.3.4 will give this extension and also show that the pushout functor is homotopical under even more general conditions than described here.

Example 11.5.10   The duals of Lemma 11.5.4 and Lemma 11.1.4 identify a large class of injective fibrations. However, there is no characterization of *all* injective fibrations as in Remark 11.5.5 because the hypothesis in the dual of Corollary 12.2.4 are never satisfied in practice. Dualizing Example 11.5.7 or by a direct lifting argument, a diagram of shape $b \xrightarrow{f} a \xleftarrow{g} c$ in which the objects are fibrant and the maps are fibrations is injectively fibrant. Hence the conclusion of Remark 11.5.9 can be dualized: in a combinatorial model category, homotopy pullbacks can be computed by replacing the objects of a pullback diagram by fibrant objects and the maps by fibrations. In fact, up to the caveats of Remark 11.5.9, homotopy pullbacks can be computed in this manner in *any* model category, combinatorial or otherwise; see Section 14.3.

Example 11.5.11   Let us prove the claim asserted in Example 6.4.8, that the homotopy colimit of a countable sequence of maps

$$X_0 \xrightarrow{f_{01}} X_1 \xrightarrow{f_{12}} X_2 \xrightarrow{f_{23}} \cdots \tag{11.5.12}$$

in a cofibrantly generated simplicial model category can be computed by replacing the generating maps by a sequence of cofibrations between cofibrant objects and then forming the ordinary colimit. In topological spaces, each map f_{ii+1} may be replaced by its **mapping cylinder**. Gluing these together, the homotopy

---

[5] The precise reason is that the pushout diagram is a Reedy category.

colimit is then the **mapping telescope**. We will see in Exercise 14.3.6 that the hypotheses on the model structure can be dropped.

Let ω denote the ordinal category $0 \to 1 \to 2 \to \cdots$ and write $F \colon \omega \to \mathcal{M}$ for the diagram (11.5.12). We will construct a projective cofibrant replacement of F by an inductive process. For the initial step, we take a cofibrant replacement $q_0 \colon Q_0 \xrightarrow{\sim} X_0$ of X_0 and form the functor $G^0 := \omega(0, -) \cdot Q_0$. By Lemma 11.5.4 and Lemma 11.1.4, this functor is projectively cofibrant; by construction, there is a natural transformation

$$
\begin{array}{ccccccccc}
G^0 & & Q_0 & = & Q_0 & = & Q_0 & = & \cdots \\
\Big\Vert & & {\scriptstyle q_0}\Big\downarrow {\scriptstyle \wr} & & \Big\vert & & \Big\vert & & \\
F & & X_0 & \xrightarrow{f_{01}} & X_1 & \xrightarrow{f_{12}} & X_2 & \xrightarrow{f_{23}} & \cdots
\end{array}
$$

in which the dotted arrows are defined to be the composites of the arrows to their left. Note that $\omega(0, -) \cdot Q_0$ is not yet a cofibrant replacement of F because only its initial component is a weak equivalence.

For the next step, we use the factorization in $\mathcal{M}$ to form a cofibrant replacement of the map $f_{01}q_0$:

$$
\begin{array}{ccc}
Q_0 & \xrightarrow{\ g_{01}\ } & Q_1 \\
{\scriptstyle q_0}\Big\downarrow {\scriptstyle \wr} & & {\scriptstyle q_1}\Big\downarrow {\scriptstyle \wr} \\
X_0 & \xrightarrow{\ f_{01}\ } & X_1
\end{array}
$$

We then form a pushout in $\mathcal{M}^\omega$:

$$
\begin{array}{ccc}
\omega(1, -) \cdot Q_0 & \longrightarrow & G^0 = \omega(0, -) \cdot Q_0 \\
{\scriptstyle g_{01}}\Big\downarrow & & \Big\downarrow \\
\omega(1, -) \cdot Q_1 & \xrightarrow[\ \ulcorner\ ]{} & G^1
\end{array}
$$

Here the top horizontal "attaching map" is adjunct to the identity at Q_0. The vertical maps are projective cofibrations by Lemmas 11.5.4 and 11.1.4. The universal property of the pushout defining the functor G^1 furnishes a natural transformation

$$
\begin{array}{ccccccccc}
G^1 & & Q_0 & \xrightarrow{\ g_{01}\ } & Q_1 & = & Q_1 & = & \cdots \\
\Big\Vert & & {\scriptstyle q_0}\Big\downarrow {\scriptstyle \wr} & & {\scriptstyle q_1}\Big\downarrow {\scriptstyle \wr} & & \Big\vert & & \\
F & & X_0 & \xrightarrow{f_{01}} & X_1 & \xrightarrow{f_{12}} & X_2 & \xrightarrow{f_{23}} & \cdots
\end{array}
$$

in which the first two components are weak equivalences.

At step n, we define a functor G^n in an analogous fashion

$$
\begin{array}{ccc}
Q_{n-1} & \xrightarrow{g_{n-1,n}} & Q_n \\
{\scriptstyle q_{n-1}} \downarrow {\scriptstyle \wr} & & {\scriptstyle q_n} \downarrow {\scriptstyle \wr} \\
X_{n-1} & \xrightarrow{f_{n-1,n}} & X_n
\end{array}
\qquad
\begin{array}{ccc}
\omega(n,-) \cdot Q_{n-1} & \longrightarrow & G^{n-1} \\
{\scriptstyle g_{n-1,n}} \downarrow & & \downarrow \\
\omega(n,-) \cdot Q_n & \xrightarrow{\ulcorner} & G^n
\end{array}
$$

By Lemma 11.1.4, the transfinite composite

$$
\emptyset \rightarrowtail G^0 \rightarrowtail G^1 \rightarrowtail G^2 \rightarrowtail \cdots \rightarrowtail \operatorname{colim}_n G^n =: G
$$

is a projectively cofibrant functor with a natural weak equivalence to F:

$$
\begin{array}{ccccccccc}
G & & Q_0 & \xrightarrow{g_{01}} & Q_1 & \xrightarrow{g_{12}} & Q_2 & \xrightarrow{g_{23}} & \cdots \\
{\scriptstyle q} \Vert {\scriptstyle \wr} & & {\scriptstyle q_0} \downarrow {\scriptstyle \wr} & & {\scriptstyle q_1} \downarrow {\scriptstyle \wr} & & {\scriptstyle q_2} \downarrow {\scriptstyle \wr} & & \\
F & & X_0 & \xrightarrow{f_{01}} & X_1 & \xrightarrow{f_{12}} & X_2 & \xrightarrow{f_{23}} & \cdots
\end{array}
$$

By Corollary 11.5.3, the colimit of G is the homotopy colimit of F.

Theorem 11.5.1 enables new proofs that certain weighted colimits compute homotopy colimits. Granting for a moment the assertion about projective cofibrancy, Theorem 6.6.1 reappears as a corollary of Theorem 11.5.1.

Corollary 11.5.13 ([36, 14.8.8]) $N(-/\mathcal{D}) \colon \mathcal{D}^{\mathrm{op}} \to \mathbf{sSet}$ *and* $N(\mathcal{D}/-) \colon$ $\mathcal{D} \to \mathbf{sSet}$ *are projective cofibrant replacements for the constant functors* $*$. *Hence, for any diagram F of shape $\mathcal{D}$ taking values in a combinatorial simplicial model category,*

$$
\operatorname{hocolim} F \cong \operatorname{colim}^{N(-/\mathcal{D})} QF \qquad and \qquad \operatorname{holim} F \cong \lim^{N(\mathcal{D}/-)} RF.
$$

Note that $N(-/\mathcal{D})^{\mathrm{op}}$ is equally a projective cofibrant replacement for the constant weight. This finally proves that any of the possible conventions discussed in Remark 7.8.3 gives results with the same homotopy type.

Corollary 11.5.14 *The homotopy colimit of a diagram of shape $\mathcal{D}$ in a combinatorial simplicial model category is computed by the weighted colimit with weight $N(-/\mathcal{D})$ or $N(-/\mathcal{D})^{\mathrm{op}}$ of a pointwise cofibrant replacement. Dually, the homotopy limit of a diagram of shape $\mathcal{D}$ is computed by the weighted limit with weight $N(\mathcal{D}/-)$ or $N(\mathcal{D}/-)^{\mathrm{op}}$ of a pointwise fibrant replacement.*

For similar reasons, the weight $N(-/K) \colon \mathcal{D}^{\mathrm{op}} \to \mathbf{sSet}$ is projectively cofibrant. Hence Ken Brown's Lemma 11.3.14 and the Reduction Theorem 8.1.8 given an immediate proof of the Homotopy Finality Theorem 8.5.6.

Alternate proof of Theorem 8.5.6 The weighted colimits of a pointwise cofibrant diagram weighted by weakly equivalent projective cofibrant weights are necessarily weakly equivalent. If $N(d/K)$ is contractible, then the natural map $N(-/K) \to N(-/\mathcal{D})$ is a weak equivalence, so the result follows. $\square$

It remains only to explain why these weights are projectively cofibrant. We will sketch the argument for $N(\mathcal{D}/-)\colon \mathcal{D} \to \mathbf{sSet}$; the other cases are similar. The proof uses the strategy outlined in Remark 11.5.5 and implemented in Example 11.5.11. We factor $\emptyset \to N(\mathcal{D}/-)$ as a composite of pushouts of coproducts of projective cofibrations, each of which is built by applying Lemma 11.5.4 to the maps $\partial\Delta^n \to \Delta^n$, which are cofibrations in $\mathbf{sSet}$. Lemma 11.1.4 then implies that $N(\mathcal{D}/-)$ is projective cofibrant. We refer to this as a **cellular decomposition** of $N(\mathcal{D}/-)$; we will have more to say about this terminology in the next chapter.

This cellular decomposition attaches a copy of $\mathcal{D}(d, -) \cdot \partial\Delta^n \to \mathcal{D}(d, -) \cdot \Delta^n$ for each n-simplex in $N\mathcal{D}$ ending at d. By the Yoneda lemma, the inclusion map

$$\mathcal{D}(d, -) \cdot \Delta^n \to N(\mathcal{D}/-)$$

corresponds to the associated n-simplex in $N(\mathcal{D}/d)$, whose last object is the identity at d. More details can be found in [36, 14.8.5].

12

Algebraic perspectives on the small object argument

The modern definition of a model category typically asks that the factorizations are **functorial**, and for good reason: the construction of derived *functors* of Quillen functors described in Chapter 2 requires a functorial cofibrant and fibrant replacement, which come for free with functorial factorizations. Depending on which examples of model categories you have in mind, this functoriality hypothesis can seem restrictive. For example, it is fairly clear how to replace a chain complex of abelian groups by a quasi-isomorphic chain complex of free abelian groups, defining a cofibrant replacement for a model structure on $\mathbf{Ch}_{\geq 0}(\mathbb{Z})$, but the naïve method of doing so is not functorial. Fortunately, there is a general argument, due to Quillen, that saves the day in many cases.

The **small object argument**, introduced by Quillen to construct factorizations for his model structure on **Top** [65, §II.3], is well treated in a number of sources [24, 36, 38, 58]. We review this construction in Section 12.2, but our focus here is somewhat different. We present a number of insights that we expect will be new to most homotopy theorists deriving from recent work of Richard Garner – the "algebraic perspectives" mentioned in the title. To whet the reader's appetite, we briefly mention one corollary: a principle for recognizing when a map constructed as a generic colimit, such as a coequalizer, of a diagram of cofibrations is again a cofibration, despite the quotienting (see Lemma 12.6.13).

The key technical preliminary to this and similar results is an alternative "algebraic" small argument that produces functorial factorizations that bear a much closer relationship to their compulsory lifting properties. Indeed, these ideas can be extended to produce functorial factorizations for model categories that are not cofibrantly generated – even in the extended senses of this term on display in Example 12.5.11 and Example 13.4.5. See, for instance, [3], which describes how to construct factorizations for Hurewicz-type model structures on topologically bicomplete categories.

190

12.1 Functorial factorizations

Before we describe the small object argument, we should be precise about what is being asked for. For a reason that will become clear eventually, we prefer to call our model category $\mathcal{K}$ in this chapter. Just as the functor category $\mathcal{K}^2$ is the category of commutative squares in $\mathcal{K}$, the functor category $\mathcal{K}^3$ is the category of pairs of composable arrows in $\mathcal{K}$ and natural transformations of such. The three injections $d^0, d^1, d^2 \colon 2 \to 3$, whose images omit the object indicated by the superscript, give rise to composition and projection functors $\mathcal{K}^3 \to \mathcal{K}^2$.

Definition 12.1.1 A **functorial factorization** is a section of the composition functor $\mathcal{K}^3 \to \mathcal{K}^2$, obtained by precomposing with $d^1 \colon 2 \to 3$.

We typically write $L, R \colon \mathcal{K}^2 \rightrightarrows \mathcal{K}^2$ for the functors that take an arrow to their left and right factors; these are the composites of the functorial factorization with the projections $d^2, d^0 \colon \mathcal{K}^3 \rightrightarrows \mathcal{K}^2$. It is also convenient to have notation, say, $E \colon \mathcal{K}^2 \to \mathcal{K}$, for the functor that sends an arrow to the object through which it factors. In this notation, a functorial factorization factors a commutative square,

$$
\begin{array}{ccc}
X & \xrightarrow{u} & W \\
{\scriptstyle f}\downarrow & & \downarrow{\scriptstyle g} \\
Y & \xrightarrow{v} & Z,
\end{array}
$$

regarded as a morphism $(u, v) \colon f \Rightarrow g$ in $\mathcal{K}^2$, as

$$
\begin{array}{ccc}
X & \xrightarrow{u} & W \\
\left(\begin{array}{c} \downarrow{\scriptstyle Lf} \\ \\ \downarrow{\scriptstyle Rf} \end{array}\right. &
\begin{array}{c} \\ Ef \xrightarrow{E(u,v)} Eg \\ \end{array} &
\left.\begin{array}{c} \downarrow{\scriptstyle Lg} \\ \\ \downarrow{\scriptstyle Rg} \end{array}\right) \\
Y & \xrightarrow{v} & Z.
\end{array}
$$

Note that L preserves domains and R preserves codomains of both objects and morphisms. The injections $d^1, d^0 \colon 1 \rightrightarrows 2$ induce functors dom, cod $\colon \mathcal{K}^2 \rightrightarrows \mathcal{K}$, and Definition 12.1.1 demands that dom $L = $ dom and cod $R = $ cod.

Remark 12.1.2   Slicing over the initial object defines a functor $\mathcal{K} \to \mathcal{K}^2$. Because L preserves domains, the restriction of L to the full subcategory of objects sliced under the initial object defines a functor $\mathcal{K} \to \mathcal{K}$. Given a functorial factorization for the (cofibration, trivial fibration) weak factorization system in a model category, this procedure defines a cofibrant replacement functor Q. The components of the functor R define a natural transformation $Q \Rightarrow 1$. Hence a functorial factorization for the model category gives rise to a left deformation into the subcategory of cofibrant objects. A right deformation into the subcategory of fibrant objects is obtained dually from a functorial factorization for the (trivial cofibration, fibration) weak factorization system.

12.2 Quillen's small object argument

Some model categories have natural occurring functorial factorizations, but for many examples, a general procedure for producing factorizations is desired. If we are not presented with a naturally occurring functorial factorization, how might we conjure one out of mid air? The key point is that, for many weak factorization systems in model structures, the fibrations or trivial fibrations are characterized by a right lifting property against a *set* (rather than a proper class) of trivial cofibrations or cofibrations. Recall that such weak factorization systems, introduced in 11.2.6, are called cofibrantly generated.

Example 12.2.1

- The Kan fibrations in **sSet** are the maps $\{\Lambda_k^n \to \Delta^n \mid n \geq 0, 0 \leq k \leq n\}^{\boxtimes}$.
- The trivial fibrations in the Quillen-type model structures on **sSet** or **Top** or $\mathbf{Ch}_{\geq 0}(R)$ are the maps $\{S^{n-1} \to D^n \mid n \geq 0\}^{\boxtimes}$ for various interpretations of spheres and disks (cf. 11.3.5, 11.3.6, 11.3.7).
- The Serre fibrations in **Top** or $\mathbf{Ch}_{\geq 0}(R)$ are the maps $\{i_0 \colon D^n \to D^n \otimes I \mid n \geq 0\}^{\boxtimes}$ for the appropriate interpretation of the unit interval and the tensor.
- The trivial fibrations in the folk model structure on **Cat** are the surjective categorical equivalences, that is, the maps $\{\emptyset \to \mathbb{1}, \mathbb{1} \sqcup \mathbb{1} \to \mathbb{2}, \mathbb{2}_2 \to \mathbb{2}\}^{\boxtimes}$, this last generator being the surjective functor mapping a parallel pair of arrows to the walking arrow.
- The pointwise trivial fibrations in the projective model structure on $\mathbf{sSet}^{\mathcal{D}}$ are the maps $\{\mathcal{D}(d, -) \cdot \partial \Delta^n \to \mathcal{D}(d, -) \cdot \Delta^n \mid d \in \mathcal{D}, n \geq 0\}^{\boxtimes}$.

Quillen's small object argument proves that a set of maps $\mathcal{J}$ in a category satisfying a certain set theoretical condition generates a cofibrantly generated weak factorization system $(^{\boxtimes}(\mathcal{J}^{\boxtimes}), \mathcal{J}^{\boxtimes})$ with a functorial factorization. An oft-forgotten corollary, Corollary 12.2.4, characterizes the left class of a cofibrantly generated weak factorization system.

Theorem 12.2.2 (Quillen's small object argument) *If $\mathcal{K}$ permits the small object argument for $\mathcal{J}$, then there exists a functorial factorization making $(^\boxtimes(\mathcal{J}^\boxtimes), \mathcal{J}^\boxtimes)$ a weak factorization system.*

What is meant by "permits the small object argument" varies depending on context; see [24, 36, 38, 58]. Here we will use this phrase as shorthand for the following assumption. An object $k \in \mathcal{K}$ is κ-**small**, where κ is a regular cardinal, if the representable $\mathcal{K}(k, -)\colon \mathcal{K} \to \mathbf{Set}$ preserves α-composites for any limit ordinal $\alpha \geq \kappa$. In topological contexts, it often suffices to consider **compact** objects, defined only with reference to colimits of countable sequences; to say k is compact is to say whenever we have a colimit diagram $\omega \to \mathcal{K}$, as displayed,

$$x_0 \to x_1 \to x_2 \to \cdots \to \operatorname*{colim}_n x_n = x_\omega$$

then $\mathcal{K}(k, \operatorname*{colim}_n x_n) \cong \operatorname*{colim}_n \mathcal{K}(k, x_n)$. This hom-set isomorphism says that any arrow $k \to x_\omega$ factors through some x_n.

To avoid getting bogged down by set theoretical technicalities, here our language and notation will tacitly suppose that we are in the compact case. There is little conceptual difference in the general setting, though each appearance of the ordinal ω will have to be replaced with a larger ordinal as appropriate. See [58, §15.1], where this issue is addressed carefully.

We say $\mathcal{K}$ permits the small object argument for $\mathcal{J}$ if $\mathcal{K}$ is cocomplete and if the domains of the objects in $\mathcal{J}$ are compact.[1]

Proof of Theorem 12.2.2   Quillen's small object argument constructs a functorial factorization in which the left factor is formed from the arrows in $\mathcal{J}$ using the colimits described in Lemma 11.1.4. It follows that the left factor is in the left class of the weak factorization system. The right factor is shown to be in the class $\mathcal{J}^\boxtimes$ using the compactness hypothesis.

Fixing an arrow f to be factored, form the coproduct, indexed first over arrows $j \in \mathcal{J}$ and then over commutative squares from j to f, of the arrows j. There is a canonical map from this coproduct to f defining the commutative square

$$\begin{array}{ccc} \cdot & \xrightarrow{\ d_f\ } & \cdot \\ {\scriptstyle \coprod_{j \in \mathcal{J}} \coprod_{\mathbf{Sq}(j,f)} j}\Big\downarrow & & \Big\downarrow{\scriptstyle f} \\ \cdot & \xrightarrow[\ c_f\ ]{} & \cdot \end{array} \qquad (12.2.3)$$

---

[1] In fact one can, and frequently does, get by with less – it suffices that $\mathcal{K}(k, -)$ commutes with colimits of sequences comprised of arrows in $^\boxtimes(\mathcal{J}^\boxtimes)$; cf., e.g., [38, 2.4.1–2].

Here d_f and c_f are shorthand for the canonical maps induced by the domain and codomain components of the commutative squares. The step-one functorial factorization (L, R) is defined by the pushout

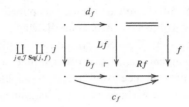

By construction, the map Lf is in $^\boxtimes(\mathcal{J}^\boxtimes)$ but the map Rf need not be in $\mathcal{J}^\boxtimes$. To remedy this, this process is repeated, but with the map Rf in place of f; thus Rf is factored as $R^2 f \cdot LRf$. The left factor in the step-two functorial factorization is the composite $LRf \cdot Lf$, which is still in the class $^\boxtimes(\mathcal{J}^\boxtimes)$ by 11.1.4; however, the step-two right factor $R^2 f$ need not be in $\mathcal{J}^\boxtimes$ yet, so the map $R^2 f$ is then factored again. Iterating produces a factorization

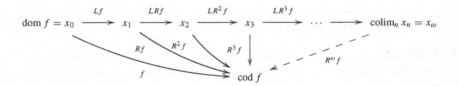

The horizontal ω-composite is still in $^\boxtimes(\mathcal{J}^\boxtimes)$ by Lemma 11.1.4 and defines the left factor in the functorial factorization. This time the right factor $R^\omega f$ is in $\mathcal{J}^\boxtimes$. By compactness, the domain component of any lifting problem factors through $x_n \to x_\omega$ for some n, as displayed:

$$
\begin{array}{ccc}
\cdot \longrightarrow x_\omega & & \cdot \xrightarrow{\;u\;} x_n \longrightarrow x_\omega \\
j\Big\downarrow \quad \Big\downarrow R^\omega f & = & j\Big\downarrow \quad \Big\downarrow R^n f \quad \Big\downarrow R^\omega f \\
\cdot \xrightarrow[\;v\;]{} \cdot & & \cdot \xrightarrow[\;v\;]{} \cdot =\!=\!= \cdot
\end{array}
$$

The left-hand square $(u, v)\colon j \Rightarrow R^n f$ in the right-hand rectangle necessarily occurs as a component of the coproduct whose pushout factors $R^n f$ as $LR^n f$ followed by $R^{n+1} f$. In particular, one leg of this pushout square defines a map

from the codomain of j to x_{n+1},

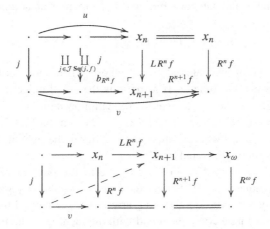

which gives the desired lift. □

We leave the proof of the following corollary as an exercise:

Corollary 12.2.4 *If $\mathcal{K}$ permits the small object argument, then $^{\boxtimes}(\mathcal{J}^{\boxtimes})$ is the smallest class of maps containing $\mathcal{J}$ and is closed under the colimits listed in Lemma 11.1.4. More specifically, any map in $^{\boxtimes}(\mathcal{J}^{\boxtimes})$ is a retract of a transfinite composite of pushouts of coproducts of maps in $\mathcal{J}$. This shows that $^{\boxtimes}(\mathcal{J}^{\boxtimes})$ is the* **weak saturation** *of $\mathcal{J}$, the smallest weakly saturated class containing $\mathcal{J}$.*

This result gives a description of both classes of the weak factorization system generated by $\mathcal{J}$. Elements of the right class are characterized by their lifting property against $\mathcal{J}$. Elements of the left class are retracts of transfinite composite of pushouts of coproducts of elements of $\mathcal{J}$.

12.3 Benefits of cofibrant generation

Recall that a model category is cofibrantly generated just when its two weak factorization systems are. Cofibrantly generated weak factorization systems are surprisingly common. One reason for their ubiquity is that cofibrantly generated weak factorization systems beget other cofibrantly generated weak factorization systems. For instance, a cofibrantly generated weak factorization system can be lifted along an adjunction $F \colon \mathcal{K} \rightleftarrows \mathcal{M} \colon U$ provided that the ambient categories permit the small object argument. If $\mathcal{J}$ generates a weak factorization system on $\mathcal{K}$, then $F\mathcal{J}$ generates a weak factorization system on $\mathcal{M}$ whose right class is created by the right adjoint U; that is, $f \in F\mathcal{J}^{\boxtimes}$ if and only if $Uf \in \mathcal{J}^{\boxtimes}$ (see Lemma 11.1.5).

For example, suppose $\mathcal{K}$ has a weak factorization system generated by $\mathcal{J}$, and let $\mathcal{D}$ be any small category. Fixing $d \in \mathcal{D}$, there is an adjunction

$$\mathcal{D}(d, -) \cdot - : \mathcal{K} \;\underset{\longleftarrow}{\overset{\longrightarrow}{\perp}}\; \mathcal{K}^{\mathcal{D}} : \mathrm{ev}_d \tag{12.3.1}$$

whose right adjoint evaluates at d and whose left adjoint forms a copower with the representable functor. If $\mathcal{K}$ permits the small object argument, then so does $\mathcal{K}^{\mathcal{D}}$. Hence there is a weak factorization system on $\mathcal{K}^{\mathcal{D}}$ generated by the set $\mathcal{D}(d, -) \cdot \mathcal{J} = \{\mathcal{D}(d, -) \cdot j \mid j \in \mathcal{J}\}$. By Lemma 11.1.5, its right class consists of those maps whose component at d lies in the right class of the weak factorization system generated by $\mathcal{J}$.

The functor category $\mathcal{K}^{\mathrm{ob}\,\mathcal{D}}$ is isomorphic to the product of the category $\mathcal{K}$ indexed by objects of $\mathcal{D}$. This category has a cofibrantly generated weak factorization system whose generating arrows are the natural transformations with $j \in \mathcal{J}$ in a single component and with the identity at the initial object in each other component. This set generates a weak factorization system on $\mathcal{K}^{\mathrm{ob}\,\mathcal{D}}$ whose left class consists of ob $\mathcal{D}$-indexed products of arrows each in the left class and whose right class consists of ob $\mathcal{D}$-indexed products of arrows in the right class.

There is a "many objects" version of the adjunction (12.3.1), in the form of an adjunction

$$\mathrm{Lan} : \mathcal{K}^{\mathrm{ob}\,\mathcal{D}} \;\underset{\longleftarrow}{\overset{\longrightarrow}{\perp}}\; \mathcal{K}^{\mathcal{D}} : \mathrm{ev}$$

whose right adjoint precomposes by the inclusion ob $\mathcal{D} \hookrightarrow \mathcal{D}$ and whose left adjoint is given by left Kan extension. The right adjoint has the effect of evaluating a functor at each object and placing the results in the associated components of the left-hand product. The category $\mathcal{K}^{\mathcal{D}}$ has a cofibrantly generated weak factorization system generated by the image under the left adjoint of the set of maps generating the pointwise defined weak factorization system on $\mathcal{K}^{\mathrm{ob}\,\mathcal{D}}$. Using Theorem 1.2.1 to compute the left Kan extension, these generating cofibrations have the form

$$\mathcal{J}_{\mathcal{D}} := \{\mathcal{D}(d, -) \cdot j \mid d \in \mathcal{D}, j \in \mathcal{J}\}$$

By Lemma 11.1.5, the right class of the weak factorization system generated by $\mathcal{J}_{\mathcal{D}}$ consists precisely of the natural transformations whose components are in the right class of the weak factorization system generated by $\mathcal{J}$ on $\mathcal{K}$.

This (nearly) proves:

Theorem 12.3.2 *For any cofibrantly generated model category $\mathcal{K}$ that permits the small object argument and any small category $\mathcal{D}$, the functor category $\mathcal{K}^{\mathcal{D}}$ admits a projective model structure.*

Proof If $\mathcal{J}$ is a set of generating trivial cofibrations and $\mathcal{I}$ is a set of generating cofibrations for the model structure on $\mathcal{K}$, then $\mathcal{J}_\mathcal{D}$ and $\mathcal{I}_\mathcal{D}$ generate weak factorization systems whose right classes consist of the pointwise fibrations and pointwise trivial fibrations. It remains only to verify that these weak factorization systems fit together to define a model structure with pointwise weak equivalences, as in Definition 11.3.1. By a well-known argument [36, 11.3.1–2] that we prove in an enriched form in Theorem 13.5.1, it remains only to check the "acyclicity condition," that is, to show that the elements in the left class of the weak factorization system generated by $\mathcal{J}_\mathcal{D}$ are pointwise weak equivalences.

Write α for an element of $^\boxtimes(\mathcal{J}_\mathcal{D}^\boxtimes)$. By adjunction, we know α lifts against any pointwise fibration. In particular, if we factor α by applying the *pointwise-defined* functorial factorization[2] produced by the small object argument for $\mathcal{J}$ on $\mathcal{K}$ (not the functorial factorization produced by $\mathcal{J}_\mathcal{D}$), then α lifts against its right factor, which is a pointwise fibration. Hence α is a retract of its left factor, which is a pointwise trivial cofibration, and, in particular, a pointwise weak equivalence. $\square$

This proof used the idea that appeared in the retract argument (Lemma 11.2.3). If f factors as $r \cdot \ell$ and f lifts against r, then f is a retract of ℓ, as displayed in (11.2.4).

Digression 12.3.3 (Bousfield localization) There is a fully developed machinery for producing new model structures on the same category, called **Bousfield localizations**, from a cofibrantly generated model structure satisfying additional set-theoretical hypotheses. A **left Bousfield localization** is a model structure with the same cofibrations whose weak equivalences contain the original weak equivalences. In particular, the identity defines a left Quillen functor from the original model structure to its left Bousfield localization.

Existence results come in a number of forms. For one, we suppose that $\mathcal{K}$ is a combinatorial simplicial model category that is also **left proper** (see 14.3.5 for a definition). Given any set S of cofibrations, there is a left Bousfield localization, again a left proper combinatorial simplicial model category, whose weak equivalences are the S-equivalences and whose fibrant objects are S-local objects that are fibrant in the original model structure. Writing $\underline{\text{Ho}}\mathcal{K}$ for the $\mathcal{H}$-enriched category defined by the proof of Theorem 10.0.1, an object $k \in \mathcal{K}$

---

[2] By postcomposition, a functorial factorization on $\mathcal{K}$ defines a functorial factorization on $\mathcal{K}^\mathcal{D}$; in particular, the functorial factorization constructed from $\mathcal{J}$ allows us to factor any natural transformation as a pointwise trivial cofibration followed by a pointwise fibration. But these left and right factors might not lift against each other because pointwise-defined lifts between their components need not assemble into a *natural* transformation. This is why the projective cofibrations are not the pointwise cofibrations. We will have more to say about this in Example 12.5.11.

is S-**local** if $\underline{\mathrm{Ho}}\mathcal{K}(b,k) \xrightarrow{f^*} \underline{\mathrm{Ho}}\mathcal{K}(a,k)$ is a homotopy equivalence for each $f: a \to b$ in S. A map $g: x \to y$ is an S-**equivalence** if $\underline{\mathrm{Ho}}\mathcal{K}(y,k) \xrightarrow{g^*} \underline{\mathrm{Ho}}\mathcal{K}(x,k)$ is a homotopy equivalence for each S-local object k. Textbook references include [36, §§4–5], [49, §A.3.7], and [58, §§19.2, 19.4–5].

12.4 Algebraic perspectives

To introduce the algebraic perspective, let us go back to Quillen's construction and try to understand why it worked. Taking artistic license, we will refer to a set of arrows $\mathcal{J}$ as the "generating trivial cofibrations" and the set $\mathcal{J}^{\boxtimes}$ as the "fibrations," though nothing we will say depends in any way on the existence of an ambient model structure. A common narrative is that the small object argument first produces a factorization whose left factor is a trivial cofibration but whose right factor is badly behaved, and the magic worked by iteration and compactness somehow corrects this deficiency. But from our perspective, and presumably from Quillen's, the small object argument is all about the fibrations from the very start.

To see what we mean by this, let us return to step zero, which takes the map f to be factored and produces the square (12.2.3). This square should be thought of as the "generic lifting problem" that tests whether f is a fibration. Indeed, a single lift

$$
\coprod_{j \in \mathcal{J}} \coprod_{\mathrm{Sq}(j,f)} j \left\downarrow \nearrow^{\phi_f} \right\downarrow f \tag{12.4.1}
$$

simultaneously solves all lifting problems, guaranteeing that $f \in \mathcal{J}^{\boxtimes}$. We might call ϕ_f a **lifting function** because this arrow specifies a solution to any lifting problem against a generating trivial cofibration.

The step-one pushout translates this generic lifting problem into another generic lifting problem whose domain component is the identity. By the universal property, a solution to the lifting problem presented by the right-hand square

$$
\coprod_{j \in \mathcal{J}} \coprod_{\mathrm{Sq}(j,f)} j \left\downarrow \quad \left\downarrow^{Lf} \nearrow \right. \left\downarrow f \right. \quad \underset{Rf}{} \tag{12.4.2}
$$

precisely specifies a lifting function. In words, we see that f is a fibration if and only if it lifts against its left factor in the canonical lifting problem defined by its step-one functorial factorization.

This turns out to be an extraordinarily useful observation. For any functorial factorization – not just for the step-one factorization constructed as part of Quillen's small object argument – the functors L and R are equipped with canonical natural transformations to and from the identity on $\mathcal{K}^2$, respectively, which we denote by $\vec{\epsilon} \colon L \Rightarrow 1$ and $\vec{\eta} \colon 1 \Rightarrow R$. The components of these natural transformations at $f \colon X \to Y$ are the squares

$$
\begin{array}{ccc}
X & \!=\!=\! & X \\
{\scriptstyle Lf}\big\downarrow & & \big\downarrow{\scriptstyle f} \\
Ef & \xrightarrow{\;\;Rf\;\;} & Y
\end{array}
\qquad
\begin{array}{ccc}
X & \xrightarrow{\;\;Lf\;\;} & Ef \\
{\scriptstyle f}\big\downarrow & & \big\downarrow{\scriptstyle Rf} \\
Y & \!=\!=\! & Y
\end{array}
\qquad (12.4.3)
$$

In other words, L and R are **pointed endofunctors** of $\mathcal{K}^2$, where we let context indicate in which direction (left or right) the functors are pointed. An **algebra** for the pointed endofunctor R is defined analogously to the notion of an algebra for a monad, except, of course, there is no associativity condition. Similarly, a **coalgebra** for the pointed endofunctor L is defined analogously to the notation of a coalgebra for a comonad. In the framework of these definitions, the retract argument (11.2.4) takes the following form:

Lemma 12.4.4 *$f \in \mathcal{K}^2$ is an R-algebra just when there exists a lift*

$$
\begin{array}{ccc}
X & \!=\!=\! & X \\
{\scriptstyle Lf}\big\downarrow & \overset{t}{\nearrow} & \big\downarrow{\scriptstyle f} \\
Ef & \xrightarrow{\;\;Rf\;\;} & Y
\end{array}
\qquad (12.4.5)
$$

Conversely, any choice of lift determines an R-algebra structure for f. Dually, $i \in \mathcal{K}^2$ is an L-coalgebra just when there exists a lift

$$
\begin{array}{ccc}
A & \xrightarrow{\;\;Li\;\;} & Ef \\
{\scriptstyle i}\big\downarrow & \overset{s}{\nearrow} & \big\downarrow{\scriptstyle Ri} \\
Y & \!=\!=\! & B
\end{array}
\qquad (12.4.6)
$$

Conversely, any choice of lift determines an L-coalgebra structure for i.

A key point, which we use later, is that any L-coalgebra lifts (canonically) against any R-algebra. In other words, a choice of solutions for the generic lifting problems (12.4.5) and (12.4.6) renders further choices unnecessary.

Lemma 12.4.7   *Any L-coalgebra (i, s) lifts canonically against any R-algebra (f, t).*

Proof   Given a lifting problem, that is, a commutative square $(u, v): i \Rightarrow f$, the functorial factorization together with the coalgebra and algebra structures define a solution:

$$
\begin{array}{ccc}
A & \xrightarrow{u} & X \\
\end{array}
$$

Specializing again, suppose (L, R) is constructed as in (12.4.2). By the discussion there, the R-algebras are precisely the elements of the set $\mathcal{J}^{\boxtimes}$. It follows from Lemma 12.4.7 that the L-coalgebras are trivial cofibrations: they lift against the R-algebras and hence the fibrations. Indeed, this is the reason why Lf is a trivial cofibration: it is a (free) L-coalgebra – but this is for later.

The reason that Quillen's small object argument continues beyond step one is that Rf is not itself an R-algebra (a fibration), precisely because R is not a monad. If we could somehow replace R by a monad without changing the algebras, this problem would be solved: objects in the image of the monad are then free algebras for the monad. It turns out that under certain set-theoretic hypotheses, this is possible.

Digression 12.4.8 (the free monad on a pointed endofunctor)   The **free monad** on a pointed endofunctor R is a monad $\mathbb{F} = (F, \vec{\eta}: 1 \Rightarrow F, \vec{\mu}: F^2 \Rightarrow F)$ together with a universal map of pointed endofunctors $R \Rightarrow F$. When the ambient category is complete and locally small, the free monad has the property that the induced functor from the category of $\mathbb{F}$-algebras to the category of R-algebras is an isomorphism commuting with the forgetful functors [45, §22]. This is the property of interest.

How might the free monad $\mathbb{F}$ be constructed? One guess would be to form the colimit

$$1 \to R \to R^2 \to \cdots \to \operatorname{colim} R^n \overset{?}{\cong} F$$

and use the defining universal property to try to construct the necessary multiplication map $\vec{\mu}$. But this does not work unless the two natural maps $\vec{\eta}R, R\vec{\eta}: R \rightrightarrows R^2$ are equal, in which case the pointed endofunctor R is called **well-pointed**.

The general construction iteratively "guesses" the value of the free monad and then "corrects" it by forming coequalizers until this process converges. The first guess is R. The second is the coequalizer of $\vec{\eta}R$ and $R\vec{\eta}$, a quotient of R^2. Further details can be found in [45].[3]

Example 12.4.9 The step-one right factor R fails to be well-pointed; indeed, this failure precisely highlights the redundancy in Quillen's small object argument. To fix ideas, suppose $\mathcal{J} = \{S^{n-1} \to D^n\}_{n \geq 0}$ includes "spheres" into "disks." A map is in $\mathcal{J}^{\boxtimes}$ just when, for every sphere in its domain which becomes contractible in its base, there is a lift of any specified contracting homotopy. The step-one factorization of $f : X \to Y$ glues disks filling all such spheres in X, producing a new space Ef. Step two of Quillen's small object argument repeats this process, gluing disks to spheres in Ef to produce a new space ERf. The map $\vec{\eta}R$ includes Ef into ERf as a component of the defining pushout. The map $R\vec{\eta}$ agrees with this inclusion on the subspace X but then maps each disk attached to some sphere in X to the corresponding disk attached to the image of this sphere in Ef.

12.5 Garner's small object argument

At this point, an alternate "algebraic" small object argument, due to Garner [30, 31], diverges from Quillen's construction. Recognizing that the algebras for the step-one right factor are precisely the fibrations, Garner's small object argument forms the free monad on the pointed endofunctor R by means of the construction described in 12.4.8. Because R is an endofunctor of $\mathcal{K}^2$ that preserves codomains, its unit map (12.4.3) defines a functorial factorization. Similarly, each stage of the free monad construction defines a functorial factorization. Hence the free monad $\mathbb{F}$ is also the right factor in a functorial factorization, whose left factor, defined by the unit map, we will call C.

The claim is that (C, F) is a functorial factorization for the weak factorization system generated by $\mathcal{J}$ – but this appears a bit mad: the functor F and hence also C was constructed through a lot of quotienting, which makes it seem unlikely that the arrows Cf are trivial cofibrations. But this is actually true and for completely general reasons: the pointed endofunctor C turns out to be a comonad. In particular, Cf is a C-coalgebra, and C-coalgebras lift against F-algebras (fibrations) by Lemma 12.4.7.

Before we explain why this works, let us state the theorem.

---

[3] The *constructibility* of the free monad will be important because it allows us to prove that the free monad associated to the right factor in a functorial factorization is again the right factor in a functorial factorization.

Theorem 12.5.1 (Garner's small object argument) *If $\mathcal{K}$ permits the small object argument, then for any small set (or even small category) of arrows $\mathcal{J}$, there exists a functorial factorization (an algebraic weak factorization system) $(\mathbb{C}, \mathbb{F})$ generated by $\mathcal{K}$ whose underlying weak factorization system is $({}^{\boxtimes}(\mathcal{J}^{\boxtimes}), \mathcal{J}^{\boxtimes})$.*

Ignore the parentheticals for now; we will return to them shortly. Here "permits the small object argument" means that either

(*) for each $k \in \mathcal{K}$, there is some regular cardinal κ_k so that $\mathcal{K}(k, -)$ preserves κ_k-filtered colimits

(†) for each $k \in \mathcal{K}$ there is some regular cardinal κ_k so that $\mathcal{K}(k, -)$ sends κ_k-filtered unions of $\mathcal{M}$-subobjects to κ_k-filtered unions of sets, where $\mathcal{M}$ is the right class of some proper, well-copowered orthogonal factorization system $(\mathcal{E}, \mathcal{M})$ on $\mathcal{K}$

Proper means that maps in the left class are epimorphisms and maps in the right class are monomorphisms, though the converses need not apply. As a consequence, both the factorizations and liftings for the weak factorization system $(\mathcal{E}, \mathcal{M})$ are unique; hence the factorization system is orthogonal. **Well-copowered** means that every object has a mere set of $\mathcal{E}$-quotients up to isomorphism – this is the case in any category of interest. For example, surjections and subspace inclusions define an orthogonal factorization system with these properties on **Top** [46, §6]. These conditions guarantee that the free monad construction described in 12.4.8 converges.

Because our original pointed endofunctor preserved codomains, the free monad does as well, defining a functorial factorization. However, we want more control over the left factor. To achieve this, Garner lifts the free monad construction to the category of functorial factorizations on $\mathcal{K}$ whose left factor is a comonad. It follows that the free monad so produced is the right factor of an **algebraic weak factorization system**, defined in 12.6.1; in particular, the left factor is a comonad, and hence, by Lemma 12.4.7, maps in its image will lift against $\mathbb{F}$-algebras and therefore be trivial cofibrations.

For this outline to make sense, the step-one left factor in the Quillen/Garner small object argument would have to be a comonad – indeed, this is the case! To see this, we return once more to step zero. Write L_0 for the functor that sends the arrow f to the left-hand side of the square (12.4.1). We associate the generating arrows with their image in $\mathcal{K}$ via a functor $\mathcal{J} \to \mathcal{K}^2$. Our favorite Theorem 1.2.1 allows us to recognize the functor L_0 as the left Kan extension of $\mathcal{J} \to \mathcal{K}^2$ along itself. By an easy formal argument, any endofunctor constructed by forming the left Kan extension of some other functor along itself is a comonad, called the **density comonad** [16, §II.1]. The counit for this comonad is the generic lifting problem displayed in (12.4.1). The step-one left factor L is just the pushout of L^0 – or, if you prefer, its reflection via the (pushout square, isomorphism-on-domain-component) orthogonal factorization system on $\mathcal{K}^2$ – and hence inherits a comonad structure as well.

Remark 12.5.2 Note that step zero of Garner's small object argument, which takes a left Kan extension of $\mathcal{J} \rightarrow \mathcal{K}^2$ along itself, works equally well if $\mathcal{J}$ is any small *category* of arrows. The remaining steps refer only to the functor L_0 so produced and hence also make sense when the generators are taken to be a small category of arrows and commutative squares. See Example 12.5.11 for an illustration.

Remark 12.5.3 Because the right functor is a monad and the left functor is a comonad, the procedure described in Remark 12.1.2 can be used to obtain a cofibrant replacement comonad and a fibrant replacement monad for any cofibrantly generated model category that permits Garner's small object argument. These functors can be used to construct point-set level derived monad and comonad resolutions for any Quillen adjunction between two such model categories (see [7]).

The upshot is that Garner's small object argument presents an alternative to Quillen's, producing functorial factorizations appropriate for a cofibrantly generated weak factorization system (or model structure), but unlike Quillen's small object argument, this procedure converges. As a result, the functorial factorization is "less redundant" and often easier to explicitly describe. Furthermore, the close relationship between the functorial factorization and the desired lifting properties means that Garner's small object argument can be more easily generalized to produce factorizations for non-cofibrantly generated weak factorization systems.

Further details about this construction can be found in the original [30, 31] or in [71, 73]. Here we prefer to develop familiarity with these ideas through examples.

Example 12.5.4 Fix a commutative ring R and write D^n for the unbounded chain complex with the ring R in degrees n and $n - 1$, with identity differential, and zeros elsewhere. The set with the right lifting property with respect to $\mathcal{J} = \{0 \rightarrow D^n\}_{n \in \mathbb{Z}}$ in $\mathbf{Ch}_\bullet(R)$ is the set of chain maps whose components are epimorphisms. These are the fibrations in the Quillen-type model structure on this category described in 11.3.7.

Let us use Garner's small object argument to factor $f \colon X_\bullet \rightarrow Y_\bullet$. Observe that the set of squares from $0 \rightarrow D^n$ to f is isomorphic to the underlying set of the R-module Y_n. Hence the step-one factorization has the form

$$
\begin{array}{ccccc}
0 & \longrightarrow & X_{\bullet} & =\!=\!= & X_{\bullet} \\
\downarrow & & \downarrow{\scriptstyle Lf} & & \downarrow{\scriptstyle f} \\
\oplus_{n,Y_n} D^n & \xrightarrow{\ \ulcorner\ } & X_{\bullet} \oplus (\oplus_{n,Y_n} D^n) & \xrightarrow[Rf]{} & Y_{\bullet}
\end{array}
\tag{12.5.5}
$$

In degree k, the chain complex $\oplus_{n,Y_n} D^n$ is isomorphic to the direct sum of the free R-modules on the underlying sets of Y_k and Y_{k+1}; the latter component is in the image of the differential from degree $k+1$.

Because the set of squares from $0 \to D^n$ to the map Rf is again isomorphic to the set Y_n, Quillen's step-two factorization has the form

$$
\begin{array}{ccccc}
0 & \longrightarrow & X_{\bullet} \oplus (\oplus_{n,Y_n} D^n) & =\!=\!= & X_{\bullet} \oplus (\oplus_{n,Y_n} D^n) \\
\downarrow & & \downarrow{\scriptstyle LRf} & & \downarrow{\scriptstyle Rf} \\
\oplus_{n,Y_n} D^n & \xrightarrow{\ \ulcorner\ } & X_{\bullet} \oplus (\oplus_{n,Y_n} D^n) \oplus (\oplus_{n,Y_n} D^n) & \xrightarrow[R^2 f]{} & Y_{\bullet}
\end{array}
$$

By contrast, Garner's step-two factorization is formed by taking the coequalizer of the two evident maps

$$
X_{\bullet} \oplus (\oplus_{n,Y_n} D^n) \rightrightarrows X_{\bullet} \oplus (\oplus_{n,Y_n} D^n) \oplus (\oplus_{n,Y_n} D^n) \dashrightarrow X_{\bullet} \oplus (\oplus_{n,Y_n} D^n)
$$

But this factorization is (L, R) again! Hence Garner's small object argument converges at step one. Indeed, observe that the map Rf is already a pointwise surjection, and therefore no further steps are needed to obtain a factorization with the desired lifting properties.

Remark 12.5.6 (simplified version of the algebraic small object argument) In general, when the left class of the weak factorization system generated by $\mathcal{J}$ is contained in the monomorphisms, Garner's small object argument admits a particularly simple description, as was noticed independently in the PhD thesis of Andrei Radulescu-Banu [66]; more details can be found there. In this case, the role of the coequalizers taken at each step after step one is precisely to avoid attaching redundant cells. Put another way, when the left class is contained in the monomorphisms, each attaching map from the domain of some $j \in \mathcal{J}$ factors (uniquely) through a minimal stage in the small object argument. At each subsequent stage, the cell that is temporarily reattached via this attaching map is identified with the first cell so attached. Hence the free monad construction is equally described as follows: At step one, attach a cell for any commutative square from j to f. After this, attach cells only for those squares that do not factor through a previous stage. In this way, it is easy to see that Garner's small object argument is just a "less redundant" form of Quillen's small object argument.

Example 12.5.7 Consider $\mathcal{J} = \{\partial \Delta^n \to \Delta^n\}_{n \geq 0}$ in **sSet**. In this case, the left class of the weak factorization system generated by $\mathcal{J}$ is the set of monomorphisms. To see this, note that any monomorphism $A \to B$ has a canonical cellular decomposition $A = A^{-1} \to A^0 \to A^1 \to A^2 \to \cdots \to A^\omega = B$, where A^n contains A and every n-simplex in B. Each $A^{k-1} \to A^k$ is a pushout of a coproduct of copies of the generator $\partial \Delta^k \to \Delta^k$, each attaching a k-simplex in B but not in A to its boundary in A^{k-1}. This shows that monomorphisms are contained in the left class of the weak factorization system generated by $\mathcal{J}$. The converse, left to the reader, makes use of Corollary 12.2.4 and the closure properties of the monomorphisms in **Set** that derive from the existence of the (monomorphism, epimorphism) weak factorization system.

In addition to Remark 12.5.6, a further simplification of the construction of the Garner small object argument, particular to this example, is possible. This is best explained from the point of view of one of the generating cofibrations, say, $\emptyset \to \Delta^0$. Fix a map $X \to Y$ to be factored. In the first step, lifting problems against $\emptyset \to \Delta^0$ induce us to attach all the vertices of Y to X. In the second step, we do not reattach these vertices; instead, we only attach new vertices appearing in the codomain. But the codomain is unchanged. So the generator $\emptyset \to \Delta^0$ makes no contribution after step one.

Now consider $\partial \Delta^1 \to \Delta^1$. In the first and second steps, it induces us to attach edges of Y to vertices appearing in the domain. But in the third and subsequent steps, there is nothing new to attach unless new vertices appear in the domain, but they will not because only $\emptyset \to \Delta^0$ produces new vertices (the other generators being identities on 0-simplices), and this generator does not contribute after step one. Indeed, if we waited until step two to attach any edges, it would suffice to consider lifting problems against $\partial \Delta^1 \to \Delta^1$ only once.

Continuing this line of reasoning, it suffices to run Garner's small object argument in countably infinitely many steps, the $(n+1)$th step attaching n-cells only.

Remark 12.5.8 Note there is still some redundancy in Garner's small object argument: it attaches cells to every sphere, even if a "filler" is already present in the domain. An algebra for the monad produced by this construction precisely exhibits the redundancy of these attached cells, mapping each attached simplex to some preexisting filler.

There are interesting weak factorization systems whose left class is not contained within the monomorphisms.

Example 12.5.9 Consider the (cofibration, trivial fibration) weak factorization system for the folk model structure on **Cat**. Its left class consists of functors that are injective on objects; its right class consists of surjective equivalences of categories. This weak factorization system is generated by a set of three functors $\{\emptyset \to \mathbb{1}, 2 \to 2, 2_2 \to 2\}$ described in 12.2.1.

By a careful[4] argument similar to the one given in the previous example, Garner's small object argument is equivalent to one that considers only lifting problems against the first generator, then those against the second, and then finally those against the third. Each process converges after a single step, which means that each functorial factorization is constructed by a single pushout of the coproduct over squares of the generator in question.

The functorial factorization produced by the small object argument is equivalent to the usual mapping cylinder construction:

$$
\begin{array}{ccc}
A & \xrightarrow{\;\;f\;\;} & B \\
\downarrow{\scriptstyle i_1} & & \downarrow \\
A \times \mathbb{I} & \xrightarrow{\quad\ulcorner\quad} & A \times \mathbb{I} \coprod_A B
\end{array}
$$

Here $\mathbb{I}$ is the "free-standing isomorphism," the category with two objects "0" and "1" and a unique isomorphism between them. Concretely, $A \times \mathbb{I} \coprod_A B$ is the unique category with objects $A_0 \coprod B_0$ such that the functor (f, id) to B is fully faithful. This cylinder object defines a functorial factorization

$$
A \xrightarrow{\;f\;} B \quad \mapsto \quad A \xrightarrow{\;i_0\;} A \times \mathbb{I} \coprod_A B \xrightarrow{\;(f,\mathrm{id})\;} B
$$

of a functor f as a cofibration followed by a trivial fibration.

Exercise 12.5.10   Compute the factorizations produced on **Set** by Garner's small object argument applied to each of the following single-element sets: $\{\emptyset \to *\}$ and $\{* \to * \sqcup *\}$.

Example 12.5.11   Suppose $\mathcal{K}$ permits the small object argument and $\mathcal{J}$ is some set[5] of arrows so that Garner's small object argument produces the functorial factorization (L, R). Now consider the diagram category $\mathcal{K}^{\mathcal{D}}$, where $\mathcal{D}$ is small, and form a new *category* of arrows isomorphic to $\mathcal{D}^{\mathrm{op}} \times \mathcal{J}$ whose objects have the form $\mathcal{D}(d, -) \cdot j$ for $d \in \mathcal{D}$ and $j \in \mathcal{J}$ and with morphisms $g^* \colon \mathcal{D}(d', -) \cdot j \to \mathcal{D}(d, -) \cdot j$ for each $g \colon d \to d'$ in $\mathcal{D}$. The objects of this category coincide with the set of generating projective cofibrations $\mathcal{J}_{\mathcal{D}}$, but the presence of morphisms will alter the result of the algebraic small object argument.

---

[4] Care must be taken because the cofibrations in this case are not monomorphisms. However, composites of pushouts of coproducts of the first two generators are monomorphisms.

[5] This discussion easily extends to the case where $\mathcal{J}$ is itself a small category (cf. [71, 4.3]) .

Let us calculate this functorial factorization. Because morphisms in $\mathcal{K}^{\mathcal{D}}$ are natural transformations, we will call the map we wish to factor α. Step zero of the small object argument defines a functor $L_0^{\mathcal{D}}: (\mathcal{K}^{\mathcal{D}})^2 \to (\mathcal{K}^{\mathcal{D}})^2$ formed by taking the left Kan extension of the inclusion $\mathcal{D}^{\mathrm{op}} \times \mathcal{J} \to (\mathcal{K}^{\mathcal{D}})^2$ along itself. By Theorem 1.2.1, this yields

$$L_0^{\mathcal{D}}\alpha \cong \int^{(d,j)\in\mathcal{D}^{\mathrm{op}}\times\mathcal{J}} \coprod_{\mathbf{Sq}(\mathcal{D}(d,-)\cdot j,\alpha)} \mathcal{D}(d,-)\cdot j$$

$$\cong \int^{(d,j)\in\mathcal{D}^{\mathrm{op}}\times\mathcal{J}} \coprod_{\mathbf{Sq}(j,\alpha_d)} \mathcal{D}(d,-)\cdot j$$

where we have used the adjunction (12.3.1) to rename the indexing set. By Fubini's theorem and the fact the left adjoints preserve coends, this is

$$\cong \int^{d\in\mathcal{D}^{\mathrm{op}}} \mathcal{D}(d,-)\cdot\left(\int^{j\in\mathcal{J}} \coprod_{\mathbf{Sq}(j,\alpha_d)} j\right)$$

$$\cong \int^{d\in\mathcal{D}^{\mathrm{op}}} \mathcal{D}(d,-)\cdot L_0\alpha_d$$

where the last isomorphism uses the definition of the step-zero comonad L_0 constructed for $\mathcal{J}$. When we evaluate this formula for $L_0^{\mathcal{D}}\alpha$ at a component $c \in \mathcal{D}$, the co-Yoneda lemma yields $(L_0^{\mathcal{D}}\alpha)_c = L_0\alpha_c$. In other words, the step zero comonad produced by $\mathcal{D}^{\mathrm{op}} \times \mathcal{J}$ on $\mathcal{K}^{\mathcal{D}}$ is just the step zero comonad produced by $\mathcal{J}$ applied pointwise. But all the other steps in the small object argument are already pointwise colimits. So the upshot is that the small object argument applied to $\mathcal{D}^{\mathrm{op}} \times \mathcal{J}$ produces the functional factorization on $\mathcal{K}^{\mathcal{D}}$ defined by pointwise application of the factorization generated by $\mathcal{J}$.

Remark 12.5.12 Write $\mathbb{R}^{\mathcal{D}}$ for the monad produced by Example 12.5.11. What does it mean to be an $\mathbb{R}^{\mathcal{D}}$-algebra? As is always the case, an $\mathbb{R}^{\mathcal{D}}$-algebra is precisely a map α that has a solution to the generic lifting problem

By the universal property of the coend defining $L_0^{\mathcal{D}}$, a solution to this lifting problem exists if and only if $\alpha: F \Rightarrow G$ lifts "coherently" against the generators $\mathcal{D}(d,-)\cdot j$. This means that we must be able to choose solutions to each lifting problem so that the triangles of lifts

$$
\begin{array}{ccc}
\cdot \xrightarrow{\;g^*\;} \cdot \longrightarrow F & & \cdot \xrightarrow{\;Fg\;} \cdot \\
\mathcal{D}(d',-)\cdot j \Bigg\downarrow \quad \mathcal{D}(d,-)\cdot j \Bigg\downarrow \quad \Bigg\downarrow \alpha & \rightsquigarrow & j \Bigg\downarrow \quad \alpha_d \quad \Bigg\downarrow \alpha_{d'} \\
\cdot \xrightarrow{\;g^*\;} \cdot \longrightarrow G & & \cdot \longrightarrow \cdot \xrightarrow{\;Gg\;} \cdot
\end{array}
$$

$$(12.5.13)$$

commute for each $g: d \to d'$ in $\mathcal{D}$.

By contrast, an algebra for the monad produced by $\mathcal{J}_{\mathcal{D}}$, the underlying set of the category $\mathcal{D}^{\mathrm{op}} \times \mathcal{J}$, is just a natural transformation whose components are in $\mathcal{J}^{\boxtimes}$; hence each component lifts against $\mathcal{J}$, but these lifts are not necessarily natural (cf. the proof of Theorem 12.3.2 and Exercise 12.6.8).

When $\mathcal{J}$ is a small category of arrows, $\mathcal{J}^{\boxtimes}$ should be interpreted to be the class of maps that lift coherently against $\mathcal{J}$ in the sense illustrated by the left-hand side of (12.5.13). As illustrated by Remark 12.5.12, this class is smaller in general than the class of maps that lift "incoherently" against the objects of the category $\mathcal{J}$.

12.6 Algebraic weak factorization systems and universal properties

The functorial factorizations produced by Garner's small object argument are part of algebraic weak factorization systems. The general philosophy is that in an *algebraic* weak factorization system, the functorial factorization is intimately related to the lifting properties that define the left and right classes.

Definition 12.6.1   An **algebraic weak factorization system** $(\mathbb{C}, \mathbb{F})$ on $\mathcal{K}$ consists of a comonad $\mathbb{C} = (C, \vec{\epsilon}, \vec{\delta})$ and a monad $\mathbb{F} = (F, \vec{\eta}, \vec{\mu})$ on $\mathcal{K}^2$ such that

- the pointed endofunctors $(C, \vec{\epsilon})$ and $(F, \vec{\eta})$ form a functorial factorization
- the canonical map $(\delta, \mu): CF \Rightarrow FC$ is a distributive law

The original definition is due to Marco Grandis and Walter Tholen under the name "natural weak factorization system" [33]. The second condition was added by Garner and holds for free in every example we know. The natural transformations δ and μ are the codomain and domain components, respectively, of $\vec{\delta}$ and $\vec{\mu}$. These maps give the arrows in the image of C and F their free coalgebra and algebra structures:

$$
\begin{array}{ccc}
\operatorname{dom} f \xrightarrow{C^2 f} ECf & \qquad & Ef === Ef \\
\end{array}
$$

An algebraic weak factorization system should be thought of as extra structure accompanying an ordinary weak factorization system, including in particular a well-behaved functorial factorization. Note that any functorial factorization (C, F) for which $Cf \mathbin{\square} Fg$ for all maps f and g – as is the case for an algebraic weak factorization system by the existence of free (co)algebra structures and Lemma 12.4.7 – determines an **underlying weak factorization system** whose left and right classes can be deduced from the retract argument. We will say more about underlying weak factorization systems momentarily.

Here the important point is that Garner's small object argument produces an algebraic weak factorization system satisfying two universal properties:

Theorem 12.6.2 ([31, 4.4]) *If $\mathcal{K}$ permits the small object argument, then any small category of arrows $\mathcal{J}$ produces an algebraic weak factorization system $(\mathbb{C}, \mathbb{F})$ such that*

- *there is a universal functor $\mathcal{J} \to \mathbb{C}$-**coalg** over $\mathcal{K}^2$*
- *there is a canonical isomorphism of categories $\mathbb{F}$-**alg** $\cong \mathcal{J}^{\square}$ over $\mathcal{K}^2$*

The functor $\mathcal{J} \to \mathbb{C}$-**coalg** assigns $\mathbb{C}$-coalgebra structures to the generating cofibrations. Its universal property is instrumental to the work of [71, 73]; we give the barest hint of its meaning in 12.8 and its applications in Section 12.9. Here we focus on the second universal property. As a first step, we must define the category $\mathcal{J}^{\square}$.

Definition 12.6.3 If $\mathcal{J} \to \mathcal{K}^2$ is some subcategory of arrows, frequently discrete, write $\mathcal{J}^{\square}$ for the category whose objects are pairs (f, ϕ_f), where $f \in \mathcal{K}^2$ and ϕ_f is a **lifting function** that specifies a solution

to any lifting problem against some $j \in \mathcal{J}$ in such a way that the specified lifts commute with morphisms in $\mathcal{J}$ in the sense illustrated by the left-hand diagram of (12.5.13). A morphism $(f, \phi_f) \to (g, \phi_g)$ is a morphism $f \Rightarrow g$ in $\mathcal{K}^2$ commuting with the chosen lifts in the sense illustrated by the right-hand diagram of (12.5.13)

We will let context disambiguate between the category and the class of maps $\mathcal{J}^{\boxtimes}$, the latter consisting of the objects in the image of the forgetful functor associated to the former; this notion most frequently refers to the class of maps. The proof of the second universal property of Theorem 12.6.2 is via the discussion of "generic lifting problems" in Section 12.4. By construction, $\mathcal{J}^{\boxtimes}$ is isomorphic to the category of algebras for the step-one right factor. Because $\mathbb{F}$ is the free monad on this pointed endofunctor, it follows that $\mathcal{J}^{\boxtimes}$ is isomorphic to the category of $\mathbb{F}$-algebras.

We say an algebraic weak factorization system is **cofibrantly generated** when it is produced by the unenriched (and later, in Chapter 13, also the enriched) version of Garner's small object argument. Our use of this terminology can be thought of as a categorification of the previous notion, which adds extra structure (the (co)monad and the category of (co)algebras) and new examples (such as Example 12.5.11, revisited in Example 12.6.7). Supporting our expansion of this terminology is the fact that the theorems of Chapter 11 admit the obvious categorifications. For example, we can extend 11.1.6:

Lemma 12.6.4   *Suppose* $F: \mathcal{M} \rightleftarrows \mathcal{N}: U$ *is an adjunction, and let* $\mathcal{J}$ *be any small category of arrows in* $\mathcal{K}$. *The following diagram of categories is a pullback:*

$$
\begin{array}{ccc}
(F\mathcal{J})^{\boxtimes} & \xrightarrow{\quad U \quad} & \mathcal{J}^{\boxtimes} \\
\downarrow & & \downarrow \\
\mathcal{N}^2 & \xrightarrow[\quad U \quad]{} & \mathcal{M}^2
\end{array}
\qquad (12.6.5)
$$

Example 12.6.6   Let $\mathcal{J}_{\mathcal{D}} = \{\mathcal{D}(d, -) \cdot j \mid j \in \mathcal{J}, d \in \mathcal{D}\}$ be the set of generators for $\mathcal{K}^{\mathcal{D}}$ defined in the proof of 12.3.2, where $\mathcal{J}$ is a set of arrows in $\mathcal{K}$. Then, by adjunction, the category $\mathcal{J}_{\mathcal{D}}^{\boxtimes}$ is isomorphic to the category of natural transformations whose components are equipped with lifting functions against $\mathcal{J}$.

Example 12.6.7   Now extend the set $\mathcal{J}_{\mathcal{D}}$ to the category $\mathcal{D}^{\mathrm{op}} \times \mathcal{J}$ defined in Example 12.5.11. The category $(\mathcal{D}^{\mathrm{op}} \times \mathcal{J})^{\boxtimes}$ is the category of natural transformations equipped with natural lifts against $\mathcal{J}$ in the sense described there.

Depending on $\mathcal{D}$ and $\mathcal{J}$, this category might have many forms. For example, consider the folk model structure on **Cat** with generating cofibrations

$\mathcal{I}$ described in Example 12.5.9 and a single generating trivial cofibration $\mathcal{J} = \{* \to \mathbb{I}\}$ as described in Example 11.3.9. The 2-category $\mathbf{Cat}^{\mathcal{D}}$ of functors, natural transformations, and modifications inherits a model structure with its fibrations and weak equivalences defined representably [48]. The constituent weak factorization systems in this model structure are proven to be non-cofibrantly generated in the classical sense, but they are in this new sense. These (algebraic) weak factorization systems are generated by $\mathcal{D}^{\mathrm{op}} \times \mathcal{J}$ and $\mathcal{D}^{\mathrm{op}} \times \mathcal{I}$.

Exercise 12.6.8 Let $\mathcal{D}$ be a small category. Find generators for an algebraic weak factorization system on $\mathbf{Set}^{\mathcal{D}}$ whose right class consists of epimorphisms admitting a section (in $\mathbf{Set}^{\mathcal{D}}$).

For technical set-theoretical reasons, weak factorization systems are more often cofibrantly generated than "fibrantly generated." For this reason, coalgebras for the comonad of an algebraic weak factorization system tend to be rather more complicated than algebras for the monad. The simplicity of the following example is not typical.

Example 12.6.9 Let $\mathbb{C}$ be the comonad generated by $\{\partial\Delta^n \to \Delta^n\}_{n\geq 0}$ on **sSet**. A map is a $\mathbb{C}$-coalgebra if and only if it is a monomorphism, in which case it admits a unique $\mathbb{C}$-coalgebra structure. Write $A \to Ef \to B$ for the factorization of a monomorphism f described in Example 12.5.7. The $\mathbb{C}$-coalgebra structure is given by a map $B \to Ef$ with the following description. This map fixes the simplicial subset A. Each vertex of B not contained in A is sent to the vertex attached in step one of the small object argument. Each 1-simplex not in A is sent the unique 1-simplex attached in step two to the vertices to which we have just mapped its boundary. And so on.

Part of the complication arises from a subtlety in the definition of the underlying weak factorization system of an algebraic weak factorization system. At issue is that in general, neither the categories of algebras for the monad or coalgebras for the comonad of an algebraic weak factorization system $(\mathbb{C}, \mathbb{F})$ are closed under retracts, as is required by the definition of 11.2.3. This has to do with Lemma 12.4.4 and Lemma 12.4.7: (co)algebra structures for the pointed endofunctors of a functorial factorization are necessary and sufficient for a map to have the appropriate lifting property.

With this in mind, the left and right classes of the **underlying weak factorization system** are defined to be the retract closures of the classes of maps admitting $\mathbb{C}$-coalgebra and $\mathbb{F}$-algebra structures. In other words, the left and right classes consist of all maps that admit $(C, \vec{\epsilon})$-coalgebra structures and $(F, \vec{\eta})$-algebra structures. When $(\mathbb{C}, \mathbb{F})$ is cofibrantly generated, $\mathbb{F}$-**alg** is closed under retracts because the category $\mathcal{J}^{\boxtimes}$ is. Thus all maps in the right class are algebras for the monad. But even in the cofibrantly generated case, it is not necessarily true that all left maps are comonad coalgebras.

Exercise 12.6.10   Describe the algebras and coalgebras for the algebraic weak factorization systems from Exercise 12.5.10. For the latter example, find another "naturally occurring" algebraic weak factorization system with the same underlying weak factorization system. (Hint: its factorization is in some sense "dual" to the factorization in the first example.)

We call an element of the left class of the underlying weak factorization system of an algebraic weak factorization system $(\mathbb{C}, \mathbb{F})$ **cellular** if it admits the structure of a $\mathbb{C}$-coalgebra. In Example 12.6.9, all left maps are cellular, but in general, this is not the case.

Example 12.6.11   The map $0 \to R$ generates an algebraic weak factorization system on $\mathbf{Mod}_R$ whose right class consists of the epimorphisms and whose left class consists of those injective maps with projective cokernel. By contrast, the cellular maps are the injective maps with free cokernel. Unlike in Example 12.6.9, coalgebra structures are not unique.

Example 12.6.12   The name "cellular" was motivated by the algebraic weak factorization system $(\mathbb{C}, \mathbb{F})$ generated by $\{S^{n-1} \to D^n\}_{n \geq 0}$ on **Top**. We prove in Theorem 12.8.4 that the $\mathbb{C}$-coalgebras are the relative cell complexes and coalgebra structures are cellular decompositions; generic elements in the left class are retracts of relative cell complexes.

It is reasonable to ask why it is worth distinguishing cellular cofibrations from generic cofibrations. Appealing to authority, we might note that Quillen's original definition of a model category differs from the modern notion because it does *not* require the cofibrations and fibrations to be closed under retracts, presumably with Example 12.6.12 in mind. Another justification is given by the closure properties expressed by the following lemma, which categorifies (and extends) Lemma 11.1.4.

Lemma 12.6.13   *Let $(\mathbb{C}, \mathbb{F})$ be any algebraic weak factorization system. The category of $\mathbb{C}$-coalgebras contains the isomorphisms in $\mathcal{K}$ and is closed under pushouts, colimits in $\mathcal{K}^2$, and (transfinite) vertical composition in the sense that each of these maps inherits a canonical $\mathbb{C}$-coalgebra structure. Furthermore, the $\mathbb{C}$-coalgebra structures assigned to pushouts, colimits, and composites are such that the pushout square, colimits cone, and canonical map from the first arrow in the composable pair to the composite are maps of $\mathbb{C}$-coalgebras.*

Proof   It is easy to see that isomorphisms admit a unique $\mathbb{C}$-coalgebra structure. We leave it as an exercise to the reader to verify that the pushout (in $\mathcal{K}$) of a $\mathbb{C}$-coalgebra is again (canonically) a $\mathbb{C}$-coalgebra. Closure under colimits in the arrow category is a consequence of the general categorical theorem that says that the forgetful functor $\mathbb{C}$-**coalg** $\to \mathcal{K}^2$ creates all colimits [50, VI.2]. Closure under vertical composition will be discussed in Section 12.7. $\square$

One of these closure properties has obvious applications in model category theory. It is well known that colimits of cofibrations are not necessarily cofibrations. However, if the (cofibration, trivial fibration) weak factorization system is an algebraic weak factorization system, and if the cofibrations in the diagram admit $\mathbb{C}$-coalgebra structures so that the maps in the diagram are maps of $\mathbb{C}$-coalgebras, then the colimit admits a canonical $\mathbb{C}$-coalgebra structure and is in particular a (cellular) cofibration. Note that these remarks hold for *all* colimits in the arrow category, including coequalizers, which seem less likely to preserve cofibrations.

Example 12.6.14 The cofibrations, displayed vertically, in the pushout diagram

$$
\begin{array}{ccccc}
D^n & \longleftarrow & S^{n-1} & \longrightarrow & * \\
\downarrow & & \downarrow & & \downarrow \\
S^n & \longleftarrow & D^n & \longrightarrow & S^n
\end{array}
$$

in Quillen's model structure on **Top** are cellular, and both commutative squares, being pushout squares, are maps of $\mathbb{C}$-coalgebras by Lemma 12.6.13. Hence the pushout is also a cofibration.

Similarly, the cofibrations in the pushout diagram

$$
\begin{array}{ccccc}
D^n & \longleftarrow & * & \longrightarrow & * \\
\| & & \downarrow & & \| \\
D^n & \longleftarrow & S^{n-1} & \longrightarrow & *
\end{array}
$$

are cellular, but in this case neither square is a map of $\mathbb{C}$-coalgebras. There are multiple cellular decompositions possible for the middle map; for example, we might attach a single n-cell to the point. But neither of the outside vertical maps are obtained by attaching an n-cell; hence the squares are not maps of coalgebras, and indeed the pushout in this case is not a cofibration.

The final closure property of Lemma 12.6.13 is worthy of closer consideration.

12.7 Composing algebras and coalgebras

The categories $\mathbb{C}$-**coalg** and $\mathbb{F}$-**alg** associated to any algebraic weak factorization system are in fact **double categories**; in particular, each is equipped with an associative "vertical" composition law for (co)algebras and (co)algebra maps. Furthermore, and somewhat unexpectedly, this vertical composition law on either $\mathbb{C}$-**coalg** or $\mathbb{F}$-**alg** completely determines the algebraic weak factorization system, allowing one to recognize examples that "occur in the wild" by examining the category of algebras or coalgebras only. In particular, this leads to an alternative form of Garner's small object argument, which we illustrate below by means of the relative cell complexes described in Example 12.6.12.

We will not go into detail about how this composition is defined in general (for this, see [71, §2.5], [73, §5.1]), contenting ourselves instead with an illustrative example.

Example 12.7.1 Suppose $(\mathbb{C}, \mathbb{F})$ is generated by a small category $\mathcal{J}$, so that $\mathbb{F}$-**alg**$\cong \mathcal{J}^{\boxtimes}$ over $\mathcal{K}^2$; recall the notation introduced in 12.6.3. The vertical composition of $\mathbb{F}$-algebras is defined as follows. If (f, ϕ_f), $(g, \phi_g) \in \mathcal{J}^{\boxtimes} \cong \mathbb{F}$-**alg** with cod $f =$ dom g, define a lifting function $\phi_g \bullet \phi_f$ for the composite gf by

$$\phi_g \bullet \phi_f(j, a, b) := \phi_f(j, a, \phi_g(j, fa, b))$$

$$(12.7.2)$$

Observe that this definition is strictly associative.

This vertical composition preserves morphisms of algebras in the following sense. If $(u, v) \colon f \Rightarrow f'$ and $(v, w) \colon g \Rightarrow g'$ are algebra maps, then so is $(u, w) \colon gf \Rightarrow g'f'$ when these composites are assigned the algebra structures just defined. This is what it means for $\mathcal{J}^{\boxtimes}$ to be a double category. Indeed, (12.6.5), already a categorification of 11.1.6, can be "doubly categorified": it is a pullback of double categories.

The following theorem describes a composition criterion due to Garner that allows one to recognize algebraic weak factorization systems "in the wild."

Theorem 12.7.3 ([71, 2.24]) *Let $\mathbb{F}$ be a monad on $\mathcal{K}^2$ that preserves codomains. If the category $\mathbb{F}$-**alg** admits a vertical composition law that gives* *this category the structure of a double category, then this structure defines an algebraic weak factorization system with monad* $\mathbb{F}$. *Conversely, the category of algebras for the monad of an algebraic weak factorization system always has this structure.*

Example 12.7.4 The Moore paths functorial factorization of [51, 56] is an algebraic weak factorization system occurring "in the wild." The maps admitting algebra structures for the pointed endofunctor underlying its right factor are precisely the Hurewicz fibrations. The category of algebras for the monad admits a vertical composition law, which composes *transitive* path lifting functions; see [3, §3].

Remark 12.7.5 As suggested by this example, the composition criterion presented by Theorem 12.7.3 is particularly useful in the non-cofibrantly generated case. The category of algebras for an algebraic weak factorization system may be pulled back along the right adjoint of an adjunction, generalizing Lemma 12.6.4; the pullback inherits a vertical composition law. This putative category of algebras is monadic if and only if its forgetful functor admits a left adjoint, in which case this category specifies an algebraic weak factorization system.

Example 12.7.6 The Hurewicz fibrations, the fibrations for the Strøm model structure on **Top** (see Example 11.3.6), are defined by a right lifting property against a proper *class* of maps (see Example 11.1.3). Indeed, this model structure is not cofibrantly generated [67]. Nevertheless, the algebraic perspective can be used to construct appropriate functorial factorizations. The interest in this construction is not for the Strøm model structure, for which other factorizations (such as the one described in Example 12.7.4) exist, but for generalized Hurewicz-type model structures on topologically bicomplete categories.

The key insight is that there is a generic lifting problem characterizing the Hurewicz fibrations. Using exponential notation for the cotensor, we can define the mapping path space associated to $f : X \to Y$ to be the pullback

$$
\begin{array}{ccc}
Nf & \xrightarrow{\chi_f} & Y^I \\
\phi_f \downarrow & \lrcorner & \downarrow p_0 \\
X & \xrightarrow{f} & Y
\end{array}
$$

Note that this homotopy limit, constructed in Example 6.5.2, makes sense in any topologically bicomplete category. It turns out that the map f is a Hurewicz fibration if and only if there exists a solution to the equivalent generic lifting problems

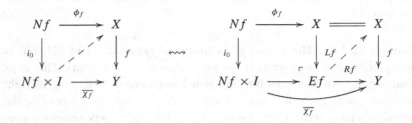

The proof of this claim stems from the observation that Nf represents the functor $\mathbf{Top}^{\mathrm{op}} \to \mathbf{Set}$ that sends a space A to the set of lifting problems between $i_0 \colon A \to A \times I$ and f.

By this observation and the Yoneda lemma, it follows that the category of algebras for the pointed endofunctor R is isomorphic to the category $\mathcal{J}^{\boxtimes}$, where $\mathcal{J}$ is the *category* whose objects are arrows $i_0 \colon A \to A \times I$ and whose morphisms are defined so that the domain-projection functor $\mathcal{J} \to \mathbf{Top}$ is an equivalence. Because $\mathbf{Top}$ satisfies a particular set theoretical hypothesis, we can form the free monad $\mathbb{F}$ on R as described in 12.4.8. On account of the isomorphism $\mathbb{F}\text{-}\mathbf{alg} \cong \mathcal{J}^{\boxtimes}$ and Example 12.7.1, the category of $\mathbb{F}$-algebras admits a vertical composition law. Hence Theorem 12.7.3 implies that this monad is the right factor of an algebraic weak factorization system. By construction, the algebras for the monad are the Hurewicz fibrations; hence the coalgebras for the comonad are trivial cofibrations in the model structure. See [3] for a complete discussion.

12.8 Algebraic cell complexes

Let (L, R) be any functorial factorization on $\mathcal{K}$. As mentioned in Remark 12.7.5, by Jon Beck's monadicity theorem, the forgetful functor from the category of algebras for the pointed endofunctor R to $\mathcal{K}^2$ is monadic if and only if it admits a left adjoint, in which case the monad of this adjunction is the free monad on R. In this way, we can think of the algebraic small object argument as a construction of an adjoint for the forgetful functor associated to the right factor in the step-one factorization. More generally, Theorem 12.7.3 allows us to recognize when a monadic or comonadic category over $\mathcal{K}^2$ encodes an algebraic weak factorization system. Depending on how the category of algebras or coalgebras is presented, the most intricate stage in the proof is likely the construction of the appropriate adjoint. In this section, we explore these ideas in an important special case.

Consider the set $\{S^{n-1} \to D^n\}_{n \geq 0}$ of generating cofibrations for Quillen's model structure on topological spaces. By the first universal property of Theorem 12.6.2, these maps are canonically coalgebras for the comonad $\mathbb{C}$ of the algebraic weak factorization system generated by this set. It follows from Lemma 12.6.13 that all relative cell complexes are cellular cofibrations: a relative cell complex structure factors a given map as a sequential composite of pushouts of coproducts of these canonical $\mathbb{C}$-coalgebras, and each of these pieces inherits a canonical $\mathbb{C}$-coalgebra structure.[6] What is less obvious is that all cellular cofibrations are relative cell complexes. This is what we show via a clever encoding of the algebraic small object argument following work of Thomas Athorne [1].

A **stratum** is a space X together with a set S of cells e_n, which are understood to come with specified attaching maps $S^{n-1} \to X$ of appropriate dimension. The space X is also referred to as the **boundary** of the stratum. Its associated **body** is the space formed by the pushout

$$
\begin{array}{ccc}
\coprod\limits_{e_n \in S} S^{n-1} & \longrightarrow & X \\
\downarrow & & \downarrow \\
\coprod\limits_{e_n \in S} D^n & \xrightarrow{\quad\ulcorner\quad} & \overline{X}
\end{array}
$$

Speaking colloquially, we may refer to the map $X \to \overline{X}$ as the stratum, but the cells and attaching maps should also be specified. We might think of $X \to \overline{X}$ as an element of $\mathbf{Top}^2$ and the stratum structure an an extra bit of coalgebraic data; indeed, we shortly make use of the forgetful functor $\mathbf{Strata} \to \mathbf{Top}^2$.

A morphism of strata $(X, S) \to (X', S')$ is a continuous map $f : X \to X'$ together with a set function $g : S \to S'$ such that the attaching map for $g(e_n)$ is the composite of f and the attaching map for e_n.

Example 12.8.1 Each $S^{n-1} \to D^n$ is canonically a stratum with a single cell and the identity attaching map.

---

[6] The sequential composite does so because it can be expressed as a colimit of a diagram of $\mathbb{C}$-coalgebras:

$$
\begin{array}{ccccccccc}
\cdot & = & \cdot & = & \cdot & = & \cdot & \cdots & \cdot \\
\Big\downarrow{\scriptstyle f_1} & & \Big\downarrow{\scriptstyle f_2 f_1} & & \Big\downarrow{\scriptstyle f_3 f_2 f_1} & & \Big\downarrow{\scriptstyle f_4 f_3 f_2 f_1} & & \Big\downarrow{\scriptstyle \mathrm{colim}=\cdots f_4 f_3 f_2 f_1} \\
\cdot & \xrightarrow{\ f_2\ } & \cdot & \xrightarrow{\ f_3\ } & \cdot & \xrightarrow{\ f_4\ } & \cdot & \cdots & \cdot
\end{array}
$$

Note that the category of strata is closed under pushouts and colimits in $\mathbf{Top}^2$ [1, 3.3–4], the same closure properties of Lemma 12.6.13 bar one. This leads us to define an **algebraic cell complex**.

Definition 12.8.2   An **algebraic cell complex** (X_k, S_k) is a countable sequence of strata that is **connected** (the body of each stratum of the sequence is the boundary of the next) and **proper** (no attaching map factors through a previous boundary).

In other words, an algebraic cell complex is a relative cell complex with a specified cellular decomposition. Morphisms of algebraic cell complexes must preserve these decompositions. More precisely, a morphism of algebraic cell complexes $(f_k, g_k): (X_k, S_k) \to (X_k', S_k')$ is a collection of morphisms of strata that are compatible in the sense that the map $f_{k+1}: X_{k+1} \to X_{k+1}'$ is the pushout of $f_k: X_k \to X_k'$ and the morphisms induced by the map of cells $g_k: S_k \to S_k'$, as displayed:

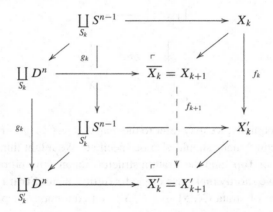

With these definitions, one can check that the category of algebraic cell complexes is closed under pushouts, colimits in $\mathbf{Top}^2$, and vertical composition [1, 4.5–8]. This last proof requires a bit of point-set topology, as becomes apparent when we try to compose a stratum (Y, T) onto an algebraic cell complex (X_k, S_k) whose body $\mathrm{colim}_k X_k$ is Y. Because the maps $X_k \to X_{k+1}$ are closed, T_1-inclusions and the boundaries of cells are compact, each attaching map $S^{n-1} \to Y$ factors uniquely through some X_k where k is minimal. This partitions the set T of cells for Y into sets $T = T_0 \sqcup T_1 \sqcup T_2 \sqcup \cdots$ of cells which could be attached to $X_0, X_1, X_2, \ldots$ The vertical composite of the algebraic cell complex (X_k, S_k) with (Y, T) is the algebraic cell complex $(Z_k, S_k \sqcup T_k)$ where $Z_0 = X_0$ and the subsequent Z_k are defined to be the obvious pushouts. The vertical composite of two algebraic cell complexes is formed by iterating this procedure for each stratum in the second cell complex.

Each algebraic cell complex is a relative cell complex, and furthermore, any relative cell complex underlies an algebraic cell complex. The algebraic small object argument reappears in the proof of the following theorem.

Theorem 12.8.3 (Athorne) *The forgetful functor U: **AlgCellCx** $\to$ **Top**^{2} from algebraic cell complexes to the arrow category for spaces has a right adjoint C.*

Proof The first step is to note that the forgetful functor U: **Strata** $\to$ **Top**^{2} has a right adjoint L. It sends a map $f: X \to Y$ to the stratum whose boundary is the space X and whose set of cells is isomorphic to the union over $\{S^{n-1} \xrightarrow{j_n} D^n\}_{n \geq 0}$ of the sets **Sq**(j_n, f). The body of Lf is obtained via the left-hand pushout

$$
\begin{array}{ccccc}
\coprod\limits_{n,\mathbf{Sq}(j_n,f)} S^{n-1} & \longrightarrow & X & =\!=\!= & X \\[2mm]
\coprod\limits_{n,\mathbf{Sq}(j_n,f)} \Big\downarrow & & \Big\downarrow {\scriptstyle Lf} & & \Big\downarrow {\scriptstyle f} \\[4mm]
\coprod\limits_{n,\mathbf{Sq}(j_n,f)} D^n & \longrightarrow & \overline{X} & \xrightarrow{\ Rf\ } & Y
\end{array}
$$

The right-hand square displays the counit of the adjunction.

To define the right adjoint to U: **AlgCellCx** $\to$ **Top**^{2}, one might try to iterate this procedure, forming an algebraic cell complex whose first stratum is Lf, whose second is L applied to the map $Rf: \overline{X} \to Y$, and so on. However, this sequence of strata is not proper unless we explicitly exclude from each step any cells whose attaching map factors through some previous boundary. This modification precisely distinguishes Garner's small object argument from Quillen's (see Remark 12.5.6) and also produces the desired result here. In summary, C takes f to the algebraic cell complex (X_k, S_k) with $X_0 = X$; $S_0 = \sqcup_n \mathbf{Sq}(j_n, f)$; X_1 equal to the body of Lf; S_1 equal to the subset of $\coprod_n \mathbf{Sq}(j_n, Rf)$ of squares whose attaching maps do not factor through X_0; and so on. $\qquad\square$

In other words, the algebraic small object argument reflects an arrow f into the category of relative cell complexes formed from the maps $S^{n-1} \to D^n$, at least when we remember the coalgebraic data determining each cellular decomposition. Considered carefully, this argument essentially establishes the first universal property of Theorem 12.6.2.

Theorem 12.8.4 (Athorne) *The adjunction U: **AlgCellCx** $\rightleftarrows$ **Top**^{2}: C is comonadic. Hence **AlgCellCx** is the category of coalgebras for the comonad of the algebraic weak factorization system generated by $\{S^{n-1} \to D^n\}_{n \geq 0}$.*

Proof   Comonadicity follows from one of Beck's theorems; see [1, §6]. As is evident from the proof of Theorem 12.8.3, the comonad is the one arising from the small object argument applied to the set $\{S^{n-1} \to D^n\}_{n\geq 0}$; hence the category of algebraic cell complexes is equivalent to the category of coalgebras for the comonad of the algebraic weak factorization system generated by this set. Alternatively, this last fact could be deduced from the dual of Theorem 12.7.3. Clearly the comonad on $\mathbf{Top}^2$ induced by the adjunction preserves domains. We have seen that its category of coalgebras, here the category of algebraic cell complexes, admits a vertical composition law. Hence the category of algebraic cell complexes encodes the data of an algebraic weak factorization system. Furthermore, $\mathbf{AlgCellCx}$ is universal among categories of coalgebras for an algebraic weak factorization system admitting a map from $\{S^{n-1} \to D^n\}_{n\geq 0}$ over $\mathbf{Top}^2$ [1, 8.4]. Hence Theorem 12.6.2 implies that this category encodes the algebraic weak factorization system generated by this set. $\square$

The following rephrasing of Theorem 12.8.4 justifies our use of "cellular" for the arrows in the left class of an algebraic weak factorization system admitting a coalgebra structure for the comonad.

Corollary 12.8.5   *The cellular cofibrations for the algebraic weak factorization system generated by $\{S^{n-1} \to D^n\}_{n\geq 0}$ on* $\mathbf{Top}$ *are precisely the relative cell complexes.*

12.9 Epilogue on algebraic model categories

It is clear that in any cofibrantly generated model category that permits the small object argument, the constituent weak factorization systems can be upgraded to algebraic weak factorization systems, which might be denoted by $(\mathbb{C}_t, \mathbb{F})$ and $(\mathbb{C}, \mathbb{F}_t)$. But we have not really explored the role that algebraic weak factorization systems play in model categories. This is a rather long story [71, 73]; here we content ourselves with just a brief taste. A main theme is that, as a consequence of a generalized version of the first universal property in Theorem 12.6.2, a number of complicated algebraic structures are determined by a simple "cellularity condition" having to do with the generating (trivial) cofibrations.

Here is a very simple example of this principle. If the generating trivial cofibrations are cellular cofibrations,[7] then there exists a canonical natural

---

[7] The generating trivial cofibrations can always be replaced by other generators for an algebraic weak factorization system with the same underlying weak factorization system so that this is the case [71, 3.8].

transformation

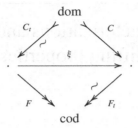

that defines a natural solution to the lifting problem between the trivial cofibration and the trivial fibration obtained by applying the two functorial factorizations to a given map. This natural transformation does much more: First, it specifies a natural solution to any lifting problem between an (algebraic) trivial cofibration and an (algebraic) trivial fibration. It also determines functors $\mathbb{C}_t$-**coalg** $\to \mathbb{C}$-**coalg** and $\mathbb{F}_t$-**alg** $\to \mathbb{F}$-**alg** that map the categories of algebraic trivial (co)fibrations into the categories of algebraic (co)fibrations. We call a model category equipped with algebraic weak factorization systems $(\mathbb{C}_t, \mathbb{F})$ and $(\mathbb{C}, \mathbb{F}_t)$ and a map of this form an **algebraic model category**.

Writing $\mathbb{Q}$ for the cofibrant replacement monad derived from $\mathbb{C}$ and $\mathbb{R}$ for the fibrant replacement monad derived from $\mathbb{F}$, there is a canonical lifting problem

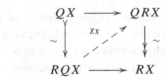

comparing the two fibrant–cofibrant replacements of an object X. In an ordinary model category, the lifting properties imply that there exists a comparison arrow from RQX to QRX. In an algebraic model category, there are canonical lifts that assemble into a natural transformation $\chi : RQ \Rightarrow QR$. This turns out to be a distributive law of the monad $\mathbb{R}$ over the comonad $\mathbb{Q}$, which implies that R lifts to a monad on the category $\mathbb{Q}$-**coalg** of **algebraic cofibrant objects**, Q lifts to a monad on the category $\mathbb{R}$-**alg** of **algebraic fibrant objects**, and the algebras for the former and coalgebras for the latter coincide, defining a category of algebraic fibrant–cofibrant objects [71, 3.5].

13

Enriched factorizations and enriched lifting properties

Often the context in which one runs the small object argument is an enriched category. These enrichments need not necessarily be topological in flavor – for instance, $\mathbf{Ch}_\bullet(R)$ is enriched over $\mathbf{Mod}_R$, provided the ring R is commutative – though we see later that the case of a simplicial model category is particularly well behaved. As defined in the previous chapter, Quillen's and Garner's small object arguments fail to produce enriched functors, but only because the initial step, which forms the coproduct of generating arrows indexed over the *sets* of commutative squares from each generator to the map to be factored, is not enriched, as illustrated by Example 3.7.17.

In this chapter, we prove that a modified version of either Garner's or Quillen's construction does produce an enriched functorial factorization. We call this the **enriched small object argument**. By design, the right factor produced by the enriched small object argument is a member of the right class of the original cofibrantly generated weak factorization system. If a certain condition is satisfied, then the left factor is a member of the left class. In this case, the unenriched factorizations of the cofibrantly generated weak factorization system can be replaced by enriched factorizations whenever it is convenient.

Perhaps more interesting is the case when the condition needed to guarantee compatibility of the left factor produced by the enriched small object argument fails. When this happens, the class of maps satisfying an **enriched lifting property** against the generating arrows $\mathcal{J}$ is strictly smaller than the class satisfying usual unenriched lifting property. In this context, the enriched Quillen small object argument is badly behaved – the maps in the image of the left factor need not lift against the maps in the image of the right factor – but the enriched Garner small object argument can be integrated with enriched lifting properties to produce **enriched weak factorization systems**.

The enriched weak factorization system $(^{\boxtimes}(\mathcal{J}^{\boxtimes}), \mathcal{J}^{\boxtimes})$ differs from the usual cofibrantly generated weak factorization system $(^{\boxtimes}(\mathcal{J}^{\boxtimes}), \mathcal{J}^{\boxtimes})$. In general, the enriched version will have a larger left class and a smaller right one. As the reader might now expect, the enriched notions of lifting are similarly well behaved with respect to adjunctions, and so on; we present enriched analogs of a number of the results from Chapters 11 and 12. In summary, we claim that the new enriched theory, while less familiar, is fundamentally no more mysterious than the classical one.

The theory of enriched weak factorization systems developed here is the key to a comparison between the two classical model structures on the category of unbounded chain complexes of modules over a commutative ring described in Example 11.3.7. Interestingly, one of these is not cofibrantly generated in the unenriched sense but becomes so in the enriched sense. We preview the comparison here and refer the reader to [2] for the full story.

13.1 Enriched arrow categories

It has been a while since we have required a serious consideration of enrichment, so we take a moment to recall a few preliminaries. Let $\mathcal{V}$ be a complete and cocomplete closed symmetric monoidal category with monoidal unit $*$. Let $\mathcal{K}$ be a complete and cocomplete category. When $\mathcal{K}$ is a $\mathcal{V}$-category, so is its arrow category $\mathcal{K}^2$. As per usual, we write $\underline{\mathcal{K}}(x, y)$ to denote the hom-object in $\mathcal{V}$ of morphisms from x to y in $\mathcal{K}$. The hom-object between $j_0 \xrightarrow{j} j_1$ and $f_0 \xrightarrow{f} f_1$ in $\underline{\mathcal{K}}^2$ is defined by the following pullback in $\mathcal{V}$:

$$
\begin{array}{ccc}
\underline{\mathbf{Sq}}(j, f) & \longrightarrow & \underline{\mathcal{K}}(j_0, f_0) \\
\downarrow & & \downarrow{\scriptstyle f_*} \\
\underline{\mathcal{K}}(j_1, f_1) & \xrightarrow{j^*} & \underline{\mathcal{K}}(j_0, f_1)
\end{array}
\tag{13.1.1}
$$

When $\mathcal{K}$ is tensored and cotensored over $\mathcal{V}$, the category $\mathcal{K}^2$ is also, with the tensors and cotensors defined pointwise (see Remark 3.8.2). These facts generalize to the category $\underline{\mathcal{K}}^{\mathcal{D}}$ of diagrams of any shape; in the general case, the hom-objects in $\underline{\mathcal{K}}^{\mathcal{D}}$ are defined using an end (7.3.2) in $\mathcal{V}$.

Recall from Definition 3.4.5 that a morphism from x to y in the underlying category of the $\mathcal{V}$-category $\underline{\mathcal{K}}$ is an arrow $f : * \to \underline{\mathcal{K}}(x, y)$ in $\mathcal{V}$. When $\mathcal{K}$ is tensored, this corresponds, by adjunction, to an arrow $f : x \cong * \otimes x \to y$ in $\mathcal{K}$. Because there is no ambiguity, we give these maps the same name.

Remark 13.1.2 Elements in the underlying set of the hom-object (13.1.1) are maps $* \to \underline{\mathbf{Sq}}(j, f)$ in $\mathcal{V}$, which correspond precisely, by the defining universal property and the Yoneda lemma, to commutative squares in $\mathcal{K}$ from j to f. That is to say,

$$\mathbf{Sq}(j, f) = \mathcal{V}(*, \underline{\mathbf{Sq}}(j, f)).$$

More precisely, assuming $\mathcal{K}$ is tensored, the map $(u, v): * \to \underline{\mathbf{Sq}}(j, f)$ in $\mathcal{V}$ classifies a commutative square $(u, v): j \Rightarrow f$ in $\mathcal{K}$ in the sense that the commutative square factors, as the composite of the component of the counit of the adjunction $- \otimes j: \mathcal{V}^2 \rightleftarrows \mathcal{K}^2: \underline{\mathbf{Sq}}(j, -)$ at f and the map defined by tensoring $* \to \underline{\mathbf{Sq}}(j, f)$ with j:

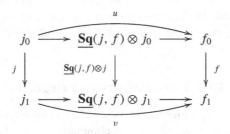

Henceforth, for convenience, we will suppose that $\mathcal{K}$ is complete and cocomplete, tensored, cotensored, and enriched. Recall from Theorem 7.5.3 that it follows that ordinary (co)limits in $\mathcal{K}$ become conical (co)limits, that is, satisfy an enriched universal property. In particular, for any small category $\mathcal{D}$, the colimit and limit functors are $\mathcal{V}$-functors $\underline{\mathcal{K}}^{\mathcal{D}} \to \underline{\mathcal{K}}$. By Corollary 7.6.4, any $\mathcal{K}$ satisfying these hypotheses admits all $\mathcal{V}$-weighted limits and colimits, constructed out of conical (co)limits and (co)tensors. This is precisely what it means to say that $\mathcal{K}$ is $\mathcal{V}$-bicomplete. We leave it to the reader to formulate the more precise statements of the results that follow required for contexts in which certain weighted limits or weighted colimits might not exist.

13.2 Enriched functorial factorizations

To simplify the statement of our main result, we sweep set-theoretic considerations under the rug and allow the meaning of "permits the small object argument" to vary as appropriate for the version of the small object argument we wish to run. The precise hypotheses are unchanged from Theorems 12.2.2 and 12.5.1.

The following folklore result is due in various forms to many people. See in particular [36, 4.3.8], [69, 6.3], and [79, 24.2]. The proof for the algebraic small argument extends to the case where the generators are taken to be a small category of arrows.

Theorem 13.2.1 (enriched small object argument) *Let $\mathcal{K}$ be a $\mathcal{V}$-bicomplete category permitting the small object argument, and let $\mathcal{J}$ be any small set of arrows. If, for all $f \in \mathcal{K}^2$,*

$(\star)$ *the map $\underline{\mathbf{Sq}}(j, f) \otimes j$ is in $^\boxtimes(\mathcal{J}^\boxtimes)$ whenever $j \in \mathcal{J}$,*

then there exist $\mathcal{V}$-enriched functorial factorizations for the weak factorization system $(^\boxtimes(\mathcal{J}^\boxtimes), \mathcal{J}^\boxtimes)$, constructed by the enriched version of either Quillen's or Garner's small object argument.

Proof The construction of each factorization is analogous. For ease of exposition, we call elements of the left class "trivial cofibrations" and elements of the right class "fibrations."

Write $L_0 \colon \underline{\mathcal{K}}^2 \to \underline{\mathcal{K}}^2$ for the $\mathcal{V}$-enriched pointwise left Kan extension of $\mathcal{J} \to \underline{\mathcal{K}}^2$ along itself defined by

$$L_0 f = \coprod_{j \in \mathcal{J}} \mathbf{Sq}(j, f) \otimes j.$$

Write (L, R) for the step-one factorization constructed by factoring the canonical square $L_0 f \Rightarrow f$, whose components are adjunct to the identity map on $\underline{\mathbf{Sq}}(j, f)$, through the indicated pushout:

$$
\begin{array}{ccccc}
\cdot & \longrightarrow & \cdot & = & \cdot \\
{\scriptstyle L_0 f} \downarrow & {\scriptstyle Lf} \downarrow & \diagup\nearrow & & \downarrow {\scriptstyle f} \\
\cdot & \longrightarrow & \cdot & \underset{Rf}{\longrightarrow} & \cdot
\end{array}
\tag{13.2.2}
$$

Because Lf and Rf are formed from tensors and conical colimits in $\mathcal{K}$, this construction defines $\mathcal{V}$-functors $L, R \colon \underline{\mathcal{K}}^2 \rightrightarrows \underline{\mathcal{K}}^2$.

We claim that if the dashed lift in (13.2.2) exists, then f is a fibration. A lift in a square $(u, v) \colon j \Rightarrow f$ is obtained by precomposing (13.2.2) with the map induced, as described in Remark 13.1.2, by $(u, v) \colon * \to \underline{\mathbf{Sq}}(j, f)$ followed by the inclusion into the appropriate component of the coproduct. Conversely, $(\star)$ implies that $L_0 f$ is a trivial cofibration, and hence by Lemma 11.1.4, the same is true of Lf. It follows that f is an algebra for the pointed endofunctor R if and only if f is a fibration.

By easy abstract nonsense involving the universal property of enriched left Kan extension, the $\mathcal{V}$-functor L_0 is a comonad; indeed, it is the enriched density comonad associated to the functor $\mathcal{J} \to \underline{\mathcal{K}}^2$ [16, §II.1]. It follows that L is also a comonad, by direct argument or by abstract nonsense involving the (pushout square, isomorphism-on-domain) orthogonal factorization system on $\mathcal{K}^2$. Because $\mathcal{K}$ permits the small object argument, the argument of

[31, 4.22] allows us to apply [31, 4.21] to conclude that the algebraic small object argument converges, producing an algebraic weak factorization system $(\mathbb{C}, \mathbb{F})$ such that the category of $\mathbb{F}$-algebras is isomorphic, over $\mathcal{K}^2$, to the category of R-algebras. The arrow Cf is a $\mathbb{C}$-coalgebra and consequently against any fibration; hence Cf is a trivial cofibration. Furthermore, the free monad on a pointed endofunctor construction, employing the $\mathcal{V}$-functors L and R and various conical colimits in $\mathcal{K}$, produces $\mathcal{V}$-functors C and F, as claimed.

Similarly, the functors in the functorial factorization produced by Quillen's small object argument, but with L_0 in place of the usual coproduct over squares at each step, are $\mathcal{V}$-functors. Write (L^ω, R^ω) for this factorization. Because each map in the image of L is a trivial cofibration, it follows that $L^\omega f$, defined to be a countable composite of such, is a trivial cofibration. It remains only to show that $R^\omega f$ is a fibration. Given any lifting problem $(a, b)\colon j \Rightarrow R^\omega f$, the map a factors through some $R^n f$ by the compactness hypothesis of Theorem 12.2.2. Hence the lifting problem $(a, b)\colon j \Rightarrow R^\omega f$ factors as

$$
\begin{array}{ccccccccc}
j_0 & \longrightarrow & \coprod_{j\in\mathcal{J}} \mathbf{Sq}(j, R^n f) \otimes j_0 & \longrightarrow & x_n & \xrightarrow{\ L R^n f\ } & x_{n+1} & \longrightarrow & x_\omega \\
\downarrow{\scriptstyle j} & & \downarrow{\scriptstyle \coprod_{j\in\mathcal{J}} \mathbf{Sq}(j,R^n f)\otimes j} & & \downarrow{\scriptstyle R^n f} & & \downarrow{\scriptstyle R^{n+1} f} & & \downarrow{\scriptstyle R^\omega f} \\
j_1 & \longrightarrow & \coprod_{j\in\mathcal{J}} \mathbf{Sq}(j, R^n f) \otimes j_1 & \longrightarrow & f_1 & =\!=\!= & f_1 & =\!=\!= & f_1
\end{array}
$$

The map $L R^n f$ is the pushout of the second vertical map along the second top horizontal map inside the second square from the left. The other leg of this pushout cone is the dotted diagonal arrow. This pushout gives rise to a solution to the lifting problem (a, b) in the usual manner. $\qquad\square$

Remark 13.2.3   Without the hypothesis $(\star)$, the argument in the proof of Theorem 13.2.1 still implies that Ff and $R^\omega f$ are fibrations. However, Cf and $L^\omega f$ might not be trivial cofibrations. We will see in Proposition 13.4.2 that regardless of whether $(\star)$ holds, the maps in the images of C and F lift against each other. However, this might not be true of the maps in the image of L^ω and R^ω, for the reasons explained in Remark 13.4.4.

We anticipate that most applications of Theorem 13.2.1 will be in the context of a model structure. For the reader's convenience, we state an immediate corollary in that language. The main point is that the condition $(\star)$ holds if tensoring with any object of $\mathcal{V}$ preserves cofibrations and trivial cofibrations. This is the case, in particular, if the tensor–cotensor–hom is a Quillen two-variable adjunction (or even if the correct two-thirds of this SM7 axiom hold), as is necessarily true, for instance, if $\mathcal{K}$ is a $\mathcal{V}$-model category and if all objects of $\mathcal{V}$ are cofibrant.

Corollary 13.2.4   *Suppose $\mathcal{K}$ is a $\mathcal{V}$-bicomplete category and a cofibrantly generated model category for which $v \otimes -$ is a left Quillen functor for each $v \in \mathcal{V}$. Then if $\mathcal{K}$ permits the small object argument, there exist $\mathcal{V}$-enriched functorial factorizations for the model structure. In particular, if $\mathcal{K}$ permits Garner's small object argument, then it has a $\mathcal{V}$-enriched cofibrant replacement comonad and fibrant replacement monad.*

An example will illustrate how the factorizations produced by the enriched small object argument differ from their unenriched analogs.

Example 13.2.5   Let R be a commutative ring, and consider the category $\mathbf{Ch}_{\bullet}(R)$ and the generators $\mathcal{J} = \{0 \to D^n\}_{n \in \mathbb{Z}}$ of Example 12.5.4. The category $\mathbf{Ch}_{\bullet}(R)$ admits many enrichments. Here we consider its enrichment over $\mathbf{Mod}_R$. The set of chain maps between two fixed chain complexes has an R-module structure with addition and scalar multiplication defined as in the codomain. This enriched category has tensors and cotensors, obtained by applying the functors $M \otimes_R -$ and $\underline{\mathbf{Mod}}_R(M, -)$ associated to an R-module M objectwise. One way to prove all these facts simultaneously is to appeal to Theorem 3.7.11: $\mathbf{Ch}_{\bullet}(R)$ is a closed monoidal category, and there is a strong monoidal adjunction $\mathbf{Mod}_R \rightleftarrows \mathbf{Ch}_{\bullet}(R)$ whose left adjoint embeds an R-module as a chain complex concentrated in degree zero and whose right adjoint takes 0-cycles.

An easy calculation shows that the R-module of commutative squares from $0 \to D^n$ to $X_{\bullet} \to Y_{\bullet}$ is Y_n; its underlying set is then the underlying set of this R-module, as noted in Example 12.5.4. The modification performed by the enriched small object argument replaces the direct sum over the underlying set of Y_n in (12.5.5) with a tensor with the R-module Y_n. Write $D^n_{Y_n}$ for the chain complex $Y_n \otimes D^n$. It has the R-module Y_n in degrees n and $n - 1$, with identity differential, and zeros elsewhere. For the reasons explained in Example 12.5.4, the enriched algebraic small object argument converges at step one to produce the functorial factorization

$$X_{\bullet} \longrightarrow X_{\bullet} \oplus (\oplus_n D^n_{Y_n}) \longrightarrow Y_{\bullet} \ . \qquad (13.2.6)$$

Unlike the rather unwieldy functorial factorization produced by Quillen's small object argument in this case, this factorization is quite familiar: it is the mapping path space factorization described for this model structure in [58, 18.3.7].

Note, however, that the condition $(\star)$ is not satisfied by non-projective R-modules. Indeed, when Y is not degreewise projective, the first map in the factorization (13.2.6) is not a trivial cofibration (quasi-isomorphism and point-wise injection with projective cokernel) in the Quillen-type model structure. Instead, (13.2.6) factors any map $X_\bullet \to Y_\bullet$ as a trivial cofibration followed by a fibration for the Hurewicz-type model structure on $\mathbf{Ch}_\bullet(R)$. We will expand on this point when we return to this example in 13.4.5, where we prove that while $\mathcal{J}$ generates the weak factorization system for Quillen-type fibrations in the usual sense, it also generates the weak factorization system for Hurewicz-type fibrations in the $\mathbf{Mod}_R$-enriched sense.

To make sense of the just-mentioned notion of an enriched weak factorization system, we must first introduce enriched lifting properties.

13.3 Enriched lifting properties

The algebraic small object argument allows us to expand the notion of a cofibrantly generated (algebraic) weak factorization system to allow the generators to be a small category, rather than simply a set, of arrows. These morphisms can be used to impose coherence conditions on the right class of maps so classified, as illustrated in 12.5.12 and 12.6.8.

We now argue that the notion of "cofibrantly generated" might also be productively expanded to allow the lifting properties of 11.2.6 to be interpreted in an enriched sense. A primary example is the weak factorization system in $\mathbf{Ch}_\bullet(R)$ just mentioned, whose right class is the class of Hurewicz fibrations. This weak factorization system is not cofibrantly generated in the classical sense, at least for the ring $\mathbb{Z}$ [12, 5.12]; Hurewicz fibrations cannot be characterized by a right lifting property against a set of arrows. However, they are precisely the class of maps satisfying a $\mathbf{Mod}_R$-enriched lifting property against a particular set of maps, indeed, the same generating set of Example 13.2.5.

To understand what this means, it is productive to think algebraically. To say a set of arrows $\mathcal{J}$ generates a weak factorization system $(\mathcal{L}, \mathcal{R})$ is to say that a map $f \colon X \to Y$ is in the right class $\mathcal{R}$ if and only if certain additional structure exists, namely, solutions to any lifting problem between $\mathcal{J}$ and f. For example, in the category of spaces, if we take $\mathcal{J} = \{\emptyset \to *\}$, then this data takes the form of a specified fiber point above each $y \in Y$.

By contrast, we say that f satisfies the enriched lifting property against $\mathcal{J}$ when this lifting data can be enriched, that is, described by appropriate morphisms in the base category $\mathcal{V}$. Here f satisfies the **Top**-enriched lifting property against $\emptyset \to *$ just when it admits a continuous section. As before, to say that a set $\mathcal{J}$ generates a weak factorization system $(\mathcal{L}, \mathcal{R})$ in the enriched sense means that $f \in \mathcal{R}$ if and only if it satisfies the enriched lifting property against $\mathcal{J}$.

We now give the formal definition.

Definition 13.3.1 Let $j \colon j_0 \to j_1$ and $f \colon f_0 \to f_1$ be arrows in a $\mathcal{V}$-category $\mathcal{K}$. We say that j and f satisfy the $\mathcal{V}$-**enriched lifting property** and write $j \boxtimes f$ to mean that there is a section to the canonical map

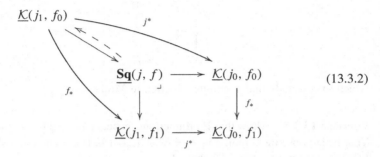

$$(13.3.2)$$

or equivalently, in the case where $\mathcal{K}$ is tensored, by Exercise 11.1.9, that there exists a solution to the generic lifting problem

$$\begin{array}{ccc}
\underline{\mathbf{Sq}}(j, f) \otimes j_0 & \longrightarrow & f_0 \\
{\scriptstyle \underline{\mathbf{Sq}}(j,f)\otimes j} \downarrow & \nearrow & \downarrow {\scriptstyle f} \\
\underline{\mathbf{Sq}}(j, f) \otimes j_1 & \longrightarrow & f_1
\end{array} \qquad (13.3.3)$$

For any set of arrows $\mathcal{J}$, we write $\mathcal{J}^{\boxtimes}$ for the (unenriched) category of arrows equipped with a lifting function in the sense of (13.3.2) or (13.3.3). As the context dictates, we also allow $\mathcal{J}^{\boxtimes}$ to denote the objects in the image of the forgetful functor $\mathcal{J}^{\boxtimes} \to \mathcal{K}^2$, that is, for the class of morphisms admitting the enriched right lifting property against each $j \in \mathcal{J}$.

Applying the underlying set functor, we see that any $f \in \mathcal{J}^{\boxtimes}$ is also in $\mathcal{J}^{\boxtimes}$; put more categorically, the forgetful functor $\mathcal{J}^{\boxtimes} \to \mathcal{K}^2$ factors through the forgetful functor $\mathcal{J}^{\boxtimes} \to \mathcal{K}^2$. However, a map satisfying the ordinary right lifting property against $\mathcal{J}$ need not satisfy the enriched right lifting property. The underlying **Set**-based version of (13.3.2) asserts that any lifting problem between j and f has a solution. The topological version, say, asserts that there is a continuous function, the dotted arrow in (13.3.2), specifying solutions to lifting problems.

Example 13.3.4 Let R be a commutative ring. The right class of the weak factorization system on $\mathbf{Mod}_R$ satisfying the right lifting property against $0 \to R$

consists of the epimorphisms. The right class satisfying the $\mathbf{Mod}_R$-enriched lifting property consists of those epimorphisms admitting a section that is a map of R-modules. This claim follows from the fact that 0 and R represent the terminal and identity $\mathbf{Mod}_R$-functors, respectively. Given an R-module homomorphism $f : M \to N$, the R-module of commutative squares from $0 \to R$ to f is just N. Hence, in this case, the diagram (13.3.3) reduces to

which says exactly that f admits a section in $\mathbf{Mod}_R$.

Exercise 13.3.5   Show that $\mathcal{V}$-adjunctions respect enriched lifting properties. That is, suppose the adjunction $F \dashv U$ of 12.6.4 is $\mathcal{V}$-enriched and show that $(F\mathcal{J})^{\boxtimes}$ is the pullback of $\mathcal{J}^{\boxtimes}$ along U.

Definition 13.3.1 can be used to explain the condition $(\star)$ appearing in the statement of the enriched small object argument: under this hypothesis, the enriched lifting property and the unenriched lifting property define the same class of maps.

Lemma 13.3.6   *Let $(\mathcal{L}, \mathcal{R})$ be a weak factorization system in a tensored $\mathcal{V}$-category $\mathcal{K}$. If*

$(\star)$ *for each $v \in \mathcal{V}$, the functor $v \otimes -$ preserves the left class,*

then every map that has the right lifting property with respect to $\mathcal{L}$ also has the $\mathcal{V}$-enriched right lifting property with respect to $\mathcal{L}$, that is, $\mathcal{L}^{\boxtimes} = \mathcal{L}^{\underline{\boxtimes}}$. In particular, $\mathcal{L}\boxtimes\mathcal{R}$.

Proof   The hypothesis implies that any $f \in \mathcal{L}^{\boxtimes}$ lifts against the left-hand arrow of (13.3.3) for any $j \in \mathcal{L}$. $\square$

Example 13.3.7   Consider $\mathcal{J} = \{\emptyset \to \Delta^0\}$ in the category of simplicial sets. By the Yoneda lemma, maps in $\mathcal{J}^{\boxtimes}$ are surjective on 0-simplices. By the simplicially enriched Yoneda lemma, maps in $\mathcal{J}^{\underline{\boxtimes}}$ are split epimorphisms. Note that the weakly saturated class generated by $\emptyset \to \Delta^0$ is not closed under tensors, so Lemma 13.3.6 does not apply.

By contrast, the weakly saturated closure of $\mathcal{I} = \{\partial\Delta^n \to \Delta^n\}_{n\geq 0}$ is closed under tensors, so Lemma 13.3.6 implies that $\mathcal{I}^{\boxtimes} = \mathcal{I}^{\underline{\boxtimes}}$. The definition (13.3.2) gives the following characterization of the trivial fibrations in Quillen's model structure on simplicial sets: a map $X \to Y$ is a trivial fibration if and only if the pullback hom

$$X^{\Delta^n} \to X^{\partial \Delta^n} \times_{Y^{\partial \Delta^n}} Y^{\Delta^n} \qquad (13.3.8)$$

is a split epimorphism for each n. Because Quillen's model structure is a simplicial model structure, the trivial fibrations satisfy an even stronger condition: each of the maps (13.3.8) is itself a trivial fibration.

As illustrated by Examples 13.3.4 and 13.3.7, when the condition $(\star)$ fails, the classes $\mathcal{J}^{\boxtimes}$ and $\mathcal{J}^{\boxminus}$ differ. The remainder of this section is devoted to comparing the enriched and unenriched lifting property in this general case.

Exercise 13.3.9 Show that in a cotensored $\mathcal{V}$-category the conditions specified by (13.3.2) and (13.3.3) are weakly saturated in the first variable, that is, closed under isomorphisms, coproducts, pushouts, retracts, and transfinite composition. Conclude that any map in the left class of the (unenriched) weak factorization system generated by $\mathcal{J}$ will also satisfy the enriched lifting property against those maps that have the enriched lifting property against $\mathcal{J}$.

Employing our symbolic notation, the conclusion of Exercise 13.3.9 is that

$$^{\boxtimes}(\mathcal{J}^{\boxtimes}) \boxminus \mathcal{J}^{\boxminus}. \qquad (13.3.10)$$

The proposition encoded by (13.3.10) implies that $^{\boxtimes}(\mathcal{J}^{\boxtimes}) \subset ^{\boxminus}(\mathcal{J}^{\boxminus})$, but the inclusion is not generally an equality. The class on the right-hand side satisfies one additional closure property not enjoyed by the left-hand side, namely, closure under tensors with objects of $\mathcal{V}$.

Lemma 13.3.11 *If $\mathcal{K}$ is a $\mathcal{V}$-category, then any class of maps of the form $^{\boxminus}\mathcal{R}$ contains the isomorphisms and is closed under retracts, finite composition, and tensors with objects of $\mathcal{V}$, whenever they exist. If, in addition, $\mathcal{K}$ is a cotensored $\mathcal{V}$-category, then $^{\boxminus}\mathcal{R}$ is also closed under coproducts, pushouts, and transfinite composition.*

In proving this lemma, which should be compared with Lemma 11.1.4, we also solve Exercise 13.3.9. Recall that when $\mathcal{K}$ is a cotensored $\mathcal{V}$-category, Theorem 7.5.3 implies that ordinary colimits in $\mathcal{K}$ satisfy an enriched universal property, expressed by an isomorphism in $\mathcal{V}$.

Proof It suffices to show, for a fixed morphism f, that the class of morphisms j satisfying $j \boxminus f$ has these closure properties. The class $^{\boxminus}\mathcal{R}$ is defined to be an intersection of conditions of this form. The arguments are enrichments of the proofs in the unenriched case. We illustrate with three examples.

Suppose that $j \boxminus f$ and that $\mathcal{K}$ admits tensors by $A \in \mathcal{V}$. We claim that the map $\underline{\mathcal{K}}(A \otimes j_1, f_0) \to \mathbf{Sq}(A \otimes j, f)$ has a section. By the universal property of the tensor in $\underline{\mathcal{K}}$ and $\underline{\mathcal{K}}^2$, the domain is naturally isomorphic to $\underline{\mathcal{V}}(A, \underline{\mathcal{K}}(j_1, f_0))$

and the codomain is naturally isomorphic to $\underline{\mathcal{V}}(A, \mathbf{Sq}(j, f))$; naturality implies that the comparison map is isomorphic to the image of $\underline{\mathcal{K}}(j_1, f_0) \to \mathbf{Sq}(j, f)$ under the $\mathcal{V}$-function $\underline{\mathcal{V}}(A, -)$. Now the desired conclusion is clear.

Now suppose that we have a composable pair of morphisms $j_0 \xrightarrow{j} j_1 \xrightarrow{k} j_2$ so that $j \boxtimes f$ and $k \boxtimes f$. We define the map $\mathbf{Sq}(kj, f) \to \underline{\mathcal{K}}(j_2, f_0)$ and leave to the reader the straightforward verification that it is the desired section. As illustrated by (12.7.2), in the dual situation, the procedure to define a lift of kj against f is to first use the lifting property of j to define a lifting problem between k and f and then appeal to the lifting property of k. Formally, we note that the pullback square defining $\mathbf{Sq}(kj, f)$ factors along $k^*: \underline{\mathcal{K}}(j_2, f_1) \to \underline{\mathcal{K}}(j_1, f_1)$ through the pullback square defining $\mathbf{Sq}(j, f)$. Hence the pullback of this k^* defines a map $\mathbf{Sq}(kj, f) \to \mathbf{Sq}(j, f)$. This map combined with the enriched lifting property of j produces the domain leg of a cone over the pullback defining $\mathbf{Sq}(k, f)$; the other leg, as we would expect from the unenriched argument, is the codomain projection from $\mathbf{Sq}(kj, f)$:

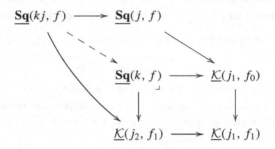

The desired section is the composite of this map $\mathbf{Sq}(kj, f) \to \mathbf{Sq}(k, f)$ with the section supplied by the enriched lifting property of k.

Now suppose $\underline{\mathcal{K}}$ is cotensored over $\mathcal{V}$ and that we are given maps $j \boxtimes f$ and $k \boxtimes f$, where j and k are no longer assumed to be composable. Making use of the $\mathcal{V}$-enriched universal property of coproducts in $\mathcal{K}$ and $\mathcal{K}^2$, we have isomorphisms commuting with the canonical maps

$$
\begin{array}{ccc}
\underline{\mathcal{K}}(j_1 \textstyle\coprod k_1, f_0) & \cong & \underline{\mathcal{K}}(j_1, f_0) \times \underline{\mathcal{K}}(k_1, f_0) \\
\downarrow & & \downarrow \\
\mathbf{Sq}(j \coprod k, f) & \cong & \mathbf{Sq}(j, f) \times \mathbf{Sq}(k, f)
\end{array}
$$

(The proof of this assertion makes use of the fact that the product of pullbacks is the pullback of products.) In particular, the right-hand map is a product of the maps for j and k; the product of the sections guaranteed by their enriched lifting properties produces the section needed to show that $(j \coprod k) \boxtimes f$. $\square$

13.4 Enriched weak factorization systems

Enriched functorial factorizations and enriched lifting properties combine to define an enriched weak factorization system. Let $\mathcal{K}$ be a $\mathcal{V}$-category.

Definition 13.4.1 A $\mathcal{V}$-**enriched weak factorization system** on $\mathcal{K}$ is a pair $(\mathcal{L}, \mathcal{R})$ of classes of maps together with an enriched functorial factorization $L, R \colon \underline{\mathcal{K}}^2 \rightrightarrows \underline{\mathcal{K}}^2$ so that

(i) the images of L and R land in $\mathcal{L}$ and $\mathcal{R}$, respectively, and
(ii) $\mathcal{L} = {}^{\boxtimes}\mathcal{R}$ and $\mathcal{R} = \mathcal{L}^{\boxtimes}$

Equivalently, in the presence of (i), (ii) can be replaced by the weaker $\mathcal{L} \boxtimes \mathcal{R}$ coupled with the requirement that both classes are closed under retracts. Using this observation, the reader should convince himself or herself that an enriched weak factorization system is in particular an ordinary weak factorization system.

The enriched algebraic small object argument described in Theorem 13.2.1 produces $\mathcal{V}$-functorial factorizations for cofibrantly generated enriched weak factorization systems, whether or not the hypothesis $(\star)$ of that theorem is satisfied.

Proposition 13.4.2 *Let $\mathcal{K}$ be a $\mathcal{V}$-bicomplete category permitting the algebraic small object argument. Then any set of maps $\mathcal{J}$ generates a $\mathcal{V}$-enriched weak factorization system $({}^{\boxtimes}(\mathcal{J}^{\boxtimes}), \mathcal{J}^{\boxtimes})$. Furthermore, this weak factorization system coincides with the ordinary weak factorization system generated by $\mathcal{J}$ if and only if the left class of the unenriched weak factorization system is closed under tensors.*

Proof Let $(\mathbb{C}, \mathbb{F})$ be the $\mathcal{V}$-enriched functorial factorization produced by the proof of Theorem 13.2.1. The monad $\mathbb{F}$ is designed so that its algebras are precisely maps g such that $j \boxtimes g$ for all $j \in \mathcal{J}$; hence Ff, as a free $\mathbb{F}$-algebra, lies in $\mathcal{J}^{\boxtimes}$. A representable version of the construction of Lemma 12.4.7 can be used to show that Cf has the enriched left lifting property against the objects of $\mathcal{J}^{\boxtimes} \cong \mathbb{F}\text{-}\mathbf{alg}$. Writing $E = \operatorname{cod} C = \operatorname{dom} F$, the enriched functorial factorization produces a map $\underline{\mathbf{Sq}}(Cf, g) \to \underline{\mathcal{K}}(ECf, Eg)$. Precomposing with the $\mathbb{C}$-coalgebra structure for $\overline{Cf}$ and postcomposing with the $\mathbb{F}$-algebra structure for g produces the desired section of (13.3.2). $\qquad\square$

Remark 13.4.3 When $\mathbb{F} = (F, \vec{\eta}, \vec{\mu})$ is produced by the $\mathcal{V}$-enriched algebraic small object argument, the entire monad of 12.6.1 is a $\mathcal{V}$-monad, that is, F is a $\mathcal{V}$-functor and $\vec{\eta}$ and $\vec{\mu}$ are $\mathcal{V}$-natural transformations. In this case, the category of $\mathbb{F}$-algebras is canonically $\mathcal{V}$-enriched [11, 2.2]. We conjecture that $\mathcal{J}^{\boxtimes}$ is also canonically a $\mathcal{V}$-category, the hom-objects $\underline{\mathcal{J}^{\boxtimes}}(f, g)$

defined to be the subobject of $\underline{\mathbf{Sq}}(f, g)$ constructed as the limit of the diagram

$$\prod_{j \in \mathcal{J}} \underline{\mathcal{V}}(\underline{\mathbf{Sq}}(j, f), \underline{\mathbf{Sq}}(j, g))$$

$$\underline{\mathbf{Sq}}(f, g) \longrightarrow \qquad \qquad \prod_{j \in \mathcal{J}} \underline{\mathcal{V}}(\underline{\mathbf{Sq}}(j, f), \underline{\mathcal{K}}(j_1, g_0))$$

$$\prod_{j \in \mathcal{J}} \underline{\mathcal{V}}(\underline{\mathcal{K}}(j_1, f_0), \underline{\mathcal{K}}(j_1, g_0))$$

Furthermore, we suspect that the isomorphism of underlying categories $\mathcal{J}^{\boxminus} \cong \mathbb{F}\text{-}\mathbf{alg}$ extends to an isomorphism of $\mathcal{V}$-categories. However, we have not checked the details.

By contrast, when the hypothesis $(\star)$ of Theorem 13.2.1 is not satisfied, that is, when the enriched and unenriched right lifting properties against $\mathcal{J}$ differ, the functorial factorization constructed by the enriched version of the Quillen small object argument is poorly behaved.

Remark 13.4.4 (on the enriched Quillen small object argument)   Let (L^{ω}, R^{ω}) denote the $\mathcal{V}$-functorial factorization constructed by the enriched Quillen small object argument. By Lemma 13.3.11, the image of the left factor L^{ω} lies in the class $^{\boxminus}(\mathcal{J}^{\boxminus})$, but when $(\star)$ is not satisfied, this class is strictly larger than $^{\boxtimes}(\mathcal{J}^{\boxtimes})$.

Consequently, to guarantee that maps in the image of L^{ω} lift against maps in the image of R^{ω}, we would have to show that each $R^{\omega} f$ satisfies the enriched lifting property against $\mathcal{J}$. However, the "smallness" hypothesis for Quillen's construction is insufficient to guarantee that this is the case. Even assuming that the *enriched* representable functors on the domains of each $j \in \mathcal{J}$ preserved sequential colimits, it does not necessarily follow that the map $\underline{\mathbf{Sq}}(j, R^{\omega} f) \to \underline{\mathcal{K}}(\mathrm{dom}\, j, \mathrm{dom}\, R^{\omega} f)$ factors through some finite stage of the colimit defining Quillen's factorization. "Smallness" is an assertion about the *elements* of the object $\underline{\mathcal{K}}(\mathrm{dom}\, j, \mathrm{dom}\, R^{\omega} f)$, not about its subobjects; the object $\underline{\mathbf{Sq}}(j, R^{\omega} f)$ is in general too big to conclude anything useful about its image.

Example 13.4.5   Let R be a commutative ring and consider the set $\mathcal{J} = \{0 \to D^n\}_{n \in \mathbb{Z}}$ of maps in $\mathbf{Ch}_{\bullet}(R)$. The unenriched weak factorization system generated by $\mathcal{J}$ is the (trivial cofibration, fibration) weak factorization system for the Quillen-type model structure on $\mathbf{Ch}_{\bullet}(R)$, whereas the $\mathbf{Mod}_R$-enriched weak factorization system is the (trivial cofibration, fibration) weak factorization system for the Hurewicz-type model structure. To prove this latter claim, note that an extension of the argument given in Example 13.3.4 proves that elements of the right class of $\mathcal{J}^{\boxminus}$ are maps admitting a section given by a map of graded R-modules. By [58, 18.3.6], this condition characterizes the class of Hurewicz fibrations. We conclude that the trivial cofibrations and fibrations in this model structure form a $\mathbf{Mod}_R$-enriched algebraic weak factorization system.

As observed in Example 13.2.5, the functorial factorization produced by the enriched algebraic small object argument is the factorization through the mapping path space, constructed using as an interval the chain complex with a single generator in degree one and two in degree zero, with boundary map taking the former generator to the difference of the latter two. A dual construction factors a map of chain complexes through the mapping cylinder [58, 18.3.7]. Because the first weak factorization system specified a $\mathbf{Mod}_R$-enriched algebraic weak factorization system, the latter also has this property. In this way, we have enriched both weak factorization systems for the Hurewicz-type model structure.

It turns out that the (cofibration, trivial fibration) weak factorization system for the Hurewicz-type model structure is also cofibrantly generated, in the sense of Proposition 13.4.2. The generating cofibrations $\mathcal{I} = \{S^{n-1} \to D^n\}_{n \in \mathbb{Z}}$ turn out to be the same maps used for the Quillen-type model structure in the unenriched case; see [2] for proof. Furthermore, the functorial factorization produced by the $\mathbf{Mod}_R$-enriched algebraic small object argument is isomorphic to the mapping cylinder factorization. The intrepid reader is encouraged to verify that the enriched algebraic small object argument converges in step two and to write down inverse isomorphic lifts comparing the two functorial factorizations.

In the next section, we explore the benefits of the observation just made.

13.5 Enriched model categories

In the context of model category theory, cofibrant generation is commonly invoked to transfer preexisting model structures along an adjunction. This technique was used to prove Theorem 12.3.2, constructing the projective model structure. As justification for our expansion of the notion of a "cofibrantly generated" weak factorization system, as encoded by Proposition 13.4.2, from its classical usage to the enriched one, we now present the enriched analog of a frequently cited theorem of Kan [36, 11.3.1–2].

Theorem 13.5.1   *Suppose* $F \colon \mathcal{M} \rightleftarrows \mathcal{N} \colon U$ *is a* $\mathcal{V}$-*adjunction between* $\mathcal{V}$-*bicomplete categories, and suppose that* $\mathcal{M}$ *has a model structure whose constituent weak factorization systems are cofibrantly generated in the* $\mathcal{V}$-*enriched sense by* $\mathcal{I}$ *and* $\mathcal{J}$. *If* $\mathcal{N}$ *permits the small object argument and if*

(†) *U takes maps in* $^{\boxtimes}(F\mathcal{J}^{\boxtimes})$ *to weak equivalences,* *then $\mathcal{N}$ has a model structure that is cofibrantly generated in the $\mathcal{V}$-enriched sense by $F\mathcal{I}$ and $F\mathcal{J}$, whose weak equivalences and fibrations are created by U.*

Because any functor preserves retracts, it suffices to restrict the condition (†) to the class $F\mathcal{J}$-cell of transfinite composites of pushouts of coproducts of tensors of maps in $F\mathcal{J}$ with objects of $\mathcal{V}$. The condition (†), which demands that certain maps are weak equivalences, is called the "acyclicity condition."

Proof By Proposition 13.4.2, $F\mathcal{I}$ and $F\mathcal{J}$ generate $\mathcal{V}$-enriched weak factorization systems $({}^{\square}(F\mathcal{I}^{\square}), F\mathcal{I}^{\square})$ and $({}^{\square}(F\mathcal{J}^{\square}), F\mathcal{J}^{\square})$ on $\mathcal{N}$. Recall that $\mathcal{V}$-enriched weak factorization systems are also ordinary weak factorization systems and hence suitable for defining a model structure. By hypothesis, the classes $\mathcal{J}^{\square}$ and $\mathcal{I}^{\square}$ define the fibrations and trivial fibrations in the model structure on $\mathcal{M}$. By Exercise 13.3.5, the classes $F\mathcal{J}^{\square}$ and $F\mathcal{I}^{\square}$ are the pullbacks of these classes along U. We call these the fibrations and trivial fibrations in $\mathcal{N}$. This is what it means to say that the right adjoint U creates the fibrations and trivial fibrations.

Let $\mathcal{W}$ be the class of weak equivalences created by U, that is, the class of maps in $\mathcal{N}$ that U sends to weak equivalences in $\mathcal{M}$. To show that $\mathcal{W}$ defines a model structure whose weak factorization systems are $({}^{\square}(F\mathcal{I}^{\square}), F\mathcal{I}^{\square})$ and $({}^{\square}(F\mathcal{J}^{\square}), F\mathcal{J}^{\square})$, it remains, by Definition 11.3.1, to show that these classes are compatible. Specifically, we must show that ${}^{\square}(F\mathcal{J}^{\square}) = {}^{\square}(F\mathcal{I}^{\square}) \cap \mathcal{W}$ and that $F\mathcal{I}^{\square} = F\mathcal{J}^{\square} \cap \mathcal{W}$.

In fact, by a standard argument [38, 2.1.19], it suffices to prove only three of the four inclusions asserted by these equalities. Suppose $f \in F\mathcal{J}^{\square} \cap \mathcal{W}$. By adjunction, $F\mathcal{J}\boxslash f$ implies that $\mathcal{J}\boxslash Uf$. So Uf is a fibration and weak equivalence, which means that $\mathcal{I}\boxslash Uf$ and hence that $F\mathcal{I}\boxslash f$. Thus $(F\mathcal{J}^{\square} \cap \mathcal{W}) \subset F\mathcal{I}^{\square}$. Conversely, if $f \in F\mathcal{I}^{\square}$, then $Uf \in \mathcal{I}^{\square}$, which is the intersection of $\mathcal{J}^{\square}$ with the class of weak equivalences in $\mathcal{M}$. Hence $f \in F\mathcal{J}^{\square} \cap \mathcal{W}$, and we have shown that $F\mathcal{I}^{\square} = F\mathcal{J}^{\square} \cap \mathcal{W}$.

It remains only to prove that ${}^{\square}(F\mathcal{J}^{\square}) \subset {}^{\square}(F\mathcal{I}^{\square}) \cap \mathcal{W}$. By hypothesis (†), we need only show that ${}^{\square}(F\mathcal{J}^{\square}) \subset {}^{\square}(F\mathcal{I}^{\square})$. As a consequence of the closure properties asserted by Lemma 13.3.11, it suffices to show that $F\mathcal{J} \subset {}^{\square}(F\mathcal{I}^{\square})$, that is, that $F\mathcal{J}\boxslash F\mathcal{I}^{\square}$. By adjunction, $f \in F\mathcal{I}^{\square}$ if and only if $Uf \in \mathcal{I}^{\square} \subset \mathcal{J}^{\square}$, because $\mathcal{I}$ and $\mathcal{J}$ define a model structure on $\mathcal{M}$. Transposing back across the $\mathcal{V}$-adjunction, we conclude that $F\mathcal{J}\boxslash f$, and hence that $F\mathcal{J}\boxslash F\mathcal{I}^{\square}$, as desired. □

Note that our notion of a model structure on a $\mathcal{V}$-category built from enriched weak factorization systems is distinct from the usual definition of a $\mathcal{V}$-model category: in particular, for this new notion, $\mathcal{V}$ does not have to be a model category itself! We should say a few words toward the comparison in the case where $\mathcal{V}$ *is* a monoidal model category, so that both Definition 11.4.7 and the definition encoded by the statement of Theorem 13.5.1 make sense. In general, neither notion implies the other.

Suppose $\mathcal{K}$ is a $\mathcal{V}$-bicomplete category, with $(\otimes, \{, \}, \underline{\mathrm{hom}})$ the tensor–cotensor–hom two-variable adjunction. Observe that the comparison map $\underline{\mathcal{K}}(j_1, f_0) \to \underline{\mathbf{Sq}}(j, f)$ of (13.3.2) is exactly the map $\widehat{\mathrm{hom}}(j, f)$ defined by 11.1.7. When $\mathcal{K}$ is a $\mathcal{V}$-model category, the map $\widehat{\mathrm{hom}}(j, f)$ is a trivial fibration whenever j and f are in the left and right classes of the same weak factorization system. If all objects of $\mathcal{V}$ are cofibrant, then this implies that $\widehat{\mathrm{hom}}(j, f)$ has a section so that $j \boxtimes f$. This can also be deduced from Lemma 13.3.6, which applies to both weak factorization systems, under the slightly weaker hypothesis that tensoring with each object of $\mathcal{V}$ defines a left Quillen functor. If the model structure is cofibrantly generated, Theorem 13.2.1 provides $\mathcal{V}$-enriched factorizations. In summary:

Theorem 13.5.2 *If tensoring with objects of $\mathcal{V}$ defines a left Quillen functor, any cofibrantly generated $\mathcal{V}$-model structure can be "fully enriched": the (trivial) cofibrations and (trivial) fibrations form $\mathcal{V}$-enriched weak factorization systems, with functorial factorizations produced by the $\mathcal{V}$-enriched Quillen or Garner small object argument.*

When some objects of $\mathcal{V}$ are not cofibrant, the fact that $\mathcal{K}$ is a $\mathcal{V}$-model category does not imply that the left and right maps in its weak factorization systems satisfy the enriched lifting property with respect to each other: the trivial fibration $\widehat{\mathrm{hom}}(j, f)$ need not be a split epimorphism. Conversely, if $\mathcal{K}$ is a model category whose weak factorization systems are $\mathcal{V}$-enriched, the comparison maps $\widehat{\mathrm{hom}}(j, f)$ are split epimorphisms whenever j is a trivial cofibration and f is a fibration or j is a cofibration and f is a trivial fibration. In particular, the map $\widehat{\mathrm{hom}}(j, f)$ lifts on the right against all objects of $\mathcal{V}$, not just cofibrant ones. However, this does not imply that $\widehat{\mathrm{hom}}(j, f)$ is a trivial fibration, so $\mathcal{K}$ need not be a $\mathcal{V}$-model category.

13.6 Enrichment as coherence

Our aim in this final short section is to describe a result of Garner that unifies the enriched lifting properties introduced in this chapter with the "coherent" lifting properties (defined with respect to a generating *category* of arrows) introduced in the previous chapter: in certain cases, enrichments can be encoded in the unenriched setting by adding morphisms to the generating category. The context for this result is a closed symmetric monoidal category $\mathcal{V}$ that has a small dense subcategory. A subcategory $\mathcal{A} \subset \mathcal{V}$ of a cocomplete category $\mathcal{V}$ is **dense** if the identity at V is a left Kan extension of the inclusion $\mathcal{A} \to V$ along itself. In this case, a corollary of 1.2.1 is that every object of V can be expressed canonically as a colimit of objects in $\mathcal{A}$. Equivalently, $\mathcal{A} \subset V$ is dense if and only if the restricted Yoneda embedding

$$V \longrightarrow \mathbf{Set}^{\mathcal{A}^{op}}$$

$$v \rightsquigarrow a \mapsto V(a, v)$$

is fully faithful. For example, the full subcategory of representables, that is, the image of the Yoneda embedding $\Delta \to \mathbf{sSet}$, is a dense subcategory of the category of simplicial sets.

Proposition 13.6.1 (Garner) *Suppose V admits a small dense subcategory $\mathcal{A}$ and suppose $\mathcal{K}$ is a V-bicomplete category that permits the small object argument. Let $\mathcal{J}$ be a small category of arrows in $\mathcal{K}$. Then the V-enriched algebraic small object applied to $\mathcal{J}$ produces the same factorization as the unenriched algebraic small object argument applied to the category $\mathcal{A} \times \mathcal{J}$ with objects $\{a \otimes j \mid a \in \mathcal{A}, j \in \mathcal{J}\}$ and morphisms given by tensor products of maps in $\mathcal{A}$ and $\mathcal{J}$.*

Proof We will show that the categories $(\mathcal{A} \times \mathcal{J})^{\boxslash}$ and $\mathcal{J}^{\boxslash}$ are isomorphic; a more precise argument shows that they are isomorphic as double categories, which suffices, by Theorem 12.7.3, to prove that the algebraic weak factorization systems these categories encode are the same. Suppose $f \in \mathcal{J}^{\boxslash}$. This means that, for each $j \in \mathcal{J}$, the map $\underline{\mathcal{K}}(j_1, f_0) \to \underline{\mathbf{Sq}}(j, f)$ admits a section, and these sections are natural in $\mathcal{J}$. Because $V \to \mathbf{Set}^{\mathcal{A}^{op}}$ is fully faithful, it follows that

$$V(a, \underline{\mathcal{K}}(j_1, f_0)) \to V(a, \underline{\mathbf{Sq}}(j, f))$$

admits a section, natural in $\mathcal{J}$ and $\mathcal{A}$. By adjunction, this means that the map

$$\mathcal{K}(a \otimes j_1, f_0) \to \mathcal{K}^2(a \otimes j, f) = \mathbf{Sq}(a \otimes j, f)$$

admits a section, natural in $\mathcal{J}$ and $\mathcal{A}$. But this says exactly that $f \in (\mathcal{A} \otimes \mathcal{J})^{\boxslash}$. □

This gives a conceptual justification for the result of 12.5.11: the construction given there is enriched over $\mathbf{Set}^{\mathcal{A}}$. Similarly, simplicial enrichments can be produced by running the unenriched algebraic small object argument with the category formed by tensoring the generating cofibrations with the category Δ.

Example 13.6.2 Consider the generating set $\mathcal{J} = \{\emptyset \to \Delta^0\}$ in the category of simplicial sets. The enriched algebraic small object argument applied to $\mathcal{J}$ produces the "co-graph" factorization $X \to X \sqcup Y \to Y$.

The category $\Delta \times \mathcal{J}$ is isomorphic to Δ; the image of the forgetful functor $\Delta \times \mathcal{J} \to \mathbf{sSet}^2$ is the full subcategory spanned by the maps $\emptyset \to \Delta^n$. By density, the left Kan extension of this functor along itself is the functor cod: $\mathbf{sSet}^2 \to \mathbf{sSet}^2$ that projects to the codomain and then slices under the initial object. This defines the step-zero functor for the algebraic small object argument. From this observation, it is an easy exercise to see that the functorial factorization produced by applying Garner's construction to the category $\Delta \times \mathcal{J}$ is again the cograph factorization.

14

A brief tour of Reedy category theory

A **Reedy category** is a category whose objects are filtered by degree together with extra structure that makes it possible to define diagrams inductively. To illustrate how this might be useful in homotopy theory, consider a simplicial space X. The objects in the indexing category Δ^{op} are naturally filtered by dimension. Write $\text{sk}_n |X|$ for the n-truncated geometric realization, that is, for the functor tensor product $\Delta^\bullet_{\leq n} \otimes_{\Delta^{\text{op}}_{\leq n}} X_{\leq n}$ of 4.4.3. As n increases, the spaces $\text{sk}_n |X|$ give better approximations to the geometric realization: a direct argument shows that the colimit of the sequence

$$\text{sk}_1 |X| \to \text{sk}_2 |X| \to \text{sk}_3 |X| \to \cdots \tag{14.0.1}$$

is $|X|$. We would like to find conditions under which it is possible to prove that this colimit is homotopy invariant.

To start, we should try to understand each map $\text{sk}_{n-1} |X| \to \text{sk}_n |X|$. If the simplicial space X were discrete, we would say the latter space is formed by gluing in a copy of the topological n-simplex $|\Delta^n|$ for each non-degenerate n-simplex of X. In general, the map $\text{sk}_{n-1} |X| \to \text{sk}_n |X|$ is a pushout of a map defined in reference to the subspace $L_n X \subset X_n$ of degenerate n-simplices.

In the topological case, the **latching space** $L_n X$ is simply the union of the images of the degeneracy maps $s_i : X_{n-1} \to X_n$. By the simplicial identities, each degeneracy map admits a retraction, and it follows that the topology on X_{n-1} coincides with the subspace topology for each of its images. The map $\text{sk}_{n-1} |X| \to \text{sk}_n |X|$ is the pushout

$$
\begin{array}{ccc}
|\Delta^n| \times L_n X \underset{|\partial \Delta^n| \times L_n X}{\coprod} |\partial \Delta^n| \times X_n & \longrightarrow & \text{sk}_{n-1} |X| \\
{\scriptstyle i_n \,\hat\times\, \ell_n} \Big\downarrow & & \Big\downarrow \\
|\Delta^n| \times X_n & \xrightarrow{\quad \ulcorner \quad} & \text{sk}_n |X|
\end{array}
$$

of the pushout-product of the inclusions $i_n \colon |\partial \Delta^n| \to |\Delta^n|$ and $\ell_n \colon L_n X \to X_n$. In words, an n-simplex is attached along its boundary for points in the complement of $L_n X$. For points in $L_n X$, these attached simplices map degenerately onto $(n-1)$-simplices previously attached to the $(n-1)$-truncated geometric realization.

Because **Top** is a simplicial model category, if each $L_n X \to X_n$ is a cofibration, then the left-hand map is a cofibration and hence its pushout is as well. In this case, the diagram (14.0.1) is projectively cofibrant by Example 11.5.11, and hence by Corollary 11.5.3, its ordinary colimit is a homotopy colimit. In this way, we have proven that geometric realization preserves weak equivalences between simplicial objects whose **latching maps** $L_n X \to X_n$ are cofibrations. Such simplicial objects are called **Reedy cofibrant**.

In Section 14.1, we continue our consideration of simplicial objects to give a geometric grounding to the formal definition of the latching and dually defined **matching objects**. In Section 14.2, we generalize the diagram shape, debuting Reedy categories and Reedy model structures. After a whirlwind introduction to these notions, we devote Sections 14.3, 14.4, and 14.5 to applications, tying up several loose ends from Parts I and II of this book. There has been a cross-pollination between this chapter and the expository article [76], which gives a much more comprehensive development of Reedy category theory from the perspective of weighted limits and colimits.

14.1 Latching and matching objects

The proof given in the introduction generalizes immediately to simplicial objects valued in any simplicial model category, provided we can work out how to define the latching objects categorically. Our definition actually makes sense for simplicial objects valued in any model category; the reason one might ask the target to be a simplicial model category is for the homotopical content of the preceding argument.

We define matching objects first because the geometric intuition is slightly simpler. The nth **matching object** of a simplicial object X is an object together with a map $m_n \colon X_n \to M_n X$, which we think of as describing the "boundary data" for the object of n-simplices. Geometrically, we think of elements of the object X_n as having the "shape" of an n-simplex; with this intuition, the image under the nth **matching map** m_n should have the shape of its boundary, given by $(n-1)$-simplices glued together appropriately.

We can make this intuition precise through a definition by means of a weighted limit. Recall that the weight for a weighted limit of a simplicial object has the form of a functor $\Delta^{\mathrm{op}} \to \mathbf{Set}$, that is, is given by a simplicial set. The limit of a simplicial object X weighted by the simplicial set Δ^n is the object X_n, by the Yoneda lemma (see Example 7.1.4).

Definition 14.1.1 The nth **matching object** $M_n X$ of a simplicial object X is the weighted limit $\lim^{\partial\Delta^n} X$. The nth **matching map** $m_n \colon X_n \to M_n X$ is the map induced by the map of weights $\partial\Delta^n \to \Delta^n$.

Example 14.1.2 When X is a simplicial set, $\lim^{\partial\Delta^n} X$ is the set of maps of simplicial sets $\partial\Delta^n \to X$ by (7.1.3). Precisely in accordance with our geometric intuition, the nth matching object of a simplicial set X is the set of $(n-1)$-spheres in X.

We can use the cocontinuity of the weighted limit bifunctor in the weight to obtain a description of $M_n X$ as an ordinary limit. This is the content of Exercise 7.2.11. A weighted limit whose weight is a colimit is equal to the limit of the limits weighted by each weight in the colimit diagram. The boundary of the n-simplex is formed by gluing together its faces, $n+1$ $(n-1)$-simplices, along their boundary $(n-2)$-simplices. These faces correspond to monomorphisms in Δ. Hence we have

$$\partial\Delta^n \cong \mathrm{colim}\left(\coprod_{[n-2]\rightarrowtail[n]} \Delta^{n-2} \xrightarrow{\;\cdots\;} \coprod_{[n-1]\rightarrowtail[n]} \Delta^{n-1} \right) \qquad (14.1.3)$$

where the colimit diagram has a morphism between the coproducts corresponding to each monomorphism $[n-2] \rightarrowtail [n-1]$. By cocontinuity of the weighted limits, we deduce that

$$M_n X \cong \lim\left(\prod_{[n-1]\rightarrowtail[n]} X_{n-1} \xrightarrow{\;\cdots\;} \prod_{[n-2]\rightarrowtail[n]} X_{n-2} \right).$$

The latching object can be defined dually as a weighted colimit; in this case, the weight should be a *covariant* functor $\Delta \to \mathbf{Set}$. We dualize (14.1.3) by exchanging Δ for Δ^{op}; this replaces contravariant representables by covariant representables and monomorphisms by epimorphisms. Define

$$\partial\Delta_n \cong \mathrm{colim}\left(\coprod_{[n]\twoheadrightarrow[n-2]} \Delta([n-2],-) \xrightarrow{\;\cdots\;} \coprod_{[n]\twoheadrightarrow[n-1]} \Delta([n-1],-) \right),$$

where the morphisms between the coproducts correspond to epimorphisms $[n-1] \twoheadrightarrow [n-2]$. Note that the covariant representable $\Delta([n],-)$ forms a canonical cocone under this colimit diagram.

Definition 14.1.4 The nth **latching object** $L_n X$ of a simplicial object X is the weighted colimit $\mathrm{colim}^{\partial\Delta_n} X$. The nth **latching map** $\ell_n \colon L_n X \to X_n$ is the map induced by the map of weights $\partial\Delta_n \to \Delta([n],-)$.

Once again, we can use the cocontinuity of the weighted colimit bifunctor to check that Definition 14.1.4 is the correct "object of degenerate n-simplices." By this cocontinuity and the co-Yoneda lemma,

$$L_n X \cong \text{colim} \left(\coprod_{[n] \twoheadrightarrow [n-2]} X_{n-2} \underset{\cdots}{\overset{\cdots}{\rightrightarrows}} \coprod_{[n] \twoheadrightarrow [n-1]} X_{n-1} \right).$$

If, as in the introduction, X is a simplicial space, then this formula says that the space $L_n X$ is formed by gluing together a copy of the space X_{n-1} for each degeneracy map $[n] \rightarrow [n-1]$ along the images of the spaces X_{n-2} corresponding to compatibly defined degeneracy maps $[n-1] \twoheadrightarrow [n-2]$.

14.2 Reedy categories and the Reedy model structures

The category $\mathbb{\Delta}^{\text{op}}$ is an example of what is now called a **Reedy category**. The eponymous model structure on simplicial objects taking values in any model category was introduced in an unpublished but nonetheless widely disseminated manuscript written by Chris Reedy [68]. Reedy notes that a dual model structure exists for cosimplicial objects, which, in the case of cosimplicial simplicial sets, coincides with a model structure introduced by Bousfield and Kan to define homotopy limits [10, §X]. The general definition, unifying these examples and many others, is due to Kan and appeared in the early drafts of the book that eventually became [22]. Various draft versions circulated in the mid-1990s and contributed to the published accounts [36, chapter 15] and [38, chapter 5]. The final [22] in turn references these sources to "review the notion of a Reedy category."

Definition 14.2.1 A **Reedy category** is a small category $\mathcal{D}$ equipped with

(i) a degree function assigning a non-negative integer to each object
(ii) a wide subcategory $\overrightarrow{\mathcal{D}}$ whose non-identity morphisms strictly raise degree
(iii) a wide subcategory $\overleftarrow{\mathcal{D}}$ whose non-identity morphisms strictly lower degree

so that every arrow factors uniquely as an arrow in $\overleftarrow{\mathcal{D}}$ followed by one in $\overrightarrow{\mathcal{D}}$.

It follows from the axioms that any map in a Reedy category factors uniquely through an object of minimal degree, and furthermore that this factorization is the **Reedy factorization**, as a map in $\overleftarrow{\mathcal{D}}$ followed by a map in $\overrightarrow{\mathcal{D}}$.

Remark 14.2.2 This definition is what some category theorists would call "evil" – it is not invariant under equivalence of categories. The axioms imply that a Reedy category must be skeletal and contain no non-trivial automorphisms. By contrast, the notion of a **generalized Reedy category** introduced by Clemens Berger and Ieke Moerdijk is preserved by categorical equivalence and includes several new examples of interest to homotopy theory, such as Segal's category Γ or Alain Connes's cyclic category Λ [5].

Exercise 14.2.3   Show that $(\overleftarrow{\mathcal{D}}, \overrightarrow{\mathcal{D}})$ defines an orthogonal factorization system. This observation forms the basis for the equivalent definition of a Reedy category given in [49, A.2.9.1].

Finite products of Reedy categories are Reedy categories, with the degree function defined by summing the degrees of the coordinates. If $\mathcal{D}$ is a Reedy category, then so is $\mathcal{D}^{\mathrm{op}}$ with the categories $\overleftarrow{\mathcal{D}}$ and $\overrightarrow{\mathcal{D}}$ exchanged, which is both a blessing and a curse. The blessing is that both Δ and Δ^{op} are Reedy categories. The curse is that we care about both examples, so there is a good possibility to mix up which morphisms are "degree decreasing" and which are "degree increasing."

Example 14.2.4   The category Δ is a Reedy category with degree function the obvious one, with $\overleftarrow{\Delta}$ the subcategory of epimorphisms, and with $\overrightarrow{\Delta}$ the subcategory of monomorphisms. Any map in Δ can be factored uniquely through an object of minimal degree; the first factor in this factorization is necessarily an epimorphism, and the second is a monomorphism.

Note that there are maps in Δ for which the source has a smaller degree than the target but which are not monomorphisms. However, we cannot include these morphisms in $\overrightarrow{\Delta}$ if the unique factorization axiom is to be satisfied.

To alleviate confusion, we will try and emphasize the role played by the degree function and deemphasize the role played by the categories $\overrightarrow{\mathcal{D}}$ and $\overleftarrow{\mathcal{D}}$. To any Reedy category there exist subcategories $\mathcal{D}_{\leq n}$ of objects of degree less than or equal to n. For $\mathcal{D} = \Delta^{\mathrm{op}}$, we have $\mathcal{D}_{\leq n} = \Delta^{\overline{\mathrm{op}}}_{\leq n}$, precisely in accordance with our previous notation. Restriction along the inclusion $\mathcal{D}_{\leq n} \hookrightarrow \mathcal{D}$ produces an n-truncation functor, defined on the category of $\mathcal{D}$-shaped diagrams valued in $\mathcal{M}$, which has left and right adjoints given by left and right Kan extension. The composite monad on $\mathcal{M}^{\mathcal{D}}$ is called sk_n, and the composite comonad is called cosk_n, as described in 1.1.9.

Let $X: \mathcal{D} \to \mathcal{M}$. Because X is covariant, we write X^d for the value of X at $d \in \mathcal{D}$, extending the usual notation for cosimplicial objects. Generalizing the definitions for simplicial objects, there are latching and matching objects corresponding to each $d \in \mathcal{D}$ that come equipped with latching and matching maps $L^d X \xrightarrow{\ell_d} X^d \xrightarrow{m_d} M^d X$. We use the direction of the latching and matching maps to help us remember how they are defined. Because the latching map points toward X^d, the latching object is defined as a colimit; dually, the matching object is a limit. Because the purpose of the Reedy category axioms are to permit us to define diagrams inductively, both $L^d X$ and $M^d X$ should be definable in reference to objects of degree strictly less than $\deg(d) = n$.

Definition 14.2.5 The dth **latching object** of $X \in \mathcal{M}^{\mathcal{D}}$ is

$$L^d X := \mathrm{colim}\left(\overrightarrow{\mathcal{D}}_{<n}/d \xrightarrow{\ U\ } \mathcal{D} \xrightarrow{\ X\ } \mathcal{M} \right)$$

and the dth **matching object** is

$$M^d X := \lim\left(d/\overleftarrow{\mathcal{D}}_{<n} \xrightarrow{\ U\ } \mathcal{D} \xrightarrow{\ X\ } \mathcal{M} \right).$$

The domains of the forgetful functors appearing in these definitions are slice categories whose objects are restricted to degree less than n and whose maps are restricted to $\overrightarrow{\mathcal{D}}$ and $\overleftarrow{\mathcal{D}}$, respectively. The canonical cones with summit X^d under the colimit and over the limit define the latching and matching maps.

Exercise 14.2.6 Show that $L^d X \cong \mathrm{colim}^{\partial\mathcal{D}(-,d)} X$ and $M^d X \cong \lim^{\partial\mathcal{D}(d,-)} X$, where $\partial\mathcal{D}(-, d)$ and $\partial\mathcal{D}(d, -)$ are the subfunctors of the representable functors consisting of the maps whose Reedy factorization is through an object of degree strictly less than n. For the proof, it might help to observe that $\overrightarrow{\mathcal{D}}_{<n}/d$ is a final subcategory of $\mathcal{D}_{<n}/d$, and dually that $d/\overleftarrow{\mathcal{D}}_{<n}$ is an initial subcategory of $d/\mathcal{D}_{<n}$.

Note that the latching object $L^d X$ and matching object $M^d X$ can be defined with respect to the $(n - 1)$-truncation of the diagram X. An important consequence of these definitions is that to extend a diagram $\mathcal{D}_{<n} \to \mathcal{M}$ to a diagram $\mathcal{D}_{\le n} \to \mathcal{M}$ is precisely to choose a factorization of $L^d X \to M^d X$ for each d of degree n. The reader is encouraged to work out (or look up) how this works. Here we are most interested in the model structure. Given a map $X \to Y$ in $\mathcal{M}^{\mathcal{D}}$, define the dth **relative latching map** and dth **relative matching map** by

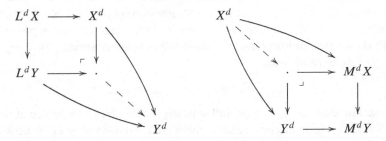

Theorem 14.2.7 ([38, 5.2.5]) *Let $\mathcal{M}$ be a model category, and let $\mathcal{D}$ be a Reedy category. There is a model structure on $\mathcal{M}^{\mathcal{D}}$ with pointwise weak*

equivalences, whose cofibrations are the maps $X \to Y$ such that each relative latching map $L^d Y \coprod_{L^d X} X^d \to Y^d$ is a cofibration in $\mathcal{M}$, and whose fibrations are the maps $X \to Y$ such that each relative matching map $X^d \to Y^d \times_{M^d Y} M^d X$ is a fibration in $\mathcal{M}$.

In particular, X is **Reedy cofibrant** if each $L^d X \to X^d$ is a cofibration. Now that this term has finally been defined, it might be a good time to revisit the proof of Lemma 5.2.1.

Example 14.2.8   The category $0 \leftarrow 1 \to 2$ indexing pushout diagrams admits multiple Reedy category structures. One of these assigns each object the degree indicated by its label. There is a single non-identity degree increasing map and a single non-identity degree decreasing map. Now let $b \xleftarrow{f} a \xrightarrow{g} c$ be a pushout diagram in a generic model category $\mathcal{M}$. The 0th and 1st latching objects are initial, while the 2nd latching object is a. Hence this diagram is Reedy cofibrant if and only if a, b, and c are cofibrant objects and $a \xrightarrow{g} c$ is a cofibration. There is another Reedy structure for which Reedy cofibrant diagrams have the form described in Example 11.5.7.

Exercise 14.2.9   The ordinal category ω generated by the graph $0 \to 1 \to 2 \to \cdots$ has a Reedy structure in which the degree of each object accords with the labels given here. Show that a countable sequence (11.5.12) is a model category is Reedy cofibrant just when each object is cofibrant and each map is a cofibration.

An inductive argument starting from the observation that the latching objects in degree zero are always initial and the matching objects in degree zero are always terminal shows that Reedy cofibrant diagrams are pointwise cofibrant and dually that Reedy fibrant diagrams are pointwise fibrant.

Lemma 14.2.10   *Any Reedy cofibrant diagram is pointwise cofibrant.*

Proof   Let $X \colon \mathcal{D} \to \mathcal{M}$ be Reedy cofibrant. It clearly suffices to show that the latching objects $L^d X$ are cofibrant for each $d \in \mathcal{D}$. We do so by induction on degree. The latching object for any degree-zero object is initial and hence automatically cofibrant. Consider an object d with degree n and suppose we have shown that all lower-degree latching objects are cofibrant. The category $\overrightarrow{\mathcal{D}}_{<n}/d$ admits a filtration

$$\mathcal{F}_0^d \to \mathcal{F}_1^d \to \cdots \to \mathcal{F}_{n-1}^d = \overrightarrow{\mathcal{D}}_{<n}/d$$

whose kth term consists of the full subcategory on objects of degree at most k. Note that $\mathcal{F}_0^d$ is discrete because there are no non-identity maps between objects of degree zero in a Reedy category. So the colimit of the diagram $\mathcal{F}_0^d \to \overrightarrow{\mathcal{D}}_{<n}/d \xrightarrow{U} \mathcal{D} \xrightarrow{F} \mathcal{M}$ is a coproduct of cofibrant objects and hence cofibrant. Our problem now reduces to a second induction in which we show that

the colimit of $\mathcal{F}_k^d \to \overrightarrow{\mathcal{D}}_{<n}/d \xrightarrow{U} \mathcal{D} \xrightarrow{F} \mathcal{M}$ is cofibrant, supposing this is true for the restriction to $\mathcal{F}_{k-1}^d$. Objects in the former category but not the latter correspond to degree-increasing maps $d' \to d$, where d' has degree k. Such a map defines an inclusion $\overrightarrow{\mathcal{D}}_{<k}/d' \to \mathcal{F}_{k-1}^d$. Indeed, $\mathcal{F}_k^d$ is the pushout

$$\begin{array}{ccc}
\coprod_{d'} \overrightarrow{\mathcal{D}}_{<k}/d' & \longrightarrow & \mathcal{F}_{k-1}^d \\
\downarrow & & \downarrow \\
\coprod_{d'} (\overrightarrow{\mathcal{D}}_{<k}/d')^{\triangleright} & \longrightarrow & \mathcal{F}_k^d
\end{array}$$

where the categories on the lower left are obtained by adjoining the identity at d', which is a terminal object. Applying the functor X and taking colimits, the left-hand vertical map is a coproduct of the latching maps $L^{d'}X \to X^{d'}$. Hence the pushout is a cofibration, proving the inductive step. □

A mild generalization of this argument shows that for any natural transformation $X \to Y$, if the relative latching maps $L^d Y \coprod_{L^d X} X^d \to Y^d$ have a left lifting property against some class of morphisms, then the maps $L^d X \to L^d Y$ and hence also the $X^d \to Y^d$ have the same lifting property.

Remark 14.2.11 A related observation, also with an inductive proof, is that a map $X \to Y$ is a Reedy trivial cofibration if and only if its relative latching maps are trivial cofibrations. There is a dual characterization of Reedy trivial fibrations. See [36, 15.3.15] or [76].

14.3 Reedy cofibrant objects and homotopy (co)limits

Our interest in Reedy model structures primarily stems from the following general theorem, which will have a number of useful corollaries. Recall from Theorem 7.6.3 that in a simplicially bicomplete category, the weighted colimit bifunctor is computed by a functor tensor product and the weighted limit bifunctor is computed by a functor cotensor product.

Theorem 14.3.1 ([36, 18.4.11]) *Let $\mathcal{D}$ be a Reedy category and let $\mathcal{M}$ be a simplicial model category. Then the functor tensor product*

$$- \otimes_{\mathcal{D}} - : \mathbf{sSet}^{\mathcal{D}^{\mathrm{op}}} \times \mathcal{M}^{\mathcal{D}} \to \mathcal{M}$$

is a left Quillen bifunctor with respect to the Reedy model structures, and the functor cotensor product

$$\{-, -\}^{\mathcal{D}} : (\mathbf{sSet}^{\mathcal{D}})^{\mathrm{op}} \times \mathcal{M}^{\mathcal{D}} \to \mathcal{M}$$

is a right Quillen bifunctor with respect to the Reedy model structures.

When a left Quillen bifunctor is evaluated at a cofibrant object in one of its variables, the result is a left Quillen functor. Similarly, when a right Quillen bifunctor is evaluated at a cofibrant object in its contravariant variable, the result is a right Quillen functor. In particular, for certain Reedy categories $\mathcal{D}$, the constant functor $* \colon \mathcal{D}^{op} \to \mathbf{sSet}$ is Reedy cofibrant. When this is the case, Ken Brown's Lemma 11.3.14 implies that colimits of pointwise weakly equivalent Reedy cofibrant diagrams are weakly equivalent and limits of pointwise weakly equivalent Reedy fibrant diagrams are weakly equivalent. For example:

Corollary 14.3.2   *Suppose given diagrams in a simplicial model category*

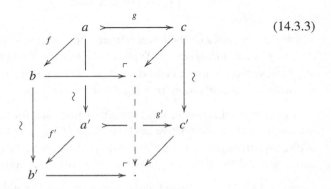

$$(14.3.3)$$

If each object is cofibrant and the maps g and g' are cofibrations, then the induced map between the pushouts is a weak equivalence.

Proof   Let $\mathcal{D}$ be the category $0 \leftarrow 1 \to 2$ with the Reedy structure described in Example 14.2.8. We claim that the constant functor $* \colon \mathcal{D}^{op} \to \mathbf{sSet}$ is Reedy cofibrant. For diagrams of shape $\mathcal{D}^{op}$, the zeroth and second latching objects are initial, whereas the first latching object is the image of the object 0. It follows that a diagram in $\mathbf{sSet}^{\mathcal{D}^{op}}$ is Reedy cofibrant if and only if its objects are cofibrant and one of its maps is a cofibration. Identities are necessarily cofibrations, and all objects in $\mathbf{sSet}$ are cofibrant, so this condition is satisfied.

For any model category $\mathcal{M}$, it follows from this observation, Theorem 14.3.1, and Lemma 11.3.14 that the functor $\mathrm{colim}_{\mathcal{D}} \colon \mathcal{M}^{\mathcal{D}} \to \mathcal{M}$ preserves weak equivalences between Reedy cofibrant diagrams. Example 14.2.8 characterizes these diagrams, completing the proof.   □

Remark 14.3.4   Recall that our notion of "homotopy pushout" required the ambient model category to be simplicial so that these objects represent homotopy coherent cones. However, in common mathematical practice, this term is expanded to include any examples where the pushout functor is homotopical. We claim that this is the case for diagrams of the form (14.3.3) in any model category, not necessarily simplicial. The proof is similar to the argument given in Remark 11.5.9, when this issue was last discussed, but with the Reedy model structure of Theorem 14.2.7 in place of the projective model structure.

There is an adjunction colim$_\mathcal{D}$: $\mathcal{M}^\mathcal{D} \rightleftarrows \mathcal{M}$: Δ between the pushout functor and the constant diagram functor. We claim this is a Quillen adjunction when $\mathcal{M}^\mathcal{D}$ is given the Reedy model structure with respect to the Reedy category structure of Example 14.2.8. The constant diagram functor manifestly preserves (pointwise) weak equivalences. To show that it is right Quillen, it therefore suffices to show that

$$
\Delta f := \quad
\begin{array}{ccccc}
x & \xleftarrow{\ 1\ } & x & \xrightarrow{\ 1\ } & x \\
f \downarrow & & f \downarrow & & \downarrow f \\
y & \xleftarrow[1]{} & y & \xrightarrow[1]{} & y
\end{array}
$$

is a Reedy fibration if f is a fibration in $\mathcal{M}$. The first and second matching objects are terminal, so the first and second matching maps are the middle and right copies of the map f. The zeroth matching objects are the images of the object 1, with the matching maps given by the horizontal arrows in the left-hand square. Hence the relative matching map is the map from the top middle x to the pullback in this square, but this square is already a pullback, so this map is an identity and hence a fibration; this implies that Δf is a Reedy fibration. By Lemma 11.3.11, it follows that colim$_\mathcal{D}$ is left Quillen and hence, by Ken Brown's Lemma 11.3.14, preserves the weak equivalences (14.3.3).

Now it follows from Theorem 2.2.8 and the existence of functorial factorizations that for any model category, a left derived functor of the pushout functor is computed by (functorially) replacing the pushout diagram by a weakly equivalent diagram whose objects are cofibrant and in which one map is a cofibration. Note that Reedy category theory is entirely self dual, so the dual conclusion applies to homotopy pullbacks.

Digression 14.3.5 (left proper model categories and the gluing lemma) Now suppose $\mathcal{M}$ is a **left proper** model category, that is, a model category with the property that a pushout of a weak equivalence along a cofibration is again a weak equivalence. The **gluing lemma** states that in such model categories, the induced map between pushouts of the form (14.3.3) is a weak equivalence, even if the objects are not cofibrant.

As in (11.5.8), it is possible to functorially replace a pushout diagram by a diagram in which each object is cofibrant and each map is a cofibration. We saw in Example 11.5.7 that such diagrams are projectively cofibrant and in Remark 11.5.9 that the colimit functor preserves weak equivalences between diagrams of this form. Hence it suffices to show that in a left proper model category, the pushout of a diagram with one arrow a cofibration is weakly equivalent to the pushout of its projective cofibrant replacement.

Choose a projective cofibrant replacement $r \overset{h}{\longleftarrow} q \overset{k}{\longrightarrow} s$ for the given diagram $b \overset{f}{\longleftarrow} a \overset{g}{\longrightarrow} c$ and form a pushout x in the square from h to f:

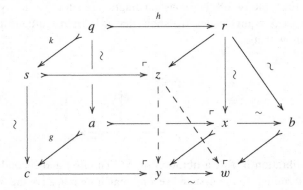

Because $\mathcal{M}$ is left proper, the map from r to x is a weak equivalence; hence so is the map from x to b. Let y be the pushout of the map from a to x along g. Composing and canceling pushouts in the faces of the cube, we see that the front face is also a pushout, and hence by left properness, the map from z to y is also a weak equivalence. By similar reasoning, the pushout of interest w factors through y along another pushout square. Left properness implies that the map from y to w is a weak equivalence, by which we see the diagonal dotted arrow is a weak equivalence, as desired.

Conversely, if a model category satisfies the gluing lemma, then it is left proper. We leave this easy exercise for the reader.

Exercise 14.3.6   Use Exercise 14.2.9, Theorem 14.3.1, and Remark 14.3.4 to generalize the conclusion of Exercise 11.5.11 to arbitrary model categories.

Further corollaries of 14.3.1 require that we identify more Reedy cofibrant objects.

Lemma 14.3.7   *Any bisimplicial set is Reedy cofibrant.*

Proof   Informally, the reason is that each latching object is, pointwise, a subset of the appropriate simplicial set, which is exactly what is meant by being Reedy cofibrant in $\mathbf{sSet}^{\Delta^{op}}$. A more precise justification has to do with the Eilenberg–Zilber lemma for simplicial sets,[1] which says that any simplex is uniquely expressible as a degenerate image of a non-degenerate simplex

---

[1] This argument generalizes to simplicial objects in a **Set**-valued functor category.

[28, II.3.1, pp. 26–27]; see also [27, 4.2.3]. By examining the weighted colimit that defines the latching object, we see that an n-simplex would have multiple preimages in the nth latching object if and only if it were degenerate in distinct ways on non-degenerate simplices. □

A similar argument enables a characterization of those cosimplicial sets (or cosimplicial objects taking values in a **Set**-valued functor category) that are Reedy cofibrant. Fixing $X \in \mathbf{Set}^\Delta$, we might say that $x \in X^n$ is **non-degenerate** if it is not in the image of any monomorphism in Δ. For degree reasons, it is clear that any $x \in X^n$ can be expressed as the image of a non-degenerate z under a monomorphism $\sigma \in \Delta$. The nth latching map is a monomorphism just when each such expression is unique. Suppose given $\sigma z = x = \sigma' z'$ and pick a left inverse τ to σ to get $z = \tau \sigma' z'$. The map $\tau \sigma'$ factors as an epimorphism followed by a monomorphism. Because z is non-degenerate, the monomorphism is the identity, and hence that $\tau \sigma'$ is an epimorphism. Repeating this argument for z', with τ' a right inverse for σ', we see that $\tau' \sigma$ is also an epimorphism. It follows that z and z' have the same degree, and thus that both epimorphisms are identities, because Δ is a Reedy category. Hence $z = z'$.

If the set of left inverses for a monomorphism uniquely characterized that monomorphism, then we could conclude that σ and σ' must be equal, and hence that such decompositions would be fully unique. This is true for nearly all monomorphisms in Δ, the only exceptions being $d^0, d^1 \colon [0] \rightrightarrows [1]$. When any degenerate simplex in X is uniquely expressible as the image of a non-degenerate simplex under a monomorphism, we say that X has the **Eilenberg–Zilber property**. By this analysis, a cosimplicial object has the Eilenberg–Zilber property if and only if it is **unaugmentable**, that is, if the equalizer of $d^0, d^1 \colon X^0 \rightrightarrows X^1$ is the initial object [27, p. 147].

A cosimplicial object in a **Set**-valued functor category has the Eilenberg–Zilber property if and only if does pointwise. This argument proves the following lemma:

Lemma 14.3.8 *If a cosimplicial object X is unaugmentable, then the latching map $L^n X \to X$ is a monomorphism. If X and Y are both unaugmentable, then any pointwise monomorphism $X \to Y$ is also a Reedy monomorphism, that is, its relative latching maps are monic.*

Here is a key example:

Example 14.3.9 The Yoneda embedding is Reedy cofibrant because the equalizer of $\Delta^0 \rightrightarrows \Delta^1$ (in **Cat** and hence in **sSet**) is empty.

This example enables the proof of another significant corollary to Theorem 14.3.1.

Corollary 14.3.10   *If $\mathcal{M}$ is a simplicial model category, $|-|: \mathcal{M}^{\Delta^{op}} \to \mathcal{M}$ is left Quillen with respect to the Reedy model structure.*

Proof   The geometric realization is the functor tensor product with the Yoneda embedding, which is Reedy cofibrant by 14.3.9.   □

The Yoneda embedding $\Delta^{\bullet}: \Delta \to \mathbf{sSet}$ is pointwise weakly equivalent to the constant functor at the terminal object. Hence it is a cofibrant replacement of the terminal diagram in the Reedy model structure $\mathbf{sSet}^{\Delta}$. In particular, it follows from Lemma 14.3.7 that the geometric realization of any bisimplicial set is its homotopy colimit. This result also has a trivial special case. Regarding a simplicial set X as a discrete bisimplicial set, the simplicial set X is isomorphic to the geometric realization of the discrete bisimplicial set. In particular, any simplicial set X is the homotopy colimit of the corresponding discrete bisimplicial set.

Other corollaries of Theorem 14.3.1 will appear in Part IV; for instance, see Propositions 17.4.8 and 18.7.1.

14.4 Localizations and completions of spaces

In [10], Bousfield and Kan introduce a completion functor $R_{\infty}: \mathbf{sSet} \to \mathbf{sSet}$, associated to a commutative ring R with unit, defined to be the totalization of a particular cosimplicial simplicial set. They show that the cosimplicial objects appearing in their construction are automatically Reedy fibrant. Hence, by 14.3.9 and Theorem 14.3.1, the totalization is its homotopy colimit, computable as the limit of a tower of fibrations dual to (14.0.1).

The simplicial sets $R_{\infty}X$ are important because they define **localizations** or **completions** of X (under certain hypotheses), in a sense inspired by previous work of Sullivan [83]. Before we introduce the Bousfield–Kan construction, let us motivate the notions of localization and completion in topology by recalling the analogous algebra. In the interest of brevity, our presentation is much less comprehensive than can be found in either of those sources or in [58].

For any set P of primes, the P-**localization** of the integers $\mathbb{Z}$ is the ring $\mathbb{Z}_P \subset \mathbb{Q}$ defined by formally inverting all primes not in P. For instance, the localization of $\mathbb{Z}$ **at the prime** p is the ring $\mathbb{Z}_{(p)}$ defined by formally inverting all primes $\ell \neq p$. The **rationalization** of $\mathbb{Z}$ is the ring $\mathbb{Z}_{(0)} \cong \mathbb{Q}$ with all primes inverted. These definitions extend to general abelian groups A via the formula

$$A_P \cong A \otimes_{\mathbb{Z}} \mathbb{Z}_P.$$

This procedure defines an exact functor from the category of abelian groups to the category of P-**local** abelian groups, in which the natural action by each prime $\ell \notin P$ is an isomorphism.

The p-**adic completion** of $\mathbb{Z}$ is the ring $\hat{\mathbb{Z}}_p$ of p-adic integers, the completion of $\mathbb{Z}$ with respect to the p-adic valuation, constructed by taking the limit of the diagram

$$\cdots \longrightarrow \mathbb{Z}/p^3 \longrightarrow \mathbb{Z}/p^2 \longrightarrow \mathbb{Z}/p.$$

The p-adic integers may be thought of as the ring of formal power series in p with coefficients in $\{0, 1, \ldots, p-1\}$.

The following pair of lemmas capture something of what is meant by the phrase "working one prime at a time."

Lemma 14.4.1 *The ring $\mathbb{Z}$ is the limit of the diagram*

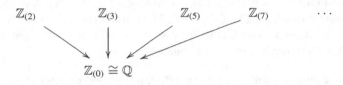

Lemma 14.4.2 *The ring $\mathbb{Z}$ is the pullback*

$$
\begin{array}{ccc}
\mathbb{Z} & \longrightarrow & \prod_p \hat{\mathbb{Z}}_p \\
\downarrow & & \downarrow \\
\mathbb{Q} \cong \mathbb{Z}_{(0)} & \longrightarrow & (\prod_p \hat{\mathbb{Z}}_p)_{(0)}
\end{array}
$$

These results, sometimes called "fracture theorems," generalize to finitely generated abelian or, more generally, nilpotent groups. Sullivan, following Serre, introduced localizations and completions of spaces with corresponding finiteness hypotheses as a way of "fracturing" homotopy types into "mod-p components."

Construction 14.4.3 ([10, §I.2]) Fix a commutative ring R with unit. Applying the free module functor $R \otimes -: \mathbf{Set} \to \mathbf{Mod}_R$ pointwise, there is an induced functor from simplicial sets to simplicial R-modules, which carries a simplicial set X to its **free simplicial R-module**. Write RX for the simplicial subset of finite R-linear combinations of simplices in which the coefficients sum to the ring unit 1. The obvious natural transformations $\eta: \mathrm{id} \Rightarrow R$ and $\mu: R^2 \Rightarrow R$ make R into a monad on $\mathbf{sSet}$.

The choice of a basepoint $* \in X$ defines an isomorphism between RX and the quotient of the free R-module on X by the submodule spanned by the basepoint, which makes RX into a simplicial R-module. The reduced R-homology of a based space X is defined by

$$\tilde{H}_*(X; R) = \pi_*(R \otimes X / R \otimes *).$$

Hence, via this isomorphism, we have

$$\pi_*(RX) \cong \tilde{H}_*(X; R), \qquad (14.4.4)$$

and moreover the unit map $X \to RX$ induces the Hurewicz homomorphism. Thus, by construction, $X \to Y$ is an R-homology isomorphism if and only if $RX \to RY$ is a weak homotopy equivalence.

Exercise 14.4.5   Show that for any simplicial set X, RX is a Kan complex. This result, analogous to [32, 3.4] or [55, 17.1], combines with Theorem 10.5.1 to imply that $X \to Y$ is an R-homology isomorphism if and only if the induced map $RX \to RY$ is a homotopy equivalence.

Given a monad such as the triple (R, η, μ) defined on **sSet** in Construction 14.4.3, the **monad resolution** is a functor $R^\bullet \colon \mathbf{sSet} \to \mathbf{sSet}^{\Delta+}$ from the base category to the category of augmented cosimplicial objects defined at a simplicial set X by the diagram

$$
X \xrightarrow{\;\;\eta\;\;} RX
\begin{array}{c}
\xleftarrow{\;\mu\;} \\[-2pt]
\xrightarrow{\;\eta R\;} \\[-2pt]
\xleftarrow{\;R\eta\;}
\end{array}
R^2 X
\begin{array}{c}
\xrightarrow{\;\eta R^2\;} \\[-2pt]
\xleftarrow{\;\mu R\;} \\[-2pt]
\xrightarrow{\;R\eta R\;} \\[-2pt]
\xleftarrow{\;R\mu\;} \\[-2pt]
\xrightarrow{\;R^2\eta\;}
\end{array}
R^3 X \cdots
$$

Exercise 14.4.6   Forgetting the augmentation, show that the cosimplicial object $R^\bullet X$ is Reedy fibrant. (Hint: see [55, p. 67] for inspiration.)

Definition 14.4.7 (R-completion)   The R-**completion** of a simplicial set X is

$$R_\infty X := \mathrm{Tot}\, R^\bullet X,$$

the totalization of the cosimplicial simplicial set $R^\bullet X$, as defined in Example 4.3.1. By Exercise 14.4.6, $R_\infty X$ is the limit of a tower of fibrations defined with respect to the totalizations of truncations of $R^\bullet X$. The augmentation $X \xrightarrow{\eta} RX$ of the cosimplicial object $R^\bullet X$ factors through the totalization, providing a natural map $X \xrightarrow{\phi} R_\infty X$.

Lemma 14.4.8   *A map $f \colon X \to Y$ of simplicial sets is an R-homology equivalence if and only if $R_\infty f \colon R_\infty X \to R_\infty Y$ is a homotopy equivalence.*

Proof   Applying R to the factorization of η through ϕ and postcomposing with the monad multiplication μ defines a retract diagram

$$
\begin{array}{ccccc}
RX & \xrightarrow{\ R\phi\ } & RR_\infty X & \longrightarrow & RX \\
{\scriptstyle Rf}\downarrow & & {\scriptstyle RR_\infty f}\downarrow & & \downarrow{\scriptstyle Rf} \\
RY & \xrightarrow[\ R\phi\ ]{} & RR_\infty Y & \longrightarrow & RY
\end{array}
$$

Taking homotopy groups, the isomorphisms (14.4.4) tell us that if $R_\infty X \to R_\infty Y$ is a homotopy equivalence, then f is an R-homology equivalence, because retracts of isomorphisms are isomorphisms.

Conversely, an R-homology equivalence $X \to Y$ induces a pointwise weak homotopy equivalence $R^\bullet X \to R^\bullet Y$ by 14.4.3. By Exercise 14.4.6 and the dual of Corollary 14.3.10, the totalization $R_\infty f : R_\infty X \to R_\infty Y$ is a weak homotopy equivalence and hence a homotopy equivalence between Kan complexes.   □

When X is nilpotent, the simplicial set $(\mathbb{Z}_P)_\infty X$ is a P-localization of X in the sense made precise by the statement of Theorem 14.4.11.

Definition 14.4.9   A space or simplicial set X is **nilpotent** if the action of $\pi_1 X$ on each $\pi_n X$, $n \geq 1$, is nilpotent, that is, if $\pi_n X$ admits a finite sequence of subgroups

$$
\pi_n X = G_1 \lhd \cdots \lhd G_k = *
$$

so that

  (i) G_{i+1} is normal in G_i and G_i / G_{i+1} is abelian
 (ii) each G_i is closed under the action of $\pi_1 X$
(iii) the induced action on G_i / G_{i+1} is trivial

A group is nilpotent if and only if its action by inner automorphisms is nilpotent. In particular, abelian groups are nilpotent. The $n = 1$ case of Definition 14.4.9 requires the fundamental group of a nilpotent space to be nilpotent.

Example 14.4.10   When X is simply connected (so that $\pi_1 X = 0$), the $\pi_1 X$-actions are trivial, and of course the higher homotopy groups are abelian. So simply connected spaces are nilpotent.

When X is nilpotent, its homotopy groups $\pi_n X$ are finitely generated for each $n \geq 1$ if and only if its integral homology groups are finitely generated, and this is the case if and only if X is weakly equivalent to a CW complex with finite skeleta [58, 4.5.2]. Nilpotent spaces can be approximated by Postnikov towers, leading to an elementary development of localization and completion and proofs of their properties; see [58, §3]. For the remainder of this section, we assume that the simplicial set X is based, connected, and nilpotent.

Theorem 14.4.11 (Bousfield–Kan) *Let P be any set of primes. If X is based, connected, and nilpotent, then $X \xrightarrow{\phi} (\mathbb{Z}_P)_\infty X$ is a $\mathbb{Z}_P$-homology equivalence, and the natural map*

$$\pi_* X \otimes_{\mathbb{Z}} \mathbb{Z}_P \to \pi_*((\mathbb{Z}_P)_\infty X)$$

is an isomorphism, that is, the homotopy groups of $(\mathbb{Z}_P)_\infty X$ are the P-localizations of the homotopy groups of X.

Even in algebra, completions are more subtle than localizations (see, e.g., [83, §1] and [58, §10]). For simplicity, we consider only finitely generated abelian groups A, for which the p-adic completion may be defined by the formula

$$\hat{A}_p \cong A \otimes_{\mathbb{Z}} \hat{\mathbb{Z}}_p.$$

In topology, this means we must restrict further to based, connected, nilpotent simplicial sets whose homotopy groups are finitely generated abelian groups. The behavior when the homotopy groups are not of finite type is also understood, but we will not go into that here.

Theorem 14.4.12 (Bousfield–Kan) *If X is based, connected, nilpotent, and has finitely generated abelian homotopy groups, then $X \xrightarrow{\phi} (\mathbb{Z}/p)_\infty X$ is a $\mathbb{Z}/p$-homology equivalence, and the natural map*

$$\pi_* X \otimes_{\mathbb{Z}} \hat{\mathbb{Z}}_p \to \pi_*((\mathbb{Z}/p)_\infty X)$$

is an isomorphism, that is, the homotopy groups of $(\mathbb{Z}/p)_\infty X$ are the p-adic completions of the homotopy groups of X.

We are now prepared to state two of the several "fracture theorems," in analogy with Lemmas 14.4.1 and 14.4.2. Given a pair of based simplicial sets, write $[W, X]$ as shorthand for the set of maps from W to X in the homotopy category $\text{Ho}(\textbf{sSet}_*)$. When X is a Kan complex, this is the usual set of homotopy classes of maps defined via the notion of homotopy provided by the simplicial tensor structure described in Example 3.7.13.

Theorem 14.4.13 ([10, V.6.2]) *Suppose that W and X are connected, based simplicial sets. If X is nilpotent and W is finite, then the set $[W, X]$ is the*

limit of the diagram

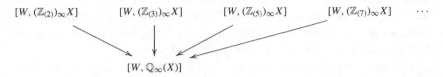

with limit cone defined by the localization maps ϕ.

Theorem 14.4.14 ([10, VI.8.1]) *Suppose that W and X are connected, based simplicial sets. If X is nilpotent, X has finitely generated homotopy groups, and W is finite, then*

$$
\begin{array}{ccc}
[W, X] & \longrightarrow & \prod_p [W, (\mathbb{Z}/p)_\infty X] \\
\downarrow & & \downarrow \\
[W, \mathbb{Q}_\infty X] & \longrightarrow & [W, \mathbb{Q}_\infty(\prod_p (\mathbb{Z}/p)_\infty X)]
\end{array}
$$

is a pullback in which the upper horizontal map is injective.

More general versions of Theorems 14.4.13 and 14.4.14 can be found in [10] or [58].

14.5 Homotopy colimits of topological spaces

Corollary 14.3.10 is essential to the proof given in Chapter 5 that our homotopy colimit formula is **homotopical**, computing a well-defined homotopy type for the homotopy colimit of pointwise weakly equivalent diagrams of the same shape. Let us briefly recall the proof. Given a small category $\mathcal{D}$, a pointwise weak equivalence $F \to F'$ between a pair of diagrams of shape $\mathcal{D}$ induces a pointwise weak equivalence

$$
B_\bullet(*, \mathcal{D}, QF) \to B_\bullet(*, \mathcal{D}, QF') \tag{14.5.1}
$$

between simplicial objects because coproducts of weak equivalences between cofibrant objects are again weak equivalences. As proven in Lemma 5.2.1, the one-sided simplicial bar construction on a pointwise cofibrant diagram is Reedy cofibrant. By Corollary 14.3.10, the geometric realization of (14.5.1) is a weak equivalence

$$
\operatorname{hocolim}_\mathcal{D} F = B(*, \mathcal{D}, QF) \to B(*, \mathcal{D}, QF') = \operatorname{hocolim}_\mathcal{D} F'.
$$

This proves that the functor $\operatorname{hocolim}_\mathcal{D}$ is homotopical.

As mentioned in Remark 6.3.4 and used throughout this text, in the convenient category of spaces, one can show that the functor $B(*, \mathcal{D}, -)$ preserves all weak equivalences, not just those between pointwise cofibrant diagrams. Our aim in this section is to outline a proof of this, referring the reader to [19, §A], our source for this material, for more details.

To begin, observe that coproducts of weak equivalences between arbitrary spaces are weak equivalences by an elementary connectedness argument. Because the natural map $q \colon QF \to F$ is a weak equivalence, it follows that $B_\bullet(*, \mathcal{D}, QF)$ and $B_\bullet(*, \mathcal{D}, F)$ are pointwise weak equivalent. The proof that $B(*, \mathcal{D}, QF)$ and $B(*, \mathcal{D}, F)$ are weak equivalent, and hence that the latter has the homotopy type of the homotopy colimit, has two remaining steps. The first is that the functor $|-| \colon \mathbf{Top}^{\Delta^{op}} \to \mathbf{Top}$ preserves weak equivalences between **split** simplicial spaces, even if they are not Reedy cofibrant. The conclusion follows because the simplicial spaces $B_\bullet(*, \mathcal{D}, F)$ are split.

Definition 14.5.2 A simplicial space $X_\bullet$ is **split** if there exist subspaces $N_n X \hookrightarrow X_n$ for each n so that the canonical map

$$\coprod_{[n] \twoheadrightarrow [k]} N_k X \to X_n$$

is an isomorphism.

The idea is that the space X_n decomposes as a direct sum of the "non-degenerate" part $N_n X$, and the remaining "degenerate" part itself decomposes as a direct sum of appropriate lower-dimensional "non-degenerate" pieces. If $X_\bullet$ is split and the spaces $N_n X$ are cofibrant, then the argument given in the proof of Lemma 5.2.1 shows that $X_\bullet$ is Reedy cofibrant.

Example 14.5.3 Given any diagram $F \colon \mathcal{D} \to \mathbf{Top}$, the simplicial space $B_\bullet(*, \mathcal{D}, F)$ is split, with $N_n B_\bullet(*, \mathcal{D}, F)$ defined to be the component of the coproduct $\coprod_{\vec{d} \colon [n] \to \mathcal{D}} F d_n$ indexed by non-degenerate n-simplices in $N\mathcal{D}$.

Recall that the geometric realization of a simplicial space is the colimit of the sequence (14.0.1). When $X_\bullet$ is split, the maps in this sequence are pushouts

$$
\begin{array}{ccc}
|\partial \Delta^n| \times N_n X & \longrightarrow & \mathrm{sk}_{n-1}|X| \\
{\scriptstyle i_n \times 1}\downarrow & & \downarrow \\
|\Delta^n| \times N_n X & \longrightarrow & \mathrm{sk}_n |X|
\end{array}
\qquad (14.5.4)
$$

An inclusion $A \hookrightarrow B$ is a **relative T_1 inclusion** if, for any open subset of A and point in $B \setminus A$, there is some open subset of B containing the open subset of A but not the point.

Exercise 14.5.5 Let N be any space. Show that the pushout

$$
\begin{array}{ccc}
|\partial\Delta^n| \times N & \longrightarrow & A \\
{\scriptstyle i_n \times 1} \downarrow & & \downarrow \\
|\Delta^n| \times N & \overset{\ulcorner}{\longrightarrow} & B
\end{array}
$$

is a relative T_1 inclusion.

The use of the term "compact" in the context of what it means to "permit the small object argument" is inspired by the following lemma:

Lemma 14.5.6 *Suppose K is compact and*

$$
Y_0 \hookrightarrow Y_1 \hookrightarrow Y_2 \hookrightarrow \cdots
$$

is a sequence of relative T_1 inclusions. Then any continuous map $f : K \to$ colim$_n Y_n$ factors through some Y_k.

Proof Suppose this is not the case and, passing to a subsequence, choose a sequence of points $k_1, k_2, \ldots \in K$ with $f(k_n) \in Y_n \backslash Y_{n-1}$. Fix $m \geq 0$ and let $U_m = Y_m$, and use the relative T_1 property to define, for each $n > m$, open subsets $U_n \subset Y_n$ that contain U_{n-1} but none of the points $f(k_j)$ for each $m < j \leq n$. The union of these subsets is an open subset $V_m \subset \text{colim}_n Y_n$. As m varies, the V_m cover colim Y and hence $f(K)$, but the construction permits no finite subcover. $\qquad\square$

Proposition 14.5.7 *If $X_\bullet \to Y_\bullet$ is a pointwise weak equivalence of split simplicial spaces, then $|X_\bullet| \to |Y_\bullet|$ is a weak equivalence.*

Proof Because $X_\bullet$ is split, $|X_\bullet|$ is a colimit of a sequence of relative T_1 inclusions by (14.5.4) and Exercise 14.5.5. Because spheres are compact, Lemma 14.5.6 gives an isomorphism

$$
\underset{n}{\text{colim}} \, \pi_k |\text{sk}_n X_\bullet| \overset{\cong}{\to} \pi_k |X_\bullet|
$$

for all $k \geq 0$. Hence it suffices to show that the maps $|\text{sk}_n X_\bullet| \to |\text{sk}_n Y_\bullet|$ are weak equivalences. By a connectedness argument, the natural transformation $X_\bullet \to Y_\bullet$ comprises weak equivalences $N_n X \to N_n Y$. The induction step is completed by applying the following lemma to the pushout squares (14.5.4). $\qquad\square$

Lemma 14.5.8 *Given arbitrary spaces and weak equivalences $A \to B$ and $X \to Y$, consider a diagram*

$$
\begin{array}{ccccc}
|\Delta^n| \times A & \longleftarrow & |\partial\Delta^n| \times A & \longrightarrow & X \\
\downarrow & & \downarrow & & \downarrow \\
|\Delta^n| \times B & \longleftarrow & |\partial\Delta^n| \times B & \longrightarrow & Y
\end{array}
$$

in which the vertical and left-hand horizontal maps are the obvious ones. Then the induced map between the pushouts is a weak equivalence.

We leave the proof as an exercise to the reader with the following hint: the key topological input is the following classical result, which also underpins the Mayer–Vietoris sequence.

Lemma 14.5.9 ([34, 16.24]) *A map $f: X \to Y$ in **Top** is a weak equivalence if there is some open cover U, V of Y such that the maps*

$$f^{-1}(U) \to U, \qquad f^{-1}(V) \to V, \qquad f^{-1}(U \cap V) \to U \cap V$$

are weak equivalences.

Remark 14.5.10 This argument does not extend to the weighted homotopy colimits discussed in Section 9.2. Unless each unit map $* \to \mathcal{D}(d, d)$ for a small topologically enriched category admits a complement, that is, unless the space $\mathcal{D}(d, d)$ decomposes into a disjoint union of the point representing the identity and its complement, the simplicial space $B_\bullet(G, \mathcal{D}, F)$ will not be split. While we do not know of an explicit example of a non-pointwise cofibrant diagram of spaces whose weighed homotopy colimit $B(QG, \mathcal{D}, QF)$ is not weakly equivalent to $B(G, \mathcal{D}, F)$, new ideas would be needed to argue that the simpler formula, without the pointwise cofibrant replacements, suffices.

Part IV

Quasi-categories

15

Preliminaries on quasi-categories

One of the fundamental invariants of algebraic topology arises when we regard a topological space as something like a category, or rather a groupoid. The points of the space become objects of the category. A path, here a continuous function from the standard unit interval, represents a morphism between its starting and ending points. More accurately, for there to be an associative composition law, we must revise this outline slightly and define a morphism to be an endpoint-preserving homotopy class of paths. This defines the fundamental groupoid of the space.

But from the topological perspective, it seems artificial to take homotopy classes of paths in pursuit of strict associativity. The more natural construction forms a (weak) ∞-groupoid with objects the points of X, 1-morphisms the paths in X, 2-morphisms the homotopies between paths, 3-morphisms the homotopies between these homotopies, and so on. With this example in mind, the **homotopy hypothesis**, a principle guiding these definitions, says that an ∞-groupoid should be the same thing as a topological space.

Continuing in this vein, mathematical structures admitting a topological enrichment assemble into $(\infty, 1)$-**categories**, loosely defined to be categories with morphisms in each dimension such that every morphism above dimension 1 is invertible. One way to encode this definition is to say that an $(\infty, 1)$-category is a category (weakly) enriched in ∞-groupoids, which are also called $(\infty, 0)$-categories. In what follows, the terms **quasi-category** and ∞-**category** are synonyms for a particular model of $(\infty, 1)$-categories for which these objects are simplicial sets with a certain lifting property.

In this chapter, we define quasi-categories and introduce the appropriate notion of equivalence between them. We then explain how quasi-categories model $(\infty, 1)$-categories by introducing the homotopy category of a quasi-category and several equivalent models for the hom-spaces between objects in a quasi-category.

Before proceeding any further, we should set notation for particular simplicial sets. As mentioned, we write Δ^n for the standard n-simplex; $\partial \Delta^n$ for its boundary, the simplicial $(n-1)$-sphere; and Λ^n_k for the subset thereof consisting of all faces containing the vertex $k \in [n]$. By convention, $\partial \Delta^0 = \emptyset$. It will be convenient to have names for the canonical inclusions: we like $j^n_k \colon \Lambda^n_k \to \Delta^n$ and $i_n \colon \partial \Delta^n \to \Delta^n$ because the js are generating trivial cofibrations and the is are generating cofibrations in Quillen's model structure of 11.3.5. Recall that weakly saturated closures of the j^n_k and the i_n are the classes of anodyne maps and cofibrations, respectively. As shorthand, we often write $*$ for the terminal object Δ^0 and I for Δ^1. This interval object gives rise to the notion of simplicial homotopy.

Using the Yoneda lemma, we write $d^i \colon \Delta^{n-1} \to \Delta^n$ for the ith face inclusion and $s^i \colon \Delta^{n+1} \to \Delta^n$ for the ith degeneracy map for each $i \in [n]$; these are the traditional names given to the corresponding morphisms in $\mathbb{\Delta}$. Again by the Yoneda lemma, precomposition with maps between representable simplicial sets defines the right action of the category $\mathbb{\Delta}$ on the graded set of simplices in a simplicial set.

We will make frequent use of the fact that the category of simplicial sets is cartesian closed. Recall, from Example 1.5.6, that an n-simplex $\Delta^n \to Y^X$ in the hom-space from X to Y is, by adjunction, a map $X \times \Delta^n \to Y$ of simplicial sets. In particular, for each m-simplex in X, we get a map $\Delta^m \times \Delta^n \to X \times \Delta^n \to Y$; conversely, the n-simplex $\Delta^n \to Y^X$ is defined by this data, chosen compatibly with faces and degeneracies. The next exercises are intended to familiarize the reader with the geometry of the simplicial set $\Delta^m \times \Delta^n$.

Exercise 15.0.1 The vertices of $\Delta^m \times \Delta^n$ are labelled by ordered pairs (i, j) with $i \in [m]$ and $j \in [n]$. Because $\Delta^m \times \Delta^n$ is the nerve of the poset $[m] \times [n]$, each simplex is uniquely determined by its vertices. Use this notation to describe the k-simplices in $\Delta^m \times \Delta^n$ and to characterize the non-degenerate simplices.

Exercise 15.0.2 (Leibniz formula) The boundary of $\Delta^m \times \Delta^n$ is the union $\partial \Delta^m \times \Delta^n \sqcup \Delta^m \times \partial \Delta^n$. Characterize the k-simplices appearing in the boundary.

Exercise 15.0.3 A top-dimensional non-degenerate simplex of $\Delta^m \times \Delta^n$ is called a **shuffle**. Explain which shuffles share codimension-one faces and use this to give a sensible partial ordering on the set of shuffles with a minimal and a maximal element.

15.1 Introducing quasi-categories

The original definition of a **weak Kan complex**, now called a quasi-category (following Joyal [40]) or an ∞-category (following Jacob Lurie [49]), is due to J. Michael Boardman and Rainer Vogt [8]. Their motivating example appears as Example 16.4.12.

Definition 15.1.1 A **quasi-category** is a simplicial set X such that $X \to *$ has the right lifting property with respect to the inner horn inclusions j_k^n for each $n \geq 2, 0 < k < n$:

$$
\begin{array}{ccc}
\Lambda_k^n & \longrightarrow & X \\
\downarrow & \nearrow & \\
\Delta^n & &
\end{array}
\tag{15.1.2}
$$

An n-simplex that extends a given horn in X is colloquially called a **filler** and is thought of as some sort of composite of the $(n-1)$-simplices in the inner horn. This intuition is clearest in the case $n = 2$. The slogan, that such composites are unique up to a contractible space of choices, will be proven as a consequence of Corollary 15.2.4.

There are two principal sources of simple examples. More sophisticated examples are produced by Lemma 16.4.10.

Example 15.1.3 Nerves of categories are quasi-categories; in fact, in this case, each lift (15.1.2) is unique.

Example 15.1.4 Tautologically, Kan complexes are quasi-categories. In particular, the total singular complex of a topological space is a Kan complex and hence a quasi-category.

Any quasi-category X has an associated **homotopy category** hX. Objects are vertices of X. A 1-simplex f represents a morphism whose source is the vertex fd^1 and whose target is the vertex fd^0. Hence 1-simplices in a quasi-category (or simplicial set) are often depicted as arrows $fd^1 \xrightarrow{f} fd^0$. The degenerate 1-simplices serve as identities in the homotopy category and are frequently depicted using an equals sign in place of the arrow.

As the name would suggest, the morphisms in hX are homotopy classes of 1-simplices, where a pair of 1-simplices f and g with common boundary are **homotopic** if there exists a 2-simplex whose boundary has any of the following forms:

$$
\nearrow \overset{f}{\cdots} \searrow \qquad \nearrow \overset{f}{\cdots} \searrow \qquad \nearrow \overset{g}{\cdots} \searrow \qquad \nearrow \overset{g}{\cdots} \searrow
$$

$$
\underset{g}{\longrightarrow} \qquad \underset{g}{\longrightarrow} \qquad \underset{f}{\longrightarrow} \qquad \underset{f}{\longrightarrow}
$$

$$(15.1.5)$$

Indeed, in a quasi-category, if any of the 2-simplices (15.1.5) exists, then there exists a 2-simplex of each type.

Exercise 15.1.6   Prove this.

Generic 2-simplices in X

$$
\overset{f}{\nearrow} \overset{\cdot}{\underset{\sim}{}} \overset{g}{\searrow}
$$
$$
\cdot \underset{h}{\longrightarrow} \cdot
$$

$$(15.1.7)$$

witness that $gf = h$ in the homotopy category. Conversely, if $h = gf$ in hX and f, g, h are any 1-simplices representing these homotopy classes, then there exists a 2-simplex (15.1.7) witnessing the composition relation. The reader who has not seen this before is encouraged to work out the details (or see [49, §1.2.3]).

Exercise 15.1.8   Show that h is the left adjoint to the nerve functor introduced in 1.5.5:

$$
\mathbf{qCat} \underset{N}{\overset{h}{\underset{\perp}{\rightleftarrows}}} \mathbf{Cat}
$$

Here we have restricted the domain of the left adjoint to $\mathbf{qCat} \subset \mathbf{sSet}$, the full subcategory of quasi-categories. The definition of the homotopy category associated to a generic simplicial set is slightly more complicated.

15.2 Closure properties

Because quasi-categories are characterized by a lifting property, they immediately inherit closure properties from Lemma 11.1.4. We call the left and right classes of the weak factorization system generated by the inner horn inclusions the **inner anodyne** maps and the **inner fibrations**, respectively. The inner fibrations are closed under products, pullbacks, retracts, and composition. It follows that quasi-categories are closed under products and retracts.

One might conjecture that if X is a quasi-category and A is a simplicial set, then X^A is a quasi-category. For one thing, this is true for Kan complexes. For another, in categorical contexts, diagram spaces tend to inherit the properties of their codomain, such as closure under certain limits or colimits. Not only is it true that X^A is a quasi-category if X is, but the proof encodes what is in some sense the key combinatorial result underlying the theory of quasi-categories.

Let us think what is being asserted by this statement. From the definition, we are asked to show that there exist extensions

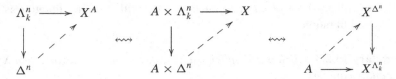

for all $n \geq 2$, $0 < k < n$. The three lifting problems correspond by the two-variable adjunction defining the cartesian closed structure on simplicial sets. To define the lift in the right-hand diagram, we must choose cylinders $\Delta^m \times \Delta^n \to X$ for each m-simplex in A in a way that is compatible with the specified horn $\Delta^m \times \Lambda^n_k \to X$ and also with previously specified cylinders $\partial \Delta^m \times \Delta^n \to X$ corresponding to the boundary of the m-simplex. In other words, inductively, we must choose extensions

The indicated lifting problems are transposes, using the fact that the map $X^{\Delta^n} \to X^{\Lambda^n_k}$ is the pullback-hom of j^n_k with $X \to *$. Such extensions always exist on account of the following result.

Proposition 15.2.1 (Joyal) *The pushout-product of a monomorphism with an inner anodyne map is inner anodyne.*

Proof Using Corollary 12.2.4, which characterizes the left class of a weak factorization system in a category permitting the small object argument as retracts of "relative cell complexes" built from the generators, it suffices to show this is true of the $i_m \hat{\times} j^n_k$s because the bifunctor $- \hat{\times} -$ preserves these colimits in each variable. A direct proof, decomposing these monomorphisms into pushouts of inner horns and thereby giving each pushout-product an inner anodyne cellular structure, is given in [20, A.1]. A non-constructive proof is given in [49, 2.3.2.4]. $\square$

Remark 15.2.2 Define a trivial fibration of simplicial sets to be a map with the right lifting property against the monomorphisms, also called the cofibrations in this context. Applying Lemma 11.1.10 to the (cofibration, trivial fibration) weak factorization system and two copies of the (inner anodyne, inner fibration) weak factorization system, Proposition 15.2.1 is equivalent to either of the following two statements:

(ii) the pullback-hom of a cofibration with an inner fibration is an inner fibration

(iii) the pullback-hom of an inner anodyne map with an inner fibration is a trivial fibration.

Corollary 15.2.3 *If A is a simplicial set and X is a quasi-category, then X^A is a quasi-category.*

Proof The pullback-hom of $\emptyset \to A$ and $X \to *$ is $X^A \to *$. $\qquad\square$

Corollary 15.2.4 *If X is a quasi-category and Λ_k^n any inner horn, then $X^{\Delta^n} \to X^{\Lambda_k^n}$ is a trivial fibration.*

In particular, by pullback stability of the trivial fibrations, the fiber over any point is a contractible Kan complex. This says that the space of fillers to a given horn in X is a contractible Kan complex. This is the common form taken by a homotopical uniqueness statement in quasi-category theory and is what is meant by saying something is "well defined up to a contractible space of choices."

Another corollary sounds more sophisticated, though the proof is no harder. By Corollary 7.6.4, the simplicial category **sSet** is complete and cocomplete in the simplicially enriched sense, admitting all weighted colimits. We write **qCat** for the full simplicial subcategory spanned by the quasi-categories. We learned this theorem from Dominic Verity.

Theorem 15.2.5 *The weighted limit of any diagram $X \colon \underline{\mathcal{D}} \to \mathbf{qCat}$ whose weight $W \colon \underline{\mathcal{D}} \to \mathbf{sSet}$ is projectively cofibrant is a quasi-category.*

When $\underline{\mathcal{D}}$ is unenriched, this notion of projectively cofibrant is exactly the one described in 11.5.5. But this result and our desired applications extend to simplicially enriched diagrams, for which case we say a simplicial functor W is projective cofibrant just when the unique map $\emptyset \to W$ is in the weakly saturated closure of the maps

$$\{\underline{\mathcal{D}}(d, -) \times \partial\Delta^n \to \underline{\mathcal{D}}(d, -) \times \Delta^n \mid n \geq 0, d \in \underline{\mathcal{D}}\}.$$

Proof By 12.2.4, W is projective cofibrant just when $\emptyset \to W$ is a retract of a (transfinite) composite of pushouts of coproducts of such maps. Note that the limit of X weighted by the empty weight is the terminal object $*$. Because the class of inner fibrations is closed under the duals of these colimits, it suffices to show that for any diagram $X\colon \mathcal{D} \to \mathbf{qCat}$, the map of weighted limits

$$\lim{}^{\mathcal{D}(d,-)\cdot\Delta^n} X \to \lim{}^{\mathcal{D}(d,-)\cdot\partial\Delta^n} X$$

is an inner fibration.

By Theorem 7.6.3 and the Yoneda lemma,

$$\lim{}^{\mathcal{D}(d,-)\cdot\Delta^n} X \cong \int_{e\in\mathcal{D}} Xe^{\mathcal{D}(d,e)\cdot\Delta^n} \cong \left(\int_{e\in\mathcal{D}} Xe^{\mathcal{D}(d,e)}\right)^{\Delta^n} = Xd^{\Delta^n}.$$

Similarly, $\lim{}^{\mathcal{D}(d,-)\cdot\partial\Delta^n} X \cong Xd^{\partial\Delta^n}$ and the map comparing the weighted limits is the map between the internal homs induced by the inclusion $\partial\Delta^n \to \Delta^n$. Because Xd is a quasi-category, 15.2.2.(ii) implies that this map is an inner fibration, as desired. $\square$

Example 15.2.6 Consider a diagram $f\colon 2 \to \mathbf{qCat}$ whose image is $f\colon X \to Y$ and the weight $N(2/-)\colon 2 \to \mathbf{sSet}$. As described at the end of Chapter 11, $N(2/-)$ is projectively cofibrant. Its image is the map $d^1\colon \Delta^0 \to \Delta^1$. Using the usual end formula, the weighted limit is the pullback

$$
\begin{array}{ccc}
\lim^{N(2/-)} f & \longrightarrow & Y^{\Delta^1} \\
\downarrow & & \downarrow{\scriptstyle d^1} \\
X & \xrightarrow{\ f\ } & Y
\end{array}
$$

In this way we see that the weighted limit the usual path space Nf defined in Example 6.5.2. Theorem 15.2.5 tells us that this space is a quasi-category. Of course, this can also be deduced directly from Proposition 15.2.2(ii) and the closure properties of the inner fibrations, but in other examples, this is much less obvious.

Indeed, it follows, essentially from Theorem 15.2.5 and Corollary 11.5.13, that quasi-categories are closed under arbitrary homotopy limits. The proof is straightforward once we settle on the correct notion of "homotopy limit" in the quasi-categorical context. We return to this topic in Section 17.7.

15.3 Toward the model structure

By a well-known theorem of Quillen, Kan complexes, which are combinatorial models for spaces, are the fibrant objects in a model structure on **sSet** whose cofibrations are the monomorphisms. We might hope that there is another model structure on **sSet** with the monomorphisms as cofibrations and whose fibrant objects are the quasi-categories, and indeed this is true. Furthermore, there is only one such model structure:

Theorem 15.3.1 ([41, E.1.10])   *The cofibrations and fibrant objects completely determine a model structure, supposing it exists.*

Proof   It suffices to prove that this data determines the weak equivalences. The cofibrations are the left class of a weak factorization system $(\mathcal{C}, \mathcal{F}_t)$ whose right class is the class of trivial fibrations. It suffices to show that this weak factorization system together with the fibrant objects determine the weak equivalences. The weak factorization system $(\mathcal{C}, \mathcal{F}_t)$ gives us a notion of cofibrant replacement for objects and maps. By the 2-of-3 property, a map is a weak equivalence if and only if its cofibrant replacement is a weak equivalence. Hence it suffices to determine the weak equivalences between cofibrant objects.

Because any model category $\mathcal{M}$ is saturated, a map is a weak equivalence if and only if it is an isomorphism in the homotopy category. By the Yoneda lemma, a map is an isomorphism if and only if the corresponding natural transformation is an isomorphism of represented functors. Because every object in the homotopy category is isomorphic to a fibrant object, a map $f : A \to B$ is a weak equivalence if and only if the maps $\mathrm{Ho}\mathcal{M}(B, X) \to \mathrm{Ho}\mathcal{M}(A, X)$ are bijections for each fibrant object in X.

We have reduced to the case where A and B are cofibrant, which allows us to exploit Quillen's construction of the hom-set from a cofibrant object to a fibrant object in $\mathrm{Ho}\mathcal{M}$. Quillen shows that $\mathrm{Ho}\mathcal{M}(A, X)$ is the quotient of $\mathcal{M}(A, X)$ by the "left-homotopy" relation defined using a cylinder object for A, which is in turn defined by the cofibration–trivial fibration factorization. In this way, the cofibrations and fibrant objects precisely determine the model structure, supposing it exists.   □

We use this result to define the weak equivalences for the hoped-for model structure for quasi-categories. A concrete description makes use of a particularly nice cylinder object. Let J be the nerve of the free-standing isomorphism $\mathbb{I}$; the name is selected because J is something like an interval. Equivalently, J is the 0-coskeletal simplicial set on the set of two vertices. This simplicial set might also be called S^∞ because it has two non-degenerate simplices in each dimension, is contractible (by the result we are about to prove), and has a natural $\mathbb{Z}/2$ action. Indeed, it is a simplicial model for the total space of the classifying space $\mathbb{R}P^\infty = K(\mathbb{Z}/2, 1) = B(*, \mathbb{Z}/2, *)$ (cf. 6.4.11).

Lemma 15.3.2   *The map $J \to *$ is a trivial fibration.*

Proof We must show that there exist solutions to lifting problems

When $n = 0$, this is true because J is non-empty. For larger n, we use the fact that $J \cong \mathrm{cosk}_0 J$. By adjunction, it suffices to show that J lifts against $\mathrm{sk}_0 \partial \Delta^n \to \mathrm{sk}_0 \Delta^n$, but for $n > 0$, the 0-skeleton of Δ^n is isomorphic to the 0-skeleton of its boundary. $\qquad \square$

For any simplicial set A, the projection $A \times J \to A$ is a pullback of $J \to *$. Hence $A \times J \to A$ is also a trivial fibration by Lemma 11.1.4. Evidently, the obvious map $A \sqcup A \to A \times J$ is a monomorphism and hence a cofibration. It follows that J can be used to define cylinder objects, that is, functorial factorizations of the form

$$A \sqcup A \rightarrowtail A \times J \xrightarrow{\sim} A$$

for any simplicial set A.

Write $[A, X]_J$ for the quotient of the set of maps from the simplicial set A to the quasi-category X by the relation generated by $f \sim g$ whenever there exists a diagram

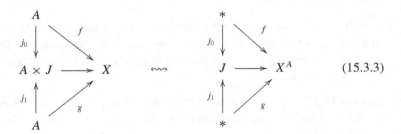

$$(15.3.3)$$

Indeed, we will eventually see that the "generated" here is unnecessary: by Lemma 17.0.2 and Lemma 17.2.5, any f and g in the same equivalence class admit such diagrams. Note that any $\emptyset \to A$ is a monomorphism; hence all simplicial sets are cofibrant. By the proof of Theorem 15.3.1, a map $f : A \to B$ of simplicial sets is a weak equivalence in the model structure for quasi-categories if and only if it induces a bijection $[B, X]_J \to [A, X]_J$ for all quasi-categories X. We follow Lurie and call these maps **categorical equivalences** or simply **equivalences** if the source and target are quasi-categories. Joyal calls these weak equivalences "weak categorical equivalences."

Example 15.3.4 By 15.2.2.(iii), if $A \to B$ is inner anodyne and X is a quasi-category, then $X^B \to X^A$ is a trivial fibration and in particular has a section $X^A \to X^B$, which can be used to show that the map $[B, X]_J \to [A, X]_J$ is surjective. The right lifting property of $X^B \to X^A$ against $* \sqcup * \to J$ can be used to prove injectivity. Hence inner anodyne maps are categorical equivalences.

Exercise 15.3.5 Show that the trivial fibrations are categorical equivalences.

We record, but do not take time to establish, the model structure. For proof, see [49, 2.2.5.1], [20, 2.13], or [70], the latter of which is an exposition of Joyal's original proof.

Theorem 15.3.6 (Joyal) *There is a left proper, cofibrantly generated, monoidal model structure on* **sSet** *whose fibrant objects are precisely the quasi-categories, whose cofibrations are monomorphisms, whose weak equivalences are the categorical equivalences just defined, and whose fibrations between fibrant objects are those maps that lift against the inner horn inclusions and also the map* $j_0 \colon * \to J$.

Remark 15.3.7 The inner horn inclusions are not the generating trivial cofibrations, though they do suffice to detect the fibrant objects. The inner horn inclusions together with j_0 suffice to detect fibrations between fibrant objects, called **isofibrations**. Set theoretical arguments can be used to show that there exists a set of generating trivial cofibrations, but no explicit description is known. See [49, §A.2.6].

In Chapter 17, we show that the definition of categorical equivalence forced on us by the model structure is a reasonable notion of equivalence for $(\infty, 1)$-categories. A corollary to the following lemma provides some preliminary supporting evidence.

Lemma 15.3.8 *The nerve and its left adjoint define a Quillen adjunction* $h \colon$ **sSet** $\rightleftarrows$ **Cat** $\colon N$ *between Joyal's model structure on* **sSet** *and the folk model structure on* **Cat**.

Proof Recall the definition of the folk model structure in 11.3.9. The homotopy category functor h sends monomorphisms to functors that are injective on objects; it remains to show that N preserves fibrations. The nerve functor is fully faithful because the counit of the adjunction $h \dashv N$ is an isomorphism. It follows from this and the fact that J is the nerve of the free-standing isomorphism that the right adjoint sends isofibrations in **Cat** to maps that lift against $j_0 \colon * \to J$. Furthermore, the nerve of any functor is an inner fibration; this is because nerves of categories have *unique* fillers for all inner horns. Hence N preserves fibrations by Example 15.1.3 and Remark 15.3.7. $\square$

Corollary 15.3.9 *If* $f \colon X \to Y$ *is a categorical equivalence, then* $hf \colon hX \to hY$ *is an equivalence of categories. If* $F \colon \mathcal{C} \to \mathcal{D}$ *is an equivalence of categories, then* $NF \colon N\mathcal{C} \to N\mathcal{D}$ *is a categorical equivalence.*

Proof This follows from the previous result and Ken Brown's lemma 11.3.14.
$\qquad\qquad\qquad\qquad\qquad\qquad\qquad\qquad\qquad\qquad\qquad\qquad\qquad$ □

Remark 15.3.10 Joyal's model structure has the same cofibrations as Quillen's and more fibrant objects; hence it has a smaller class of weak equivalences. This means that the Quillen model structure is a left Bousfield localization of the Joyal model structure (see 12.3.3). In particular, a categorical equivalence is necessarily a weak homotopy equivalence.

As a special case of a general result for left Bousfield localizations, a weak homotopy equivalence between Kan complexes is a categorical equivalence, and indeed an equivalence of quasi-categories [36, 3.2.13]. But in general, weak homotopy equivalences between quasi-categories need not be equivalence; for instance, the monomorphism $\Delta^1 \to J$ is a weak homotopy equivalence but not a categorical equivalence because $2 \to \mathbb{I}$ is not an equivalence of categories.

We see in Chapter 17 that an equivalence $X \to Y$ of quasi-categories always admits an inverse equivalence $Y \to X$ together with an "invertible homotopy equivalence" (17.2.6) using the notion of homotopy on display in (15.3.3).

15.4 Mapping spaces

For a quasi-category to model an $(\infty, 1)$-category, there must be hom-spaces between its objects (the vertices) representing well-defined homotopy types. These homotopy types, elements in the homotopy category of spaces $\mathcal{H}$, should have the property that their underlying sets, computed by applying the functor $\pi_0 \colon \mathcal{H} \to \mathbf{Set}$, coincide with the hom-sets in the homotopy category hX associated to the quasi-category. In this section, we take several stabs at the definition and prove that our guesses are all categorically equivalent. Furthermore, it is easy to show that our three candidate hom-spaces are all quasi-categories – indeed, we see in 17.2.2 that they are Kan complexes.

Exploiting the cartesian closure of simplicial sets, for any quasi-category X, we have a quasi-category X^{Δ^1} whose vertices are 1-simplices in X and whose n-simplices are cylinders $\Delta^n \times \Delta^1 \to X$. To form the mapping space between two fixed vertices $x, y \in X$, we might form the pullback

$$
\begin{array}{ccc}
\mathrm{Hom}_X(x, y) & \longrightarrow & X^{\Delta^1} \\
\downarrow & \lrcorner & \downarrow \\
* & \xrightarrow[\;(x,y)\;]{} & X \times X \cong X^{\partial\Delta^1}
\end{array}
$$

By 15.2.2.(ii), $\mathrm{Hom}_X(x, y)$ is a quasi-category. An n-simplex in $\mathrm{Hom}_X(x, y)$ is a map $\Delta^n \times \Delta^1 \to X$ such that the image of $\Delta^n \times \{0\}$ is degenerate at x and the image of $\Delta^n \times \{1\}$ is degenerate at y. In particular, 1-simplices look like

$$
\begin{array}{ccc}
x & \xrightarrow{\ f\ } & y \\[2pt]
\Big\| & \diagdown \sim & \Big\| \\[2pt]
x & \xrightarrow[g]{} & y
\end{array}
\qquad\qquad (15.4.1)
$$

from which we see that $\pi_0\mathrm{Hom}_X(x, y)$ is the hom-set from x to y in hX.

A less symmetric but more efficient construction is also possible. Let $\mathrm{Hom}_X^R(x, y)$ be the simplicial set whose 0-simplices are 1-simplices in X from x to y, whose 1-simplices are 2-simplices of the form

$$
\begin{array}{ccc}
 & x & \\
 \diagup\!\diagup \ \sim & & \searrow \\
x & \xrightarrow{\hspace{2cm}} & y
\end{array}
$$

and whose n-simplices are $(n + 1)$-simplices whose last vertex is y and whose $(n + 1)$th face is degenerate at x. Dually, $\mathrm{Hom}_X^L(x, y)$ is the simplicial set whose n-simplices are $(n + 1)$-simplices in X whose first vertex is x and whose zeroth face is degenerate at y. Once again, note that $\pi_0\mathrm{Hom}_X^L(x, y) = \pi_0\mathrm{Hom}_X^R(x, y) = hX(x, y)$.

Remark 15.4.2   The simplicial sets $\mathrm{Hom}_X^L(x, y)$ and $\mathrm{Hom}_X^R(x, y)$ are dual in the sense that $\mathrm{Hom}_X^L(x, y) = (\mathrm{Hom}_{X^{\mathrm{op}}}^R(y, x))^{\mathrm{op}}$. The annoying fact, from the perspective of homotopy (co)limits, that a simplicial set is not isomorphic to its opposite, in which the conventions on ordering of vertices in a simplex are reversed, is technically convenient here.

Exercise 15.4.3   Show that $\mathrm{Hom}_X^R(x, y)$ is a quasi-category, or at least prove that $\mathrm{Hom}_X^R(x, y)$ has fillers for horns $\Lambda_1^2 \to \Delta^2$.

To prove that $\mathrm{Hom}_X^L(x, y)$, $\mathrm{Hom}_X^R(x, y)$, and $\mathrm{Hom}_X(x, y)$ are categorically equivalent when X is a quasi-category, let us think geometrically about the difference. This discussion, and some of our notation, follows [20], with modifications due to Verity. Each quasi-category has the same zero simplices. An n-simplex in $\mathrm{Hom}_X^L(x, y)$ or $\mathrm{Hom}_X^R(x, y)$ is an $(n + 1)$-simplex in X, one of whose faces is degenerate. The shapes are given by the quotients

$$
\begin{array}{ccc}
\Delta^n & \longrightarrow & \Delta^0 \\
d^0 \downarrow & & \downarrow \\
\Delta^{n+1} & \xrightarrow{\ \ulcorner\ } & \Delta^{n+1}_{0|1}
\end{array}
\qquad\qquad
\begin{array}{ccc}
\Delta^n & \longrightarrow & \Delta^0 \\
d^{n+1} \downarrow & & \downarrow \\
\Delta^{n+1} & \xrightarrow{\ \ulcorner\ } & \Delta^{n+1}_{n|n+1}
\end{array}
$$

Let us explain the notation. Surjections $\Delta^n \twoheadrightarrow \Delta^1$ correspond to integers $0 \le i < n$, which partition the vertices of Δ^n into the fiber $[0 \cdots i]$ over 0 and the fiber $[i + 1 \cdots n]$ over 1. We write $\Delta^n_{i|i+1}$ for the quotient of Δ^n, which collapses the face spanned by the vertices $[0 \cdots i]$ to a point and the face spanned by the vertices $[i + 1 \cdots n]$ to a point.[1] This simplicial set has two vertices and has a non-degenerate k-simplex for each non-degenerate k-simplex of Δ^n whose image surjects onto Δ^1.

Similarly, the shape of an n-simplex in $\mathrm{Hom}_X(x, y)$ is given by

$$
\begin{array}{ccc}
\Delta^n \times \partial\Delta^1 & \xrightarrow{\ \mathrm{proj}_2\ } & \partial\Delta^1 \cong * \sqcup * \\
1\times i_1 \downarrow & & \downarrow \\
\Delta^n \times \Delta^1 & \xrightarrow{\ \ulcorner\ } & C^n_{\mathrm{cyl}}
\end{array}
$$

We have canonical maps

$$
C^n_L := \Delta^{n+1}_{0|1} \xleftarrow{\ r_L\ } C^n_{\mathrm{cyl}} \xrightarrow{\ r_R\ } \Delta^{n+1}_{n|n+1} =: C^n_R \tag{15.4.4}
$$

The horizontal maps are surjections: respectively, the quotients of the unique retractions $r_L, r_R \colon \Delta^n \times \Delta^1 \rightrightarrows \Delta^{n+1}$ defined on vertices by $r_L(i, 0) = 0$, $r_L(i, 1) = i + 1$, $r_R(i, 0) = i$, and $r_R(i, 1) = n + 1$.

Now write C^n_L for $\Delta^{n+1}_{0|1}$ and C^n_R for $\Delta^{n+1}_{n|n+1}$. This notation emphasizes that these constructions define three cosimplicial objects $C^\bullet_L$, $C^\bullet_{\mathrm{cyl}}$, $C^\bullet_R$, taking values in the category of simplicial sets and maps preserving two chosen basepoints. The target of these cosimplicial objects is the slice category $\partial\Delta^1/\mathbf{sSet}$, which we denote by $\mathbf{sSet}_{*,*}$. The quasi-category X with chosen vertices x, y becomes an object of $\mathbf{sSet}_{*,*}$. The geometric role played by the cosimplicial objects

---

[1] Of course, the "$i + 1$" subscript is redundant and is omitted in [20]; however, we do not want to confuse this simplicial set with the set of i-simplices in Δ^n.

$C_L^\bullet$, $C_{\text{cyl}}^\bullet$, $C_R^\bullet$ in our candidate hom-spaces is captured by the following equalities, which define the hom-spaces using the hom-sets of $\mathbf{sSet}_{*,*}$:

$$\text{Hom}_X^L(x, y) = \mathbf{sSet}_{*,*}(C_L^\bullet, X),$$

$$\text{Hom}_X(x, y) = \mathbf{sSet}_{*,*}(C_{\text{cyl}}^\bullet, X),$$

$$\text{Hom}_X^R(x, y) = \mathbf{sSet}_{*,*}(C_R^\bullet, X).$$

The natural maps $\text{Hom}_X^L(x, y) \to \text{Hom}_X(x, y) \leftarrow \text{Hom}_X^R(x, y)$ come from the maps (15.4.4) between the cosimplicial objects. We would like to show that these are categorical equivalences. Morally, this follows because $C_L^\bullet$, $C_{\text{cyl}}^\bullet$, and $C_R^\bullet$ are cofibrant resolutions of Δ^1 in the Joyal model structure. But we will prove this in a way that does not appeal to a black box. To begin:

Remark 15.4.5 The category $\mathbf{sSet}_{*,*}$, defined as a slice category, inherits a model structure from the Joyal model structure on $\mathbf{sSet}$: a map of twice-based simplicial sets is a cofibration, fibration, or weak equivalence just when the underlying map of simplicial sets is one. Fibrant objects are quasi-categories with chosen basepoints. An object is cofibrant if and only if its two chosen basepoints are distinct.

Lemma 15.4.6 $C_R^\bullet$, $C_L^\bullet$, $C_{\text{cyl}}^\bullet$ *are Reedy cofibrant.*

Proof Using Lemma 14.3.8, it suffices to show that these cosimplicial objects are unaugmentable, that is, that the equalizer of the face maps d^0, d^1 from the zeroth object to the first is initial in $\mathbf{sSet}_{*,*}$. All of the proofs are similar. For $C_{\text{cyl}}^\bullet$, the maps $C_{\text{cyl}}^0 \rightrightarrows C_{\text{cyl}}^1$ include $C_{\text{cyl}}^0 = \Delta^1$ at the top and bottom of the simplicial set of shape (15.4.1). Hence the equalizer is $\partial\Delta^1$, as desired. $\square$

The geometrical heart of the proof that our candidate hom-spaces are equivalent is in the proof of the following result.

Proposition 15.4.7 *The canonical maps $C_L^\bullet \leftarrow C_{\text{cyl}}^\bullet \to C_R^\bullet$ are pointwise categorical equivalences.*

Proof Lemmas 15.4.10 and 15.4.11, to be proven shortly, show that the natural maps $C_L^n \to \Delta^1$, $C_{\text{cyl}}^n \to \Delta^1$, and $C_R^n \to \Delta^1$ are categorical equivalences. The result then follows from the 2-of-3 property. $\square$

Having deferred the combinatorics, let us complete the proof. Given $A, X \in \mathbf{sSet}_{*,*}$ define their mapping space, the simplicial set of basepoint preserving maps, via the pullback

$$
\begin{array}{ccc}
\underline{\mathrm{hom}}(A, X) & \longrightarrow & X^A \\
\downarrow & & \downarrow {\scriptstyle X^{(a,b)}} \\
* & \xrightarrow{\;(x,y)\;} & X^{\partial \Delta^1}
\end{array}
\tag{15.4.8}
$$

in the category simplicial sets, where the pullback diagram is formed using the inclusions of the basepoints $a, b \in A$ and $x, y \in X$. When X is a quasi-category, the functor $X^{(-)}\colon \mathbf{sSet}^{\mathrm{op}} \to \mathbf{sSet}$ is right Quillen with respect to the Joyal model structure, as a consequence of the assertion in Theorem 15.3.6 that this model structure is monoidal. Alternatively, a direct proof of this assertion uses 15.2.2, Remark 15.3.7, plus one additional ingredient – that the map $X^J \to X$ is a trivial fibration.[2] Given $A \to B$ in $\mathbf{sSet}_{*,*}$, the top square of the cube

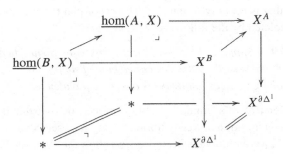

is a pullback because the bottom square and back and front faces are. This is a consequence of composition and cancellation lemmas for pullbacks appearing in two adjacent squares and their composite rectangle. In particular, $\underline{\mathrm{hom}}(-, X)$ is a pullback of $X^{(-)}$ and hence defines a right Quillen functor $\underline{\mathrm{hom}}(-, X)\colon \mathbf{sSet}^{\mathrm{op}}_{*,*} \to \mathbf{sSet}$ because pullbacks inherit right lifting properties. By Ken Brown's lemma 11.3.14, it follows that this functor preserves categorical equivalences between objects with distinct basepoints.

Consider a cosimplicial object $C^\bullet\colon \Delta \to \mathbf{sSet}_{*,*}$ and recall the definitions of latching and matching objects from 14.1.1 and 14.2.5. Applying the functor $\underline{\mathrm{hom}}(-, X)$, we have

$$
M_n \underline{\mathrm{hom}}(C^\bullet, X) \cong \lim{}^{\partial \Delta^n} \underline{\mathrm{hom}}(C^\bullet, X) \cong \underline{\mathrm{hom}}(\mathrm{colim}^{\partial \Delta^n} C^\bullet, X)
$$

$$
\cong \underline{\mathrm{hom}}(L^n C^\bullet, X).
$$

[2] The point is somewhat subtle; see [20, A.4] for a hint of what is involved.

If $C^\bullet$ is Reedy cofibrant, the maps $L^n C^\bullet \to C^n$ are cofibrations; hence the maps

$$\underline{\mathrm{hom}}(C^n, X) \to \underline{\mathrm{hom}}(L^n C^\bullet, X) \cong M_n \underline{\mathrm{hom}}(C^\bullet, X)$$

are fibrations because $\underline{\mathrm{hom}}(-, X)$ is right Quillen. This says that $\underline{\mathrm{hom}}(C^\bullet, X)$ is Reedy fibrant with respect to the Joyal model structure. Applying this result to the cosimplicial objects $C_L^\bullet$, $C_{\mathrm{cyl}}^\bullet$, $C_R^\bullet$, we see that we have pointwise equivalences between Reedy fibrant objects

$$\underline{\mathrm{hom}}(C_L^\bullet, X) \to \underline{\mathrm{hom}}(C_{\mathrm{cyl}}^\bullet, X) \leftarrow \underline{\mathrm{hom}}(C_R^\bullet, X)$$

in the category of bisimplicial sets. Recall from Lemma 14.2.10 that Reedy fibrant objects are pointwise fibrant, although in this particular case, the fact that the $\underline{\mathrm{hom}}(C^n, X)$ are quasi-categories is obvious from the definition (15.4.8).

Remembering only the vertices of each simplicial set appearing in these simplicial spaces – a process which might be called "taking vertices pointwise" – we are left with the diagram of simplicial sets $\mathrm{Hom}_X^L(x, y) \to \mathrm{Hom}_X(x, y) \leftarrow \mathrm{Hom}_X^R(x, y)$ that is actually of interest. The proof that these maps are equivalences is completed by the following lemma:

Lemma 15.4.9 *Suppose $f : X \to Y$ is a weak equivalence between Reedy fibrant bisimplicial sets. Then the associated map of simplicial sets $X_{\bullet,0} \to Y_{\bullet,0}$ obtained by taking vertices pointwise is a weak equivalence.*

Proof By Ken Brown's lemma 11.3.14, it suffices to prove that if $f : X \to Y$ is a Reedy trivial fibration of Reedy fibrant bisimplicial sets, then the associated map $X_{\bullet,0} \to Y_{\bullet,0}$ is an equivalence. Indeed, we show that $X_{\bullet,0} \to Y_{\bullet,0}$ is a trivial fibration. By Remark 14.2.11, to say that f is a Reedy trivial fibration is to say that each relative matching map $X_n \to Y_n \times_{M_n Y} M_n X$ is a trivial fibration in **sSet**.

The dimension zero content of this assertion is that the map on vertices $X_{n,0} \to (Y_n \times_{M_n Y} M_n X)_0 = Y_{n,0} \times_{(M_n Y)_0} (M_n X)_0$ is a surjection in **Set**. Limits in any complete diagram category are computed pointwise; in particular, "taking vertices pointwise" commutes with the weighted limit defining the matching objects. It follows from Definition 14.1.1 and Example 14.1.2 that $(M_n X)_0$ is the set of maps $\partial\Delta^n \to X_{\bullet,0}$. Combining this with the Yoneda lemma, we see that surjectivity of $X_{n,0} \to Y_{n,0} \times_{(M_n Y)_0} (M_n X)_0$ says exactly that any lifting problem

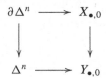

has a solution. □

It remains only to slog through the combinatorics that shows that the maps $C_L^n \to \Delta^1$, $C_{\text{cyl}}^n \to \Delta^1$, and $C_R^n \to \Delta^1$ are categorical equivalences.

Lemma 15.4.10 ([20, 9.3]) *For each $0 \le i < n$, the surjection $\Delta_{i|i+1}^n \to \Delta^1$ is a categorical equivalence.*

Proof We identify non-degenerate simplices in Δ^n with their vertices $[v_0 \ldots v_k]$. Simplices with $v_0 \le i$ and $v_k > i$ correspond bijectively to the non-degenerate simplices of $\Delta_{i|i+1}^n$, so we assign the same labeling to the latter. We start with the case $i = 0$ and prove that the section $\Delta^1 \xrightarrow{[01]} \Delta_{0|1}^n$ is inner anodyne. More specifically, we factor this map as

$$\Delta^1 = X^1 \hookrightarrow X^2 \hookrightarrow \cdots \hookrightarrow X^n = \Delta_{0|1}^n,$$

with each $X^j \to X^{j+1}$ inner anodyne. Define

$$X^2 = \bigcup_{1 < i \le n} [01i], \quad X^3 = \bigcup_{1 < i < j \le n} [01ij], \quad X^4 = \bigcup_{1 < i < j < k \le n} [01ijk],$$

and so on. In words, X^j contains all non-degenerate j-simplices in $\Delta_{0|1}^n$ that contain the edge $[01]$. It is clear that each X^j is a pushout of a coproduct of horns $\Lambda_1^j \to \Delta^j$ along a map that sends the zeroth face of each horn to a degenerate simplex on a point. For example, the new 3-simplices in X^3 are attached along Λ_1^3 horns with images

The third and second faces of this horn were attached to form X^2; the zeroth face, being degenerate, was preexisting. The conclusion follows, and $\Delta_{0|1}^n \to \Delta^1$ is thus a categorical equivalence by the 2-of-3 property. A symmetric argument proves the same result for $\Delta_{n-1|n}^n$.

For $0 < i < n - 1$, we use induction. The map $d^0 \colon \Delta^{n-1} \to \Delta^n$ induces a map $\Delta_{i-1|i}^{n-1} \to \Delta_{i|i+1}^n$ over Δ^1. We suppose, by induction, that the projection from the former to Δ^1 is a categorical equivalence and reach our desired conclusion by proving that $\Delta_{i-1|i}^{n-1} \to \Delta_{i|i+1}^n$ is inner anodyne. As before, we factor this map as

$$\Delta_{i-1|i}^{n-1} = X^1 \hookrightarrow X^2 \hookrightarrow \cdots \hookrightarrow X^n = \Delta_{i|i+1}^n$$

and show each step is inner anodyne. Let

$$X^2 = X^1 \cup \bigcup_{i < j \le n} [01j], \quad X^3 = X^2 \cup \bigcup_{1 < j < k \le n, i < k} [01jk],$$

$$X^4 = X^3 \cup \bigcup_{1 < j < k < l \le n, i < l} [01jkl], \dots$$

Each X^j is obtained by attaching the non-degenerate j-simplices that contain the edge [01] to X^{j-1}. As in the previous case, the attaching maps are Λ_1^j horns. Note that, in this case, if some of the intermediate vertices are less than or equal to i, the last few faces of the Λ_1^j horn will be degenerate at 0. The conclusion follows from the argument given previously. $\qquad\square$

This lemma shows that the maps $C_R^n \to \Delta^1$ and $C_L^n \to \Delta^1$ are categorical equivalences. Only one case remains.

Lemma 15.4.11 ([20, 9.4])   *The maps $C_{\mathrm{cyl}}^n \to \Delta^1$ are categorical equivalences.*

Proof   To condense notation, write $\{0, 1, \dots, n\}$ and $\{0', 1', \dots, n'\}$ for the vertices spanned by $\Delta^n \times \{0\}$ and $\Delta^n \times \{1\}$ in $\Delta^n \times \Delta^1$, respectively, using the prime as shorthand for the second coordinate. Non-degenerate simplices in $\Delta^n \times \Delta^1$ correspond to sequences $[v_0 \cdots v_j v_{j+1}' \cdots v_k']$ of elements of $[n]$ such that the v_i and v_i' are strictly increasing and with $v_j \le v_{j+1}'$; compare with 15.0.1. Non-degenerate simplices containing at least one primed vertex and one unprimed one correspond bijectively to non-degenerate simplices in the quotient C_{cyl}^n.

Let σ_i be the $(n+1)$-simplex $[01 \cdots (i-1)ii'(i+1)' \cdots n']$; that is, let σ_i be the ith shuffle. Its quotient in C_{cyl}^n is $\Delta_{i|i+1}^{n+1}$. We show that the categorical equivalence $C_L^n \to \Delta^1$ factors through $C_{\mathrm{cyl}}^n \to \Delta^1$ along a filtration

$$C_L^n = \Delta_{0|1}^{n+1} = X^0 \hookrightarrow X^1 \hookrightarrow \cdots \hookrightarrow X^n = C_{\mathrm{cyl}}^n.$$

Here each X^{i+1} is obtained from X^i by attaching the next shuffle, corresponding to $\Delta_{i+1|i+2}^{n+1}$. Note that the intersection $X^i \cap \Delta_{i+1|i+2}^{n+1}$ is $\Delta_{i|i+1}^n$; that is, this intersection is the n-simplex spanning $[01 \cdots i(i+1)' \cdots n']$. Hence

$$
\begin{array}{ccc}
\Delta_{i|i+1}^n & \longrightarrow & X^i \\
\downarrow & & \downarrow \\
\Delta_{i+1|i+2}^{n+1} & \longrightarrow & X^{i+1}
\end{array}
$$

is a pushout, and the result follows from the proof of Lemma 15.4.10 and the 2-of-3 property. $\qquad\square$

Thus we have proven:

Theorem 15.4.12 *The natural maps* $\operatorname{Hom}_X^L(x, y) \to \operatorname{Hom}_X(x, y) \leftarrow$ $\operatorname{Hom}^R(x, y)$ *are categorical equivalences of quasi-categories.*

Remark 15.4.13 The maps $C_L^{\bullet} \leftarrow C_{\text{cyl}}^{\bullet} \to C_R^{\bullet}$ have pointwise-defined sections, though the sections do not assemble into maps of cosimplicial objects. The maps $C_L^n \to C_{\text{cyl}}^n \leftarrow C_R^n$ are induced by the inclusions of $\Delta^{n+1} \rightrightarrows \Delta^n \times \Delta^1$ as the first and last shuffles, respectively.[3]

Remark 15.4.14 Because categorical equivalences are weak homotopy equivalences, the objects $\operatorname{Hom}_X^L(x, y)$, $\operatorname{Hom}_X(x, y)$, and $\operatorname{Hom}_X^R(x, y)$ define weakly equivalent simplicial sets whose set of path components is the hom-set $hX(x, y)$. We would like to conclude that the homotopy category hX is thereby enriched over the homotopy category of spaces, however, there is no natural composition law definable in **sSet** using any of these mapping spaces.

These considerations motivate the introduction of a fourth candidate mapping space, which is not typically a quasi-category, but which is weak homotopy equivalent (although not categorically equivalent) to these models. This new construction associates a simplicially enriched category to each simplicial set. The simplicial categories constructed in this manner are cofibrant in the model structure introduced in Example 11.3.10. It follows that they can be used to define a homotopically well-behaved notion of homotopy coherent diagrams. We turn to this subject now.

---

[3] Recall 15.0.3: simplices in $\Delta^n \times \Delta^m$ correspond bijectively to totally ordered collections of vertices (i, j) with $i \in [n]$ and $j \in [m]$. Simplices of maximal dimension are called shuffles. The first shuffle is the unique one containing the vertices $(0, 0), \ldots, (n, 0), \ldots (n, m)$. The last is the unique one containing the vertices $(0, 0), \ldots, (0, m), \ldots, (n, m)$.

16

Simplicial categories and homotopy coherence

We have seen three equivalent procedures for associating mapping spaces to pairs of objects in a quasi-category, but without a composition map, we do not yet have a way to define the ($\mathcal{H}$-enriched) homotopy category of a quasi-category. Shooting for the moon, we might try to construct a functor **sSet** → **sCat** from the category of simplicial sets to the category of simplicially enriched categories directly. A fundamental feature of the category of simplicial sets is that each object is a colimit of a diagram of standard simplices (see Example 7.2.8). With this in mind, we might hope that the functor associating a simplicial category to a simplicial set also preserved colimits. But now, because **sCat** is cocomplete, it follows from 1.5.1 that such a functor, and indeed an adjoint pair, is determined by any cosimplicial object $\Delta \to$ **sCat**. The main task of this chapter is to motivate the correct choice of cosimplicial object and explore the adjunction $\mathfrak{C} \colon$ **sSet** $\rightleftarrows$ **sCat** $\colon \mathfrak{N}$ so produced.

The right adjoint $\mathfrak{N}$ is called the **homotopy coherent nerve**. It records homotopy coherent sequences of composable arrows in a simplicial category. The left adjoint $\mathfrak{C}$ will produce simplicial categories $\mathfrak{C}X$ associated to each simplicial set X so that when X is a quasi-category, the hom-space $\mathfrak{C}X(x, y)$ is weak homotopy equivalent to the hom-spaces constructed in this previous chapter. We give an explicit description of these hom-spaces later, but first let us explore the model for $(\infty, 1)$-categories presented by simplicial categories.

16.1 Topological and simplicial categories

By a theorem of Quillen, the adjoint pair

$$\textbf{sSet} \; \underset{S}{\overset{|-|}{\underset{\perp}{\rightleftarrows}}} \; \textbf{Top}$$

is a Quillen equivalence (see 11.3.15) with respect to the model structures of Example 11.3.5 and Example 11.3.6. Because all simplicial sets are cofibrant and all spaces are fibrant, this means that the unit $K \to S|K|$ and counit $|SX| \to X$ are weak homotopy equivalences for all simplicial sets K and spaces X. Furthermore, the total derived functors form an adjoint equivalence, defining the **homotopy category of spaces** $\mathcal{H}$. For definiteness, and to recall the classical nature of this object, we might define $\mathcal{H}$ to be the category of CW complexes and homotopy classes of maps.

By Lemma 6.1.6, both the left and right adjoints preserve finite products; by (10.4.2), the localization functors to $\mathcal{H}$ are also lax monoidal. It follows that there is an induced change of base adjunction

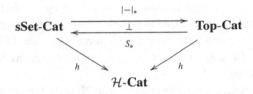

between the category of simplicial categories and the category of topological categories commuting with the change of base functors to the category of $\mathcal{H}$-enriched categories [15, 85]. Here the functor S_* turns a topological category into a simplicial category with the same objects but with hom-spaces defined to be the total singular complex of the corresponding hom-space in the simplicial category.

The localization $[-]\colon$ **Top** $\to \mathcal{H}$ can be defined to be the functor which replaces a space by a CW complex equipped with a weak equivalence $[X] \xrightarrow{\sim} X$. This procedure is well defined up to homotopy, hence defining the indicated functor. If definiteness is desired, we might set $[X] = |SX|$, but this makes no difference. In accordance with the notation introduced in Section 10.4, we write $h\underline{\mathcal{C}}$ for the $\mathcal{H}$-enriched category associated to a topological (or simplicial) category $\underline{\mathcal{C}}$ with underlying category $\mathcal{C}$. The underlying category of $h\underline{\mathcal{C}}$, denoted by $h\mathcal{C}$, is defined by applying π_0 to the hom-spaces of $\underline{\mathcal{C}}$.

We are interested in using topological or simplicial categories to model $(\infty, 1)$-categories, so it is natural to declare an enriched functor $F\colon \underline{\mathcal{C}} \to \underline{\mathcal{D}}$ between topological or simplicial categories to be a weak equivalence if and only if the induced functor $hF\colon h\underline{\mathcal{C}} \to h\underline{\mathcal{D}}$ is an equivalence of $\mathcal{H}$-enriched categories as defined in 3.5.13. That is, the functor of underlying homotopy categories must be essentially surjective (or indeed an equivalence of categories), and the maps of hom-spaces $\underline{\mathcal{C}}(x, y) \to \underline{\mathcal{D}}(Fx, Fy)$ must be weak homotopy equivalences, that is, isomorphisms in $\mathcal{H}$. In the simplicially enriched context, these are exactly the DK-equivalences introduced in Section 3.5.

Remark 16.1.1 ([49, 1.1.3.8]) The homotopy category $h\underline{\mathcal{C}}$ does not determine the weak equivalence class of $\underline{\mathcal{C}}$ even though it suffices to detect weak equivalences between topological categories. This is similar to the fact that the homotopy groups of a CW complex do not determine its homotopy type, though they suffice to detect homotopy equivalences between CW complexes.

For conciseness, we only discuss simplicial categories henceforth. A simplicial category is **locally Kan** if each of its hom-spaces is a Kan complex. In this case, an easy argument shows that two vertices in a hom-space are in the same path component if and only if there exist 1-simplices from the one to the other in both directions. From the perspective of the underlying unenriched category, we might say that maps in a locally Kan simplicial category are homotopic if and only if there exist simplicial homotopies exhibiting this fact. By a theorem of Bergner [6], mildly generalized in [49, A.3.2.4], there is a cofibrantly generated model structure on **sCat** whose weak equivalences are simplicial functors that descend to $\mathcal{H}$-equivalences and whose fibrant objects are the locally Kan simplicial categories.

To define the generating cofibrations, write $2[A]$ for the simplicial category with two objects 0,1 and which has hom-spaces $2[A](0,0) = 2[A](1,1) = *$, $2[A](1,0) = \emptyset$, $2[A](0,1) = A$. A simplicial functor has the right lifting property with respect to the functor $2[A] \to 2[B]$ if and only if each constituent map of hom-spaces has the right lifting property against $A \to B$ in **sSet**.

Theorem 16.1.2 (Bergner) *There is a cofibrantly generated model structure on* **sCat** *whose weak equivalences are simplicial functors* $F \colon \underline{\mathcal{C}} \to \underline{\mathcal{D}}$ *so that* $hF \colon h\underline{\mathcal{C}} \to h\underline{\mathcal{D}}$ *is an $\mathcal{H}$-equivalence; whose fibrant objects are the locally Kan simplicial categories; and whose cofibrations are generated by*

$$\{\emptyset \to *\} \cup \{2[\partial\Delta^n] \to 2[\Delta^n]\}_{n \geq 0}.$$

16.2 Cofibrant simplicial categories are simplicial computads

There is a more concrete description of the cofibrant objects in Bergner's model structure that we learned from Verity. Recall from Section 3.6 that any simplicial category can be encoded as a simplicial object $C \colon \Delta^{op} \to \mathbf{Cat}$ in which each category has the same objects and each functor is the identity on objects.

Given a simplicial category $\mathcal{C} \colon \Delta^{op} \to \mathbf{Cat}$, we refer to an arrow in $\mathcal{C}_n$ as an *n*-**arrow**. An *n*-arrow $f \colon a \to b$ is precisely an *n*-simplex in the simplicial set $\underline{\mathcal{C}}(a, b)$. By the Eilenberg–Zilber lemma, any *n*-simplex f is uniquely expressible as $f' \cdot \alpha$ where $\alpha : [n] \to [m]$ is an epimorphism and f' is a non-degenerate m-arrow. We say f' is the **unique non-degenerate quotient** of f. In this case, we say f has **dimension** m. Note the dimension of an n-arrow is at most n, its **simplicial degree**. The dimension is defined to be the simplicial degree of the unique non-degenerate quotient.

An arrow in an unenriched category is **atomic** if it admits no non-trivial factorizations. A category is freely generated by a reflexive directed graph,[1] abbreviated to "freely generated" or "free" below, if and only if each of its arrows may be uniquely expressed as a composite of atomic arrows. In this case, the generating graph is precisely the subgraph of atomic arrows.

Definition 16.2.1 A simplicial category $\mathcal{C} : \mathbb{\Delta}^{op} \to$ **Cat** is a **simplicial computad** if and only if

- each $\mathcal{C}_n$ is freely generated and
- for each surjection $\alpha : [n] \to [m]$ and atomic arrow $f \in \mathcal{C}_m$, the arrow $f \cdot \alpha$ is atomic in $\mathcal{C}_n$

In words, a simplicial computad is a simplicial object in **Cat**, each of whose categories is freely generated on a set of generating arrows that includes the degenerate images of all lower-dimensional generators. For instance, the discrete simplicial category on a free category is a simplicial computad. We will see less trivial examples soon.

Simplicial computads are built in an obvious (and essentially unique) way from the given generating cofibrations, making them cellular cofibrant objects in the Bergner model structure.

Lemma 16.2.2 *The simplicial computads are the cellular cofibrant objects in* **sCat**. *Furthermore, every cofibrant object is cellular and hence a simplicial computad.*

Proof We only prove the hard direction: that any cofibrant simplicial category is a simplicial computad. By Corollary 12.2.4, it suffices to show that a retract of a simplicial computad is a simplicial computad. The first step is to show that the retract of a free category is a free category. A retract $\mathcal{B} \hookrightarrow \mathcal{C} \twoheadrightarrow \mathcal{B}$ in **Cat** can be characterized as the subcategory fixed by the corresponding idempotent on $\mathcal{C}$. In a free category, every morphism is both monic and epic. It follows that if $h = gf$ in $\mathcal{C}$ and any two of these are in $\mathcal{B}$, so is the third. Hence, by induction, any arrow in $\mathcal{B}$ is uniquely decomposable into the shortest composites of atomic arrows of $\mathcal{C}$ that lie in $\mathcal{B}$.

---

[1] The category **rDirGph** of reflexive directed graphs is equivalent to the category of 1-skeletal simplicial sets. The forgetful functor $U : $ **Cat** $\to$ **rDirGph** remembers domains, codomains, and identities but forgets the composition law.

We have shown that at each level, a retract of a simplicial computad is a free category. It remains only to argue that the degenerate images of atomic arrows in $\mathcal{B}_n$ are atomic in $\mathcal{B}_{n+1}$. This is clear for those atomic arrows in $\mathcal{B}_n$ that are also atomic in $\mathcal{C}_n$. To that end, suppose a degenerate image of some atomic arrow in $\mathcal{B}_n$ factors non-trivially as gf in $\mathcal{B}_{n+1}$. Because $\mathcal{C}$ is a simplicial computad, this atomic arrow in $\mathcal{B}_n$ must also factor non-trivially as $g'f'$ in $\mathcal{C}_n$ with g' mapping to g and f' mapping to f. But applying one of the face maps that serves as a retraction of the degeneracy, we see that either g or f must map to an identity in $\mathcal{B}_n$. This contradicts the fact that g' and f' are assumed to be non-identities.   □

Example 16.2.3   There is a free-forgetful adjunction

$$F : \mathbf{rDirGph} \xrightarrow{\;\;\perp\;\;} \mathbf{Cat} : U$$

between small categories and reflexive directed graphs inducing a comonad FU on **Cat**. We claim that the comonad resolution associated to a small category $\mathcal{A}$ is a simplicial computad $FU_{\bullet}\mathcal{A}$,

$$FU\mathcal{A} \; \underset{\underset{\longleftarrow{FU\epsilon}}{\xrightarrow{\;F\eta U\;}}}{\overset{\longleftarrow{\epsilon FU}}{\longrightarrow}} \; FUFU\mathcal{A} \; \underset{\underset{\xrightarrow{\;FU F\eta U\;}}{\overset{\longleftarrow{FU\epsilon FU}}{\xrightarrow{\;F\eta UFU\;}}}}{\overset{\longleftarrow{\epsilon FUFU}}{\underset{\longleftarrow{FUFU\epsilon}}{}}} \; FUFUFU\mathcal{A} \quad \cdots ,$$

and hence a cofibrant simplicial category.

The category $FU\mathcal{A}$ is the free category on the underlying reflexive directed graph of $\mathcal{A}$. Its arrows are strings of composable non-identity arrows of $\mathcal{A}$; the atomic 0-arrows are the non-identity arrows of $\mathcal{A}$. An n-arrow is a string of composable arrows in $\mathcal{A}$ with each arrow in the string enclosed in exactly n pairs of parentheses. The atomic arrows are those enclosed in precisely one pair of parentheses on the outside. The face maps $(FU)^k\epsilon(FU)^j$ remove the parentheses that are contained in exactly k others; $FU \cdots FU\epsilon$ composes the morphisms inside the innermost parentheses. The degeneracy maps $F(UF)^k\eta(UF)^jU$ double up the parentheses that are contained in exactly k others; $F \cdots UF\eta U$ inserts parentheses around each individual morphism.

These examples reappear as the shape of homotopy coherent diagrams, which we now introduce.

16.3 Homotopy coherence

It remains to define the cosimplicial object that produces the adjunction defining the homotopy coherent nerve. A naïve choice for the cosimplicial object

$\mathfrak{C}\Delta^{\bullet}$: $\Delta \to \mathbf{sCat}$ might be simply to regard the ordinals $[n]$ as discrete simplicial categories $0 \to 1 \to \cdots \to n$. But the right adjoint specified by this cosimplicial object would simply be the ordinary nerve of the unenriched category underlying a simplicial category, which is not what we want. The idea is that $\mathfrak{C}\Delta^n$ is a simplicial category that encodes a "homotopy coherent" diagram of shape $[n]$ in a sense that we now make precise.

A good theory of homotopy coherent diagrams taking values in a locally Kan simplicial category has been developed by Jean-Marc Cordier and Tim Porter, Vogt, and others. Let $\underline{C}$ be a simplicial category with underlying category C. A **homotopy commutative diagram** of shape $\mathcal{A}$ is a map of reflexive[2] directed graphs $U\mathcal{A} \to UC$ that defines a functor $\mathcal{A} \to hC$. Recall that the category hC underlying the $\mathcal{H}$-enriched category $\underline{h}\underline{C}$ is obtained by taking path components of each hom-space. Thus a diagram $F: U\mathcal{A} \to UC$ is homotopy commutative if, whenever $h = gf$ in $\mathcal{A}$, the vertices Fh and $Fg \cdot Ff$ lie in the same path component of the hom-space from the domain of these maps to the codomain. If $\underline{C}$ is locally Kan, this is the case just when there exist 1-simplices $Fh \to Fg \cdot Ff$ and $Fg \cdot Ff \to Fh$.

A **homotopy coherent diagram** of shape $\mathcal{A}$ is a simplicial functor $FU_{\bullet}\mathcal{A} \to \underline{C}$. The map $U\mathcal{A} \to UFU\mathcal{A}$ of reflexive directed graphs[3] can be used to define a homotopy commutative diagram $U\mathcal{A} \to UFU\mathcal{A} \to UC$. The proof is by applying the path components functor $\pi_0: \mathbf{sSet} \to \mathbf{Set}$ to the hom-spaces, thereby extracting the functor $\mathcal{A} \to hC$ from the simplicial functor $FU_{\bullet}\mathcal{A} \to \underline{C}$. A **homotopy coherent natural transformation** is a homotopy coherent diagram of shape $\mathcal{A} \times 2$, that is, a simplicial functor $FU_{\bullet}(\mathcal{A} \times 2) \to \underline{C}$.

The theory of homotopy coherent diagrams was developed to describe situations for which there are affirmative answers to certain classical problems. For instance; given a commutative diagram $F: \mathcal{A} \to \underline{C}$, is it possible to form a new diagram in which each object is replaced by a (specified) homotopy equivalent one? Or given a natural transformation $\alpha: F \Rightarrow G$, it is possible to replace the maps α_a with homotopic ones? If these questions are meant strictly, the answer is no. However:

Proposition 16.3.1 (Cordier–Porter) *Given a homotopy coherent diagram $F: \mathcal{A} \to \underline{C}$ in a locally Kan simplicial category and a family of homotopy equivalences $Fa \to Ga$, this data extends to a homotopy coherent diagram $G: \mathcal{A} \to \underline{C}$ and homotopy coherent map $F \Rightarrow G$.*

---

[2] Here, for simplicity, we ignore the identities; a more careful treatment would allow homotopies between identities in the target category and the images of identities in the domain.

[3] This map is *not* a functor, since it picks out the preexisting composites in the image, not the freely added composites.

Proposition 16.3.2 (Cordier–Porter)   *Given a homotopy coherent map* $\alpha : F \Rightarrow G$ *of homotopy coherent diagrams* $F, G : \mathcal{A} \rightrightarrows \mathcal{C}$, *where* $\underline{\mathcal{C}}$ *is locally Kan, and homotopies* $\beta_a \simeq \alpha_a$, *this data extends to a homotopy coherent map* $\beta : F \Rightarrow G$ *and a coherent homotopy of homotopy coherent maps.*

The proofs are by describing and then filling horns, making use of the lifting properties guaranteed by Lemma 16.4.10 and the results of Chapter 17. We leave the details for [14], but let us examine what this means in a specific example.

Example 16.3.3   Consider the category [3] and label its non-identity morphisms as indicated:

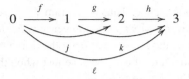

Let us describe the hom-space in $FU_\bullet[3]$ from the initial object to the terminal one. The vertices of this simplicial set are the paths of edges ℓ, kf, hj, hgf. The 1-simplices are once parenthesized strings of composable morphisms that are non-degenerate when there is more than one arrow inside some pair of parentheses. There are five such with boundary 0-simplices illustrated here:

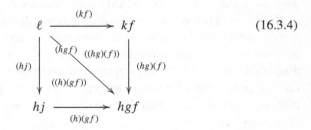   (16.3.4)

There are only two non-degenerate 2-simplices whose boundaries are depicted. Hence $FU_\bullet[3](0, 3) = \Delta^1 \times \Delta^1$. We conclude that a homotopy coherent diagram of shape [3] consists of a cube in the hom-space from the first object to the last in which certain 0-simplices and 1-simplices are specified composites of data from the other hom-spaces. One way to think about these homotopies is that they are the data one obtains if there is a lift, up to isomorphism, of a functor [3] $\rightarrow h\mathcal{C}$ to a functor landing in $\mathcal{C}$ (see [49, §1.2.6]).

With this example in mind, we define $\mathfrak{C}\Delta^n$ to be the simplicial category $FU_\bullet[n]$ so that a map $\mathfrak{C}\Delta^n \rightarrow \mathcal{C}$, by adjunction an n-simplex in $\mathfrak{N}\mathcal{C}$, is a homotopy coherent diagram of shape $[n]$. More explicitly, $\mathfrak{C}\Delta^n$ is the simplicial category with objets $0, 1, \ldots, n$ and whose hom-spaces are cubes of varying dimension

$$\mathfrak{C}\Delta^n(i, j) = \begin{cases} (\Delta^1)^{j-i-1} & \text{when } j > i, \\ \Delta^0 & \text{when } j = i, \\ \emptyset & \text{when } j < i. \end{cases} \qquad (16.3.5)$$

It is common to define these hom-spaces as nerves of posets, $\mathfrak{C}\Delta^n(i, j) = N P_{i,j}$. If $j < i$, then $P_{i,j}$, and hence $\mathfrak{C}\Delta^n(i, j)$ is empty. Otherwise $P_{i,j}$ is the poset of subsets of the interval $\{k \mid i \le k \le j\} \subset [n]$ that contain both endpoints. For $j = i$ and $j = i + 1$, this poset is the terminal category. For $j > i + 1$, a quick calculation shows that $P_{i,j}$ is isomorphic to the product of the category $\mathbf{2} = [1]$ with itself $j - i - 1$ times, hence the preceding description. An element in $P_{i,j}$ specifies a path of edges in the category $[n]$ from i to j through each of the listed vertices. Maps in the poset are "refinements" of these paths, obtained by factoring some of the edges through others.

Theorem 16.4.7 proves that these two descriptions of $\mathfrak{C}\Delta^n$ coincide; the reader who is unconvinced should adopt the geometric description (16.3.5) for now.

Remark 16.3.6 The ordinary nerve can be defined from the homotopy coherent nerve in the following manner. The path components functor $\pi_0 : \mathbf{sSet} \to \mathbf{Set}$ is left adjoint to the inclusion; both functors preserve products and hence induce a change of base adjunction. The composite adjunction

$$\mathbf{sSet} \underset{\mathfrak{N}}{\overset{\mathfrak{C}}{\rightleftarrows}} \mathbf{sCat} = \mathbf{sSet\text{-}Cat} \underset{\mathrm{incl}_*}{\overset{(\pi_0)_*}{\rightleftarrows}} \mathbf{Set\text{-}Cat} = \mathbf{Cat}$$

is the adjunction $h \dashv N$ of Example 1.5.5. In particular, the homotopy coherent nerve of a discrete simplicial category is just the nerve of the ordinary category.

The slogan that quasi-categories and locally Kan simplicial categories both model $(\infty, 1)$-categories is made precise by the following theorem, whose proof is the subject of [49, §2.2] and the paper [20].

Theorem 16.3.7 *The pair $\mathfrak{C} \dashv \mathfrak{N}$ forms a Quillen equivalence between Joyal's model structure for quasi-categories and Bergner's model structure for simplicial categories.*

In Lemma 16.4.10, we give a direct proof that the homotopy coherent nerve of a locally Kan simplicial category is a quasi-category, a fertile source of examples.

Remark 16.3.8   It is not clear from the geometric description that the $\mathfrak{C}\Delta^n$ are cofibrant simplicial categories, a necessary ingredient for the Quillen adjunction $\mathfrak{C} \dashv \mathfrak{N}$. In 16.4.5, we show that these simplicial categories are simplicial computads. Alternately, this follows from the comparison in Theorem 16.4.7 and the description of the comonad resolution given in Example 16.2.3.

Remark 16.3.9   Somewhat disingenuously (because this result is one of the ingredients in the proof of Theorem 16.3.7), we can use Theorem 16.3.7 to deduce that a map $X \to Y$ of simplicial sets is a categorical equivalence if and only if the induced functor $h\mathfrak{C}X \to h\mathfrak{C}Y$ is an equivalence of $\mathcal{H}$-enriched categories. This is Lurie's definition of the weak equivalences in the Joyal model structure.

16.4 Understanding the mapping spaces $\mathfrak{C}X(x, y)$

The functor $\mathfrak{C}\colon \mathbf{sSet} \to \mathbf{sCat}$ is defined to be the left Kan extension of the cosimplicial object $\mathfrak{C}\Delta^\bullet \colon \Delta \to \mathbf{sCat}$ along the Yoneda embedding. Hence, by definition,

$$\mathfrak{C}X = \int^{[n]\in\Delta} \coprod_{X_n} \mathfrak{C}\Delta^n = \mathrm{coeq}\left(\coprod_{[m]\to[n]} \coprod_{X_n} \mathfrak{C}\Delta^m \rightrightarrows \coprod_{[n]} \coprod_{X_n} \mathfrak{C}\Delta^n \right).$$

Informally speaking, it suffices to restrict the interior coproducts to the non-degenerate simplices of X and the left outer coproduct to the generating coface maps $d^i \colon [n-1] \to [n]$. Write $\tilde{X}_n$ for the non-degenerate n-simplices of X. Then[4]

$$\mathfrak{C}X = \mathrm{colim}\left(\begin{array}{c} \coprod_{\tilde{X}_1} \mathfrak{C}\Delta^0 \underset{d_1}{\overset{d_0}{\rightrightarrows}} \coprod_{X_0} \mathfrak{C}\Delta^0 \\[2ex] d^0 \Big\Vert d^1 \\[2ex] \coprod_{\tilde{X}_2} \mathfrak{C}\Delta^1 \Rrightarrow \coprod_{\tilde{X}_1} \mathfrak{C}\Delta^1 \\[2ex] \Biggl\Downarrow \\[1ex] \cdots\ \coprod_{\tilde{X}_2} \mathfrak{C}\Delta^2 \end{array} \right)$$

---

[4] Non-degenerate n-simplices may have degenerate $(n-1)$-simplices as faces (e.g., as in the simplicial model for S^2 formed by attaching Δ^2 to a point), so we cannot technically restrict the

The objects of $\mathfrak{C}X$ are the vertices of X. The simplicial categories $\mathfrak{C}\Delta^0$ and $\mathfrak{C}\Delta^1$ are the free simplicial categories on the poset categories [0] and [1], and the free simplicial category functor is a left adjoint and so commutes with colimits. Hence, if X is 1-skeletal so that $\tilde{X}_n = \emptyset$ for all $n > 1$, then $\mathfrak{C}X$ is the free simplicial category on the graph with vertex set X_0 and edge set $\tilde{X}_1$. Concretely, this means that the hom-spaces $\mathfrak{C}X(x, y)$ are discrete simplicial sets containing a vertex for each path of edges from x to y in X. Because each 0-arrow in $\mathfrak{C}\Delta^n$ is in the image of a 0-arrow in $\mathfrak{C}\Delta^1$ under an appropriate face map, we have completely described the 0-arrows in $\mathfrak{C}X$.

In general, for each 2-simplex of X with boundary as displayed,

$$
\begin{array}{ccc}
 & z & \\
 f \nearrow & {\scriptstyle\sim} & \searrow g \\
x & \xrightarrow[\quad j \quad]{} & w
\end{array}
$$

there exists a 1-simplex from the vertex j to the vertex gf in $\mathfrak{C}X(x, w)$. Furthermore, for each vertex in some hom-space representing a sequence of paths containing j, there is a 1-simplex connecting it to the vertex representing the same sequence, except with gf in place of j.

However, the 2-skeleton of X does not determine the 1-skeleta of the hom-spaces. For example, for each 3-simplex σ of X, as depicted,

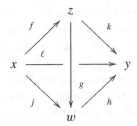

there is an edge from ℓ to hgf in $\mathfrak{C}X(x, y)$. In general, there is an edge between the vertices represented by paths $p_1 \ldots p_r$ and $q_1 \ldots q_r$ of edges from x to y in X if and only if each edge p in the first path that does not appear in the second is replaced by a sequence of n-edges that appear as the spine of some n-simplex of X with p as its diagonal. Here the **spine** of an n-simplex is the sequence of edges between consecutive vertices, using the usual ordering of the vertices, and the **diagonal** is the edge from the initial vertex to the final one.

In this way, each edge of $\mathfrak{C}X(x, y)$ corresponds to a **necklace**

$$
\Delta^{n_1} \vee \cdots \vee \Delta^{n_r} \to X
$$

---

face maps to maps $d_i : \tilde{X}_n \to \tilde{X}_{n-1}$. Instead, one must attach each degenerate face to the unique lower-dimensional non-degenerate simplex it represents. But this technicality will not affect our intuition-building discussion.

in X. By $\Delta^n \vee \Delta^k$, we always mean that the final vertex of the n-simplex is identified with the initial vertex of the k-simplex. A necklace is a sequence of **beads**, the Δ^{n_i}, that are strung together along the **joins**, defined to be the union of the initial and final vertices of each bead. We regard a necklace as an object in $\mathbf{sSet}_{*,*}$ by specifying the initial and terminal vertices as its basepoints.

We have seen that a map in $\mathbf{sSet}_{*,*}$ from a necklace to a simplicial set X with basepoints x and y determines a 1-simplex in the hom-space $\mathfrak{C}X(x, y)$. By a theorem of Dugger and David Spivak, necklaces can be used to characterize the higher-dimensional simplices of the hom-spaces $\mathfrak{C}X(x, y)$ as well, provided we keep track of additional vertex data.

Theorem 16.4.1 ([21, 4.8])   *An n-simplex in $\mathfrak{C}X(x, y)$ is uniquely represented by a triple $(T, f, \vec{T})$, where T is a necklace; $f : T \to X$ is a map in $\mathbf{sSet}_{*,*}$ that sends each bead of T to a non-degenerate simplex of X; and $\vec{T}$ is a nested sequence of sets of vertices V_T of T*

$$J_T = T^0 \subset T^1 \subset T^2 \subset \cdots \subset T^{n-1} \subset T^n = V_T, \qquad (16.4.2)$$

where J_T is the set of joins of T.

Necklaces $f : T \to X$ with the property described in the theorem statement are called **totally non-degenerate**. Note that the map f need not be injective. If $x = y$ is a vertex with a non-degenerate edge $e : x \to y$, the map $e : \Delta^1 \to X$ defines a totally non-degenerate necklace in X. Dugger and Spivak prefer to characterize the simplices of $\mathfrak{C}X(x, y)$ as equivalence classes of triples $(T, f, \vec{T})$, which are not necessarily totally non-degenerate [21, 4.4-4.5]. However, by the Eilenberg–Zilber lemma, it is always possible to replace an arbitrary triple $(T, f, \vec{T})$ by its unique totally non-degenerate quotient.

Lemma 16.4.3 ([21, 4.7])   *Given $(T, f, \vec{T})$, there is a unique quotient $(\overline{T}, \overline{f}, \vec{\overline{T}})$ such that f factors through $\overline{f}$ via a surjection $T \twoheadrightarrow \overline{T}$ and $\overline{f}$ is totally non-degenerate.*

Proof   By the Eilenberg–Zilber lemma [28, II.3.1, pp. 26–27], any simplex $\sigma \in X_n$ can be written uniquely as $\sigma'\epsilon$, where $\sigma' \in X_m$ is non-degenerate, with $m \leq n$, and $\epsilon : [n] \to [m]$ is a surjection in Δ. The necklace $\overline{T}$ agrees with T at each bead whose image is non-degenerate. If σ is a degenerate n-simplex in the image of f, then to form $\overline{T}$, we replace the bead Δ^n of T corresponding to σ by the bead Δ^m, where m is determined by the Eilenberg–Zilber decomposition of σ. Define $\overline{f} : \overline{T} \to X$ restricted to this Δ^m to equal σ'. The morphism ϵ defines an epimorphism of simplicial sets $\Delta^n \to \Delta^m$, which defines the quotient map $T \twoheadrightarrow \overline{T}$ at this bead in such a way that f factors through $\overline{f}$ along this map.

Let $\vec{\overline{T}}$ be the nested sequence of sets of vertices of $\overline{T}$ given by the direct image of $\vec{T}$ under $T \twoheadrightarrow \overline{T}$. The resulting triple $(\overline{T}, \overline{f}, \vec{\overline{T}})$ is totally non-degenerate and unique such that f factors through $\overline{f}$. $\square$

In dimensions 0 and 1, the vertex data $\vec{T}$ is completely determined by the necklace T, so we may represent the simplices of $\mathfrak{C}X(x, y)$ by necklaces alone. In particular, this result generalizes the informal description of the low-dimensional simplices in $\mathfrak{C}X$ described previously.

Remark 16.4.4 ([21, 4.6,4.9]) Importantly, with the aid of Lemma 16.4.3, the face maps $d_i : \mathfrak{C}X(x, y)_n \to \mathfrak{C}X(x, y)_{n-1}$ can also be described in the language of necklaces and vertex data. The simplicial operators d^i and s^i act on the collection of nested sequences of subsets of vertices of varying length in an obvious way – by duplicating or deleting sets of vertices. The degeneracy and inner face operations on n-arrows are defined in this way; in particular, the underlying necklace is unchanged.

The outer face maps are only marginally more difficult to describe. The chain $\vec{T}d^n$ likely has fewer vertices then $\vec{T}$. Thus the nth face of $(T, f, \vec{T})$ is the unique totally non-degenerate quotient of the restriction $(T|_{T^{n-1}}, f|_{T^{n-1}}, \vec{T}d^n)$ to these vertices. Similarly, the chain $\vec{T}d^0$ likely has more joins. Thus the zeroth face is the unique totally non-degenerate quotient of the necklace obtained by "breaking up" beads of T along the vertices in the set T^1. The image of this necklace in X has the same vertices but in general contains simplices of smaller dimension, which were faces of simplices in the necklace T. We call this new necklace the T^1-**splitting** of T. Each bead of T is replaced by a necklace with the same spine whose beads are each faces of the original bead. The vertices of each new bead will be a consecutive subset of vertices of the bead of T with initial and final vertices in T^1. The sum of the dimensions of these new beads will equal the dimension of the original bead.

Remark 16.4.5 From this description, it is easy to see that the atomic arrows in $\mathfrak{C}X$ are those whose necklace consists of a single bead. In particular, $\mathfrak{C}X$ is a simplicial computad, proving by Lemma 16.2.2 that these simplicial categories are cofibrant.

Example 16.4.6 We can use necklaces to recompute the hom-space $\mathfrak{C}\Delta^3(0, 3)$ described in Example 16.3.3. There are nine totally non-degenerate necklaces from 0 to 3 in Δ^3. Four of these contain only 1-simplices as beads and hence represent 0-arrows; the other five are the 1-arrows depicted as follows:

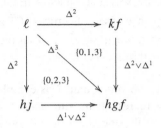

The top and left 1-arrows correspond to the second and first faces of the 3-simplex, respectively. The only two non-degenerate 2-arrows correspond to the necklace Δ^3 together with the depicted vertex data. This diagram is arranged in correspondence with (16.3.4).

Observe that the atomic k-arrows in the simplicial computad $\mathfrak{C}\Delta^3$ are those k-simplices that contain the initial vertex in the usual partial ordering for the cube defined by (16.3.5) to be the appropriate hom-space. This description holds for general $\mathfrak{C}\Delta^n$.

Using Theorem 16.4.1, we can show that the left adjoint to the homotopy coherent nerve always maps the nerve of an unenriched category to the simplicial category produced by the comonad resolution of 16.2.3.

Theorem 16.4.7 *Let $\mathcal{A}$ be any small category. The simplicial category $\mathfrak{C}N\mathcal{A}$ is isomorphic to the simplicial category $FU_\bullet\mathcal{A}$.*

Proof The objects of both simplicial categories are the objects of $\mathcal{A}$. It remains to show that the hom-spaces coincide. A necklace in the nerve of a category is uniquely determined by its spine and the set of joins; that is, a necklace is a sequence of composable non-identity morphisms each contained in one set of parentheses, indicating which morphisms are grouped together to form a bead.

An n-simplex in a hom-object of the standard free simplicial resolution is a sequence of composable non-identity morphisms, each contained within n sets of parentheses. The morphisms in the sequence describe the spine of a necklace, and the location of the outermost parentheses describes the joins. The other $(n-1)$ layers of parentheses determine the sets $T^1 \subset \cdots \subset T^{n-1}$. By Theorem 16.4.1, this exactly specifies an n-simplex in the corresponding hom-object of $\mathfrak{C}NA$. $\square$

Exercise 16.4.8 Compute $\mathfrak{C}\Lambda^n_k(0, n)$ for $0 < k < n$.

Exercise 16.4.9 Demonstrate by means of an example that $\mathfrak{C}$ does not preserve products.

Using Theorem 16.4.1, we can also show that the homotopy coherent nerve of a locally Kan simplicial category is a quasi-category; compare with [49, 1.1.5.10].

Lemma 16.4.10   *If $\underline{C}$ is a locally Kan simplicial category, then $\mathfrak{N}\underline{C}$ is a quasi-category.*

Proof   By adjunction, it suffices to prove that simplicial functors $\mathfrak{C}\Lambda_k^n \to \underline{C}$ extend to $\mathfrak{C}\Delta^n \to \underline{C}$ for all $0 < k < n$, $n \geq 2$. The categories $\mathfrak{C}\Lambda_k^n$ and $\mathfrak{C}\Delta^n$ have the same objects $0, 1, \ldots, n$ and nearly the same hom-spaces. By the necklace description, it is easy to see that $\mathfrak{C}\Lambda_k^n(i, j) = \mathfrak{C}\Delta^n(i, j)$ unless $i = 0$ and $j = n$; this is because the only simplices missing in Λ_k^n are the n-simplex and its kth face, both of which are beads from 0 to n that cannot appear in necklaces between any other pair of vertices. Thus it suffices to show that there is an extension

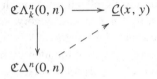

By our hypothesis on $\underline{C}$, it suffices to show that the left-hand map is anodyne. The bottom space is the cube $(\Delta^1)^{n-1}$. The top space, by Exercise 16.4.8, again using the necklace description, is a subset missing the interior and one of the codimension-one faces. The inclusion is obviously a monomorphism and weak homotopy equivalence, and thus lifts against any Kan complex. $\square$

Example 16.4.11   In a simplicial model category $\mathcal{M}$, the hom-spaces assemble into a right Quillen bifunctor $\mathcal{M}^{\mathrm{op}} \times \mathcal{M} \to \mathbf{sSet}$. As a direct consequence of Definition 11.4.4, the hom-space between any pair of fibrant–cofibrant objects is a Kan complex. Hence the full simplicial subcategory $\mathcal{M}_{QR}$ spanned by the fibrant–cofibrant objects is locally Kan. Its homotopy coherent nerve $\mathfrak{N}\mathcal{M}_{QR}$ is the quasi-category associated to the simplicial model category.

The reader is encouraged to verify that the homotopy category of the quasi-category $\mathfrak{N}\mathcal{M}_{QR}$ is equivalent to the homotopy category of $\mathcal{M}$.

Example 16.4.12   Using Theorem 16.4.7 and Lemma 16.4.10, we can see that the homotopy coherent diagrams and natural transformations introduced in Section 16.3 assemble into a quasi-category. This was the example that motivated the original definition by Boardman and Vogt (see [8, 4.9]). Given a small category $\mathcal{A}$ and a locally Kan simplicial category $\underline{C}$, define a simplicial set whose n-simplices are diagrams $\mathfrak{C}N(\mathcal{A} \times [n]) \to \underline{C}$. The action by the simplicial operators is defined by precomposing with the maps in the image of the cosimplicial object $\Delta \to \mathbf{Cat}$.

By Theorem 16.4.7, the vertices and edges are homotopy coherent diagrams and homotopy coherent natural transformations from $\mathcal{A}$ to $\underline{C}$. By adjunction, the n-simplices are isomorphic to maps of simplicial sets $N(\mathcal{A} \times [n]) \cong N\mathcal{A} \times \Delta^n \to \mathfrak{N}\underline{C}$. Again, by adjunction, the n-simplices are isomorphic to maps $\Delta^n \to \mathfrak{N}\underline{C}^{N\mathcal{A}}$. It follows from the Yoneda lemma that our simplicial set is isomorphic to $\mathfrak{N}\underline{C}^{N\mathcal{A}}$, which is a quasi-category by Lemma 16.4.10 and Corollary 15.2.3.

Remark 16.4.13   The spaces $\mathfrak{C}X(x, y)$ are essentially never Kan complexes or even quasi-categories, though in some sense they are close. Even when X is a Kan complex, it is easy to see that there are obstructions for filling Λ_0^2 horns and Λ_2^2 horns; there are also obstructions in dimensions 3 and 4. But all higher-dimensional horns admit unique fillers owing to the surprising, yet easy to prove, fact that for any simplicial set X, each hom-space $\mathfrak{C}X(x, y)$ is 3-coskeletal. Furthermore, if X is a quasi-category, any Λ_1^2 horn in $\mathfrak{C}X(x, y)$ can be filled. However, the simplicial category $\mathfrak{C}X$ is "locally quasi" if and only if X is the nerve of a category in which the identity morphisms do not admit any non-trivial factorizations. See [72] for proofs.

16.5 A gesture toward the comparison

Before concluding this chapter, we return to the problem that motivated its introduction: defining an enrichment of the homotopy category of a quasi-category X over the homotopy category of spaces. Unlike the models introduced in Section 15.4, the hom-spaces in $\mathfrak{C}X$ admit a (strictly) associative composition law. It remains to show that the spaces $\mathfrak{C}X(x, y)$ have the same homotopy type as the hom-spaces introduced previously.

Our proof follows [49, §§2.2.2, 2.2.4]: when X is a quasi-category, there exists a zigzag of weak homotopy equivalences $\mathrm{Hom}_X^R(x, y) \leftarrow Z \to \mathfrak{C}X(x, y)$ that we now describe. The simplicial set Z is defined to be the image of $\mathrm{Hom}_X^R(x, y)$ under the "straightening over a point" functor $\mathbf{sSet} \to \mathbf{sSet}$. This functor is a left adjoint and consequently determined by a cosimplicial object $Q^\bullet \colon \Delta \to \mathbf{sSet}$, which we now describe. Recall the cosimplicial object $C_R^\bullet \colon \Delta \to \mathbf{sSet}_{*,*}$ used to define $\mathrm{Hom}_X^R(x, y)$; an n-simplex in $\mathrm{Hom}_X^R(x, y)$ is precisely a map $C_R^n \to X$ in $\mathbf{sSet}_{*,*}$.[5] Write 0 and n for the two vertices of C_R^n, and define Q^n to be $\mathfrak{C}C_R^n(0, n)$; note that the other hom-spaces in the simplicial category $\mathfrak{C}C_R^n$ are either empty or terminal.

---

[5] The simplicial set C_R^n is called J^n in [49].

Using Theorem 16.4.1, we can obtain an explicit description of the $Q^n = \mathfrak{C}C_R^n(0, n)$, at least in low dimensions. It is easy to see that $Q^0 = *$ and $Q^1 = \Delta^1$. Let us also describe Q^2. Returning to the notation introduced in the proof of Lemma 15.4.10, which identified the non-degenerate simplices in C_R^2 with their preimages in Δ^3, the simplicial set C_R^2 has vertices 0, 3; edges [03], [13], [23]; 2-simplices [013], [123], [023]; and a 3-simplex [0123]. Thus Q^2 has three 0-simplices [03], [13], [23] and four 1-simplices, depicted in the following diagram, that we label with the single bead in their corresponding necklaces. There exist no totally non-degenerate necklaces in C_R^2 with multiple beads. There are two distinct 2-simplices corresponding to the only two possible non-degenerate vertex data for the necklace [0123]: $\Delta^3 \to C_R^2$. Their boundary is depicted as follows:

$$
\begin{array}{ccc}
[03] & \xrightarrow{\;\;[013]\;\;} & [13] \\[2pt]
{\scriptstyle[023]}\Big\downarrow & \searrow^{\,\sim}_{[0123]} & \Big\downarrow{\scriptstyle[123]} \\[2pt]
[23] & \overset{\sim}{=\!=\!=} & [23]
\end{array}
$$

Note that Q^2 is not the same as the simplicial set $\Delta^1 \times \Delta^1$ because it has only three distinct vertices and one of the edges of one of the 2-simplices is degenerate.

By Construction 1.5.1, the simplicial set Z is defined by the formula

$$
Z = \int^{[n]\in\Delta} \coprod_{\mathrm{Hom}_X^R(x,y)_n} \mathfrak{C}C_R^n(0, n) = \int^{[n]\in\Delta} \coprod_{\mathbf{sSet}_{*,*}(C_R^n, X)} \mathfrak{C}C_R^n(0, n)
$$

Each map $C_R^n \to X$ induces a map $\mathfrak{C}C_R^n(0, n) \to \mathfrak{C}X(x, y)$ by sending a necklace in C_R^n to the corresponding necklace in X. These maps define a cone under the coend, inducing a map $Z \to \mathfrak{C}X(x, y)$. By [49, 2.2.4.1], this map is a weak homotopy equivalence when X is a quasi-category.

At the same time, there is a map of cosimplicial objects $Q^\bullet \to \Delta^\bullet$ induced by the last vertex map of Example 1.5.7; Q^n can also be described as a quotient of $\mathrm{sd}_n \Delta^n$. This map of cosimplicial objects determines a map of coends

$$
\int^{[n]\in\Delta} \coprod_{\mathrm{Hom}_X^R(x,y)_n} \mathfrak{C}C_R^n(0, n) \to \int^{[n]\in\Delta} \coprod_{\mathrm{Hom}_X^R(x,y)_n} \Delta^n = \mathrm{Hom}_X^R(x, y).
$$

A clever inductive argument in [49, 2.2.2.7] shows that this is a weak homotopy equivalence for any simplicial set.

17

Isomorphisms in quasi-categories

A 1-simplex in a quasi-category X is an **isomorphism** if and only if its image in the homotopy category hX is an isomorphism. It follows from the description of the homotopy category associated to a quasi-category in Section 15.1 that for any isomorphism $f : x \to y$ in a quasi-category, there exists an inverse isomorphism $g : y \to x$ together with 2-simplices

$$
\begin{array}{ccc}
 & y & \\
f \nearrow & \sim & \searrow g \\
x & =\!=\!=\!= & x
\end{array}
\qquad
\begin{array}{ccc}
 & x & \\
g \nearrow & \sim & \searrow f \\
y & =\!=\!=\!= & y
\end{array}
\qquad (17.0.1)
$$

Hence a 1-simplex f is an isomorphism if and only if it can be extended to this data. This assertion is extended to higher-dimensional simplices in Lemma 17.2.5. In the presence of (17.0.1), we say that the objects x and y are **isomorphic**.

Lemma 17.0.2 *Two objects in a quasi-category are isomorphic if and only if there exists an isomorphism between them.*

Proof Objects in a quasi-category X are isomorphic just when they are in the same isomorphism class in hX, which means there is an isomorphism between them in the homotopy category. Any arrow in the homotopy category is represented by a 1-simplex in X; any isomorphism is represented by a 1-simplex that is an isomorphism. □

This is not the standard terminology: isomorphisms in a quasi-category are typically called "equivalences." We prefer our convention on account of our philosophy that every quasi-categorical term should be given the name of the categorical concept that it generalizes. The use of "isomorphism" to describe these 1-simplices helps distinguish this notion from the other "equivalences" in quasi-category theory.[1] Also, no confusion is possible: the only stricter comparison between 1-simplices is identity, witnessed by a degenerate 1-simplex (which is, in particular, an isomorphism).

A basic observation is that isomorphisms are preserved by any map between quasi-categories: one proof is that an isomorphism is witnessed equationally by the pair of 2-simplices (17.0.1) whose form is preserved by any simplicial map. In this chapter, we investigate the behavior of isomorphisms in quasi-categories. A main theme explored in Sections 17.2 and 17.6 is that the isomorphisms govern the relationship between quasi-categories and Kan complexes: we show that a Kan complex is a "groupoidal quasi-category."

Ancillary notions introduced to prove theorems about isomorphisms have other important applications. The slice and join, introduced in 17.1 to investigate lifting properties enjoyed by outer horns with invertible edges, are used to express the universal property of limits and colimits in a quasi-category and reappear in Chapter 18. Marked simplicial sets, introduced in Section 17.4 to prove that diagrams of isomorphisms can be freely inverted, are also used to define a simplicial model category of quasi-categories. For this reason, they play a role in the formation of homotopy limits of diagrams of quasi-categories, the subject of Section 17.7. The results that follow are due to several people. Our proofs follow [40], [49], and private communication with Verity.

17.1 Join and slice

The main result of the paper [40] is that the isomorphisms in a quasi-category admit the following characterization, which has a number of pleasing consequences.

Theorem 17.1.1 (Joyal) $f: \Delta^1 \to X$ *is an isomorphism if and only if any horn* Λ_0^n *with initial edge* f *extends to* Δ^n. *In particular, quasi-categories admit fillers for any* Λ_0^n *horn whose initial edge is an isomorphism or any* Λ_n^n *horn whose terminal edge is an isomorphism.*

The implication ($\Leftarrow$) is an easy exercise. The proof of the converse requires that we introduce some ancillary notions.

The category Δ_+ of finite (possibly empty) ordinals and order preserving maps has a symmetric monoidal structure given by ordinal sum; in simplicial notation, $[n] \oplus [m] = [n + m + 1]$ for all $n, m \geq -1$. This operation induces a monoidal structure, called the **join** and denoted $\star$, on augmented simplicial

---

[1] The skeptical reader is encouraged to translate Remark 18.1.5 or Remark 18.6.5 into the standard terminology.

sets by **Day convolution**: the join of two augmented simplicial sets is defined
to be the left Kan extension of the ordinal sum along the Yoneda embedding

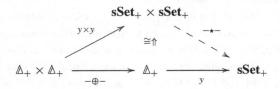

By the formula of Theorem 1.2.1,

$$X \star Y = \int^{[k],[m] \in \Delta_+ \times \Delta_+} \coprod_{X_k \times Y_m} \Delta^{k+m+1},$$

from which we deduce that

$$(X \star Y)_n = \int^{[k],[m] \in \Delta_+ \times \Delta_+} \coprod_{\Delta_+([n],[k] \oplus [m])} X_k \times Y_m,$$

which says that $X \star Y$ is the left Kan extension of the external product $\Delta_+^{\mathrm{op}} \times \Delta_+^{\mathrm{op}} \xrightarrow{X \times Y} \mathbf{Set} \times \mathbf{Set} \xrightarrow{- \times -} \mathbf{Set}$ along the opposite of the ordinal sum functor $\oplus \colon \Delta_+ \times \Delta_+ \to \Delta_+$.

One can check that $(X \star Y)_{-1} = X_{-1} \times Y_{-1}$. In particular, if X and Y are simplicial sets, given a trivial augmentation, $X \star Y$ is trivially augmented. In this way, we obtain a join bifunctor

$$- \star - \colon \mathbf{sSet} \times \mathbf{sSet} \to \mathbf{sSet}$$

on simplicial sets by restriction. By direct computation, one sees that

$$(X \star Y)_n = X_n \cup Y_n \cup \bigcup_{i+j=n-1} X_i \times Y_j.$$

In particular, observe that there are natural inclusions $X \to X \star Y \leftarrow Y$.

Example 17.1.2   The reader is encouraged to verify the preceding formula and the following computations:

- $\Delta^n \star \Delta^m \cong \Delta^{n+m+1}$
- $\partial \Delta^n \star \Delta^0 \cong \Lambda_{n+1}^{n+1}$
- $\Lambda_0^2 \star \Delta^0 \cong \Delta^1 \times \Delta^1$
- $X \star \emptyset \cong \emptyset \star X \cong X$

The last isomorphism tells us that $- \star K$ does not preserve colimits. The reason for this failure is that we have chosen to augment simplicial sets using the right adjoint to the forgetful functor $\mathbf{sSet}_+ \to \mathbf{sSet}$ and not the left adjoint, which augments a simplicial set with its set of path components (see 1.1.10).

A modified version of the join bifunctor making use of the path components augmentation is used to define the décalage functor in Section 17.5.

However, the functor $- \star K$ is cocontinuous if we reinterpret its target as the slice category under K. By an adjoint functor theorem, we have an adjunction

$$- \star K : \mathbf{sSet} \underset{\longleftarrow}{\overset{\perp}{\longrightarrow}} K/\mathbf{sSet} : (-)_{/-} \qquad (17.1.3)$$

where the right adjoint, the notation for which, following Lurie, conceals a K dependence, carries $p : K \to X$ to a simplicial set $X_{/p}$. By the Yoneda lemma, an n-simplex $\Delta^n \to X_{/p}$ corresponds to a map $\Delta^n \star K \to X$ under K. By pre- and postcomposition, a map $j : K \to L$ induces natural transformations with components $A \star K \to A \star L$ and $X_{/q} \to X_{/qj}$ at $A \in \mathbf{sSet}$ and $q : L \to X \in L/\mathbf{sSet}$.

The right adjoint of (17.1.3), called the **slice**, is worthy of further considera-tion. Given a composable triple of maps of simplicial sets, as displayed on the left-hand side of the following, we get the natural maps between the slices on the right-hand side:

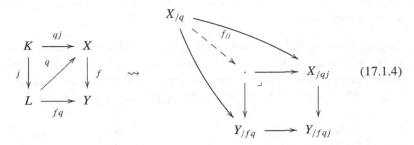

$$(17.1.4)$$

These can be thought of as some encoding of the bifunctoriality of the slice construction. When $K = \emptyset$, $X_{/qj} \cong X$, and the map $X_{/q} \to X$ is the obvious "forgetful functor."

Furthermore, on account of a verification left to the reader or [40], $i \boxtimes f_{/j}$ if and only if $i \hat{\star} j \boxtimes f$ on account of a natural bijective correspondence between commutative squares

$$
\begin{array}{ccc}
A \star L \coprod_{A \star K} B \star K \longrightarrow X & \qquad & A \longrightarrow X_{/q} \\
{\scriptstyle i \hat{\star} j} \downarrow \qquad\qquad \downarrow {\scriptstyle f} & \longleftrightarrow & {\scriptstyle i} \downarrow \qquad\qquad \downarrow {\scriptstyle f_{/j}} \\
B \star L \longrightarrow Y & \qquad & B \longrightarrow X_{/qj} \times_{Y_{/fqj}} Y_{/fq}
\end{array}
$$

$$(17.1.5)$$

Now the proof of Theorem 17.1.1 hinges on a mild combinatorial lemma and a few observations concerning the behavior of the isomorphisms with respect to the right fibrations. Recall a monomorphism is **anodyne**, **left anodyne**, **right**

anodyne, or **inner anodyne** if it is, respectively, in the weak saturation of the horn inclusions, left horn inclusions (excluding Λ_n^n), right horn inclusions (excluding Λ_0^n), or inner horn inclusions (excluding both Λ_n^n and Λ_0^n). The right classes of these weak factorization systems are the **Kan**, **left**, **right**, and **inner fibrations**.

Lemma 17.1.6 ([49, 2.1.2.3])   *Let $i\colon A \to B$ and $j\colon K \to L$ be monomorphisms. If i is right anodyne or j is left anodyne, then $i \hat{\star} j$ is inner anodyne.*

Proof   By the argument given in Proposition 15.2.1, it suffices to prove this for the generators. It turns out $j_k^n \hat{\star} i_m$ is j_k^{n+m+1} and $i_m \hat{\star} j_k^n = j_{k+m+1}^{n+m+1}$, so this is obvious. $\square$

We leave the observations concerning the behavior of isomorphisms with respect to right fibrations as an exercise for the reader.

Exercise 17.1.7   Show that right fibrations between quasi-categories reflect isomorphisms. Show furthermore that given a right fibration between quasi-categories $p\colon X \to Y$ and an isomorphism $p(x) \to y$ in Y, this 1-simplex lifts to an isomorphism $x \to x'$ in X.

Proof of Theorem 17.1.1   As observed in the proof of Lemma 17.1.6, the horn inclusion $j_0^n\colon \Lambda_0^n \to \Delta^n$ is $j_0^1 \hat{\star} i_{n-2}$, where $j_0^1 = d^1\colon \Delta^0 \to \Delta^1$. Let σ denote the first face of the image of the horn Λ_0^n, represented in the domain of $j_0^1 \hat{\star} i_{n-2}$ as $\Delta^0 \star \Delta^{n-2}$. Via (17.1.5), the desired lifting problem is adjunct to

where the right-hand vertical map is defined as in (17.1.4) with respect to the composable triple $\partial \Delta^{n-2} \xrightarrow{i_{n-2}} \Delta^{n-2}$, $\Delta^{n-2} \xrightarrow{\sigma} X$, $X \to *$.

Applying Lemma 17.1.6 to the pushout-join of a right anodyne map with the maps $\emptyset \to \partial\Delta^{n-2}$ and $\partial\Delta^{n-2} \to \Delta^{n-2}$, we conclude, after taking adjuncts, that the maps $X_{/\sigma i_{n-2}} \to X$ and $X_{/\sigma} \to X_{/\sigma i_{n-2}}$ are right fibrations; in particular, their domains and codomains are quasi-categories when X is. By Exercise 17.1.7, the bottom 1-simplex is an isomorphism that therefore lifts to $X_{/\sigma}$ by 17.1.7 again. The transpose of this lift solves the desired extension problem $\square$

This proof also shows that horns $\Lambda_0^n \to \Delta^n$ or $\Lambda_n^n \to \Delta^n$ lift against any inner fibration provided that the image of the initial or terminal edge, respectively, is an isomorphism in its domain.

17.2 Isomorphisms and Kan complexes

Several important results are immediate consequences of Theorem 17.1.1.

Corollary 17.2.1 ([49, 1.2.5.1]) *Let X be any simplicial set. The following are equivalent:*

  (i) X is a quasi-category and hX is a groupoid.
  (ii) $X \to *$ is a left fibration.
  (iii) $X \to *$ is a right fibration.
  (iv) X is a Kan complex.

In other words, Kan complexes are quasi-categories in which every edge is an isomorphism.

Corollary 17.2.2 *The simplicial sets $\mathrm{Hom}^L_X(x, y)$, $\mathrm{Hom}_X(x, y)$, and $\mathrm{Hom}^R_X(x, y)$ are Kan complexes.*

Proof The pullback

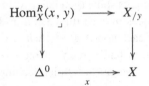

defines $\mathrm{Hom}^R_X(x, y)$. The right-hand map is the forgetful functor associated to the slice over $y \colon \Delta^0 \to X$. Applying Lemma 17.1.6 to the pushout-join of a right anodyne map with $j = \emptyset \to \Delta^0$, we conclude from (17.1.5) that the right-hand map is a right fibration. It follows from 17.2.1 (iii) $\Leftrightarrow$ (iv) that $\mathrm{Hom}^R_X(x, y)$ is a Kan complex. But we have shown that this hom-space is equivalent to the others, so it follows from 15.3.9 that their homotopy categories are also groupoids. The conclusion then follows from 17.2.1 (i) $\Leftrightarrow$ (iv). $\square$

Corollary 17.2.3 *There is a right adjoint $\iota \colon \mathbf{qCat} \to \mathbf{Kan}$ to the inclusion of full subcategories $\mathbf{Kan} \hookrightarrow \mathbf{qCat}$ that takes a quasi-category to the simplicial subset spanned by the isomorphisms.*

Proof Isomorphisms in (homotopy) categories, and hence isomorphisms in quasi-categories, are closed under composition. It follows easily that ιX is again a quasi-category. By 17.2.1 (i) $\Leftrightarrow$ (iv), it is a Kan complex. Maps of simplicial sets preserve isomorphisms because they preserve the 2-simplices that witness composition identities in the homotopy categories. Hence a map from a Kan complex K to a quasi-category X lands in ιX, and the proof is complete. $\square$

Note that the maximal sub Kan complex ιX of a quasi-category X contains all the vertices. When X is not a Kan complex, Corollary 17.2.1 implies that ιX contains fewer simplices in each positive dimension.

Remark 17.2.4 We would not expect a left adjoint to the inclusion **Kan** $\hookrightarrow$ **qCat**. If such existed, then 1-simplices in a Kan complex would have to correspond bijectively to diagrams from some Kan complex formed by applying the left adjoint to Δ^1. The natural choice for the image of Δ^1 is J, but a generic isomorphism in a Kan complex has many possible extensions to a diagram of shape J.

As a consequence of Corollary 17.2.3, the data depicted in (17.0.1) witnessing that two 1-simplices are inverse isomorphisms can be greatly extended.

Lemma 17.2.5 $f : \Delta^1 \to X$ *is an isomorphism in a quasi-category if and only if there exists an extension to* $J = N(\bullet \cong \bullet)$.

Proof Only one direction is non-obvious. For this, we make use of the following observation: an n-simplex in the nerve of a category is degenerate if and only if one of the edges along its spine is an identity. In particular, there are only two non-degenerate simplices in each dimension in J, and furthermore, only the zeroth and nth faces of each non-degenerate n-simplex are non-degenerate.

By hypothesis, the map f lands in ιX; it therefore suffices to show that $\Delta^1 \to J$ is anodyne. In fact, we give a cellular decomposition of this inclusion, building J by attaching a sequence of outer horns. Abusing terminology, we denote the non-degenerate 1-simplex of Δ^1 by f. The first attaching map $\Lambda^2_2 \to \Delta^1$ has zeroth face f and first face an identity. Call the non-degenerate 1-simplex attached by the pushout

$$
\begin{array}{ccc}
\Lambda^2_2 & \longrightarrow & \Delta^1 = J^1 \\
\downarrow & & \downarrow \\
\Delta^2 & \xrightarrow{\ulcorner} & J^2
\end{array}
$$

g. The simplicial set J^2 also contains a non-degenerate 2-simplex whose spine is fg. Next we use the Λ^3_3 horn whose boundary is depicted

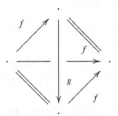

to attach a 3-simplex to J^2, defining a simplicial set J^3 that contains a non-degenerate 2-simplex with spine gf and a non-degenerate 3-simplex with spine fgf. Next attach a Λ_4^4 horn to J^3, and so on. □

Lemma 17.2.5 provides a welcome opportunity to revisit the definition of an equivalence of quasi-categories introduced in Section 15.3. Recall that $[A, X]_J$ was defined to the set of equivalence classes of maps $A \to X$, with respect to the equivalence relation generated by diagrams (15.3.3). As X and hence X^A is a quasi-category, we see from Lemma 17.2.5 that $f \sim g$ if and only if f and g are isomorphic as vertices of X^A. In other words, $[A, X]_J = \pi_0\iota(X^A)$.

Now consider an equivalence $f : X \to Y$ between quasi-categories. By definition, f induces isomorphisms between the sets of isomorphism classes of objects

$$[Y, X]_J \xrightarrow{f^*} [X, X]_J \qquad\qquad [Y, Y]_J \xrightarrow{f^*} [X, Y]_J.$$

Considering the first of these, we conclude that the identity on X is isomorphic in the quasi-category X^X to a vertex in the image of f^*. By Lemma 17.0.2 and Lemma 17.2.5, this isomorphism is represented by a map as displayed on the left:

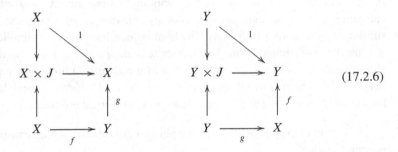

$$(17.2.6)$$

Postcomposing the isomorphism with f, we see that f and fgf are isomorphic in Y^X. From the second bijection, it follows that fg is isomorphic to the identity on Y via a map as displayed on the right.

Remark 17.2.7 The right adjoint ι preserves products. Thus $\iota_* \mathbf{qCat}$, the simplicial category obtained by applying the functor ι to each of the hom-quasi-categories, is a locally Kan simplicial category. This allows us to define the **quasi-category of quasi-categories**: it is the (large) quasi-category obtained as the homotopy coherent nerve of $\iota_* \mathbf{qCat}$.

By inspection, and the fact that $\iota_* \mathbf{qCat}$ is locally Kan, the homotopy category of $\mathfrak{N}\iota_* \mathbf{qCat}$ is $(\pi_0)_* \iota_* \mathbf{qCat}$. We call the homotopy category of the quasi-category of quasi-categories the **homotopy category of quasi-categories**.

Unpacking the definition, its objects are the quasi-categories and its morphisms are isomorphism classes of vertices in the hom-quasi-categories of **qCat**. A map $f: X \to Y$ becomes an isomorphism in the homotopy category of quasi-categories just when there exists $g: Y \to X$ so that the composites fg and gf are the same isomorphism classes as the respective identities. From (17.2.6), we see that f is an isomorphism in the homotopy category of quasi-categories, or, indeed, in the quasi-category of quasi-categories, exactly when f is an equivalence of quasi-categories.

Applying the adjunction of Corollary 17.2.3, the maps f and g of an equivalence (17.2.6) restrict to maps between the maximal sub Kan complexes of these quasi-categories. Because J is a Kan complex, the homotopies (17.2.6) restrict to maps $\iota X \times J \to \iota X$ and $\iota Y \times J \to \iota Y$ that witness, in particular, that ιf and ιg form a categorical equivalence between ιX and ιY. Thus we have proven one half of the following proposition.

Proposition 17.2.8 *If $f: X \to Y$ is an equivalence of quasi-categories, then $\iota f: \iota X \to \iota Y$ is a categorical equivalence of Kan complexes. Conversely, a weak homotopy equivalence between Kan complexes is a categorical equivalence.*

Proof For the converse, any weak homotopy equivalence $X \to Y$ of Kan complexes extends to a simplicial homotopy equivalence by a classical result subsumed by Theorem 10.5.1(iii). The homotopies, themselves 1-simplices in X^X and Y^Y, are isomorphisms because these hom-spaces are Kan complexes. Hence Lemma 17.2.5 implies that the data of the simplicial homotopy equivalence can be extended to a diagram of the form (17.2.6). Hence a weak homotopy equivalence between Kan complexes is a categorical equivalence. □

On account of Proposition 17.2.8, the phrase **equivalence of Kan complexes** is unambiguous.

17.3 Inverting simplices

We think of a 2-simplex

in a quasi-category as a witness that the composite of the 1-simplices g and f is homotopic to h. If f, g, h are isomorphisms, we might expect there to be a

2-simplex witnessing that the composite of f^{-1} with g^{-1} is homotopic to h^{-1}, for any choice of inverse isomorphisms. Indeed, this and more is true.

Theorem 17.3.1 *Consider $\Delta^n \to X$ such that the image of each edge of Δ^n is an isomorphism.[2] Then this map extends through $\tilde{\Delta}^n$, the nerve of the groupoid of $n + 1$ uniquely isomorphic objects.*

There is an easy proof[3] of this result using model category theory: the map $\Delta^n \to \tilde{\Delta}^n$ is a cofibration between cofibrant objects. By hypothesis, the n-simplex lives in ιX, so the desired extension will also land in this Kan complex, a fibrant object in Quillen's model structure. To show that a map from a cofibrant object to a fibrant object extends along a given cofibration, it suffices to replace the cofibration by any weakly equivalent map and solve the composite lifting problem [49, A.2.3.1]. We can replace the map $\Delta^n \to \tilde{\Delta}^n$ by the identity at Δ^0, which is weak homotopy equivalent (though not categorically equivalent) to both spaces, completing the proof.

In what follows, we give a different proof that allows us to draw a more refined conclusion: given any diagram in a quasi-category, that is, given any map of simplicial sets $K \to X$ whose edges are isomorphisms, each simplex in the diagram can be "freely inverted" in a sense we will make precise later. One reason we like our proof is that it gives us an opportunity to introduce a new tool, marked simplicial sets, for the study of quasi-categories that is particularly suitable for proving results about the isomorphisms or a more general class of edges satisfying certain properties.

17.4 Marked simplicial sets

Write $\mathbf{sSet}^+$ for the category of **marked simplicial sets**, that is, simplicial sets X with a specified collection of 1-simplices $X_e \subset X_1$ called **marked edges** containing the degenerate simplices. A map of marked simplicial sets is a map of simplicial sets that preserves marked edges. Any simplicial set X admits a minimal marking $X^\flat$, with only the degenerate edges marked, and a maximal marking $X^\sharp$ with all 1-simplices marked. When X is a quasi-category, the natural choice, denoted $X^\natural$, is to mark the isomorphisms.

We develop familiarity with the category of marked simplicial sets through the proof of the following proposition.

Proposition 17.4.1 $\mathbf{sSet}^+$ *is cartesian closed.*

---

[2] It suffices that the spine lands in the isomorphisms.
[3] Thanks to Gijs Heuts for pointing this out.

Proof   Write $\mathbb{A}^+$ for the category obtained by freely adjoining an object e to $\mathbb{A}$ through which the map

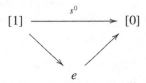

factors. The maps to and from e induce the following isomorphisms of hom-sets

$$\mathbb{A}^+(e, [n]) \cong \mathbb{A}(0, [n]) \qquad \mathbb{A}^+([n], e) \cong \mathbb{A}([n], [1]).$$

The inclusion functor $\mathbf{sSet}^+ \hookrightarrow \mathbf{Set}^{(\mathbb{A}^+)^{op}}$, which takes a marked simplicial set to the obvious presheaf on $\mathbb{A}^+$, has a left adjoint: factor the unique map $X_e \to X_1$ as an epimorphism followed by a monomorphism, and take the resulting subobject of X_1 to be the set of marked edges. The counit of this adjunction is readily seen to be an isomorphism. Hence we may regard $\mathbf{sSet}^+$ as a reflective full subcategory of the presheaf category. In the notation just introduced, the representable functors $\mathbb{A}^+(-, [n])$ correspond to the marked simplicial sets $(\Delta^n)^\flat$. The representable $\mathbb{A}^+(-, e)$ corresponds to the marked edge $(\Delta^1)^\sharp$.

The product and internal hom on $\mathbf{sSet}^+$ are the restrictions of these structures on the category of presheaves (cf. 6.1.9). In particular, an edge in a product of marked simplicial sets is marked if and only if both components are; this is necessary if the projection maps $X \leftarrow X \times Y \to Y$ are to be maps of marked simplicial sets. Write $\mathrm{Map}(X, Y)$ for the internal hom in $\mathbf{sSet}^+$. It is the simplicial set whose n-simplices are, by adjunction, maps of marked simplicial sets $\alpha \colon X \times (\Delta^n)^\flat \to Y$. The assertion that α is a map of marked simplicial sets corresponds to the condition that the horizontal edges of the cylinder $\Delta^1 \times \Delta^n \to Y$ associated to a marked edge in X are marked in Y. This is illustrated in the following in the case $n = 2$:

$$f \in X_e \;\rightsquigarrow\; (\Delta^1)^\sharp \times (\Delta^2)^\flat \xrightarrow{f \times -} X \times (\Delta^2)^\flat \to Y$$

A 1-simplex $X \times (\Delta^1)^\flat \to Y$ in $\mathrm{Map}(X, Y)$ is marked just when the diagonal 1-simplex in each cylinder $\Delta^1 \times \Delta^1 \to Y$ associated to a marked edge of X is also marked in Y:

$$f \in X_e \quad \rightsquigarrow \quad (\Delta^1)^\sharp \times (\Delta^1)^\flat \xrightarrow{f \times -} X \times (\Delta^1)^\flat \to Y \qquad \begin{array}{c} \cdot \xrightarrow{\sim} \cdot \\ \downarrow \searrow \downarrow \\ \cdot \xrightarrow{\sim} \cdot \end{array}$$

$$(17.4.2)$$

Applying this observation to the degenerate 1-simplices in X, we conclude that the vertical edges of the diagram (17.4.2) corresponding to a marked 1-simplex are also marked. Hence marked 1-simplices in $\mathrm{Map}(X, Y)$ are precisely maps $X \times (\Delta^1)^\sharp \to Y$. □

We have a string of adjunctions

$$\mathbf{sSet} \; \underset{\substack{\longrightarrow \\ \perp \\ \longleftarrow \\ \perp \\ \longrightarrow}}{\overset{\substack{\flat \\ \perp}}{}} \; \mathbf{sSet}^+$$

If K is a simplicial set, $K^\flat$ is the minimally marked simplicial set, and $K^\sharp$ is the maximally marked one. The maps $(-)^\flat, (-)^\sharp \colon \mathbf{sSet}^+ \rightrightarrows \mathbf{sSet}$ might also be called U and ι, respectively. If (X, X_e) is a marked simplicial set, $U(X, X_e) = X$ and $\iota(X, X_e)$ is the full simplicial subset spanned by the marked edges; if X is a quasi-category and X_e is the set of isomorphisms, then $\iota(X, X_e) = \iota X$. We prefer the flat–sharp notation for these forgetful functors, particularly when applied to the internal hom in $\mathbf{sSet}^+$. In particular, $\mathrm{Map}^\flat(X, Y)$ is the simplicial set underlying the internal hom, while $\mathrm{Map}^\sharp(X, Y)$ is the simplicial subset whose n-simplices are the maps $X \times \Delta^n \to Y$ whose edges

$$X \times \Delta^1 \to X \times \Delta^n \to Y$$

are marked, meaning that they extend along the map $X \times (\Delta^1)^\flat \to X \times (\Delta^1)^\sharp$.

Because both $(-)^\flat, (-)^\sharp \colon \mathbf{sSet} \rightrightarrows \mathbf{sSet}^+$ are strong monoidal, by Theorem 3.7.11, we have two choices of simplicial enrichment, tensor, and cotensor structure for $\mathbf{sSet}^+$. The one with the flats is suitable for modeling $(\infty, 2)$-categories. Here we instead use the sharps to obtain a model for $(\infty, 1)$-categories. The tensor structures pass through the left adjoint $(-)^\sharp \colon \mathbf{sSet} \to \mathbf{sSet}^+$, and the hom-spaces are $\mathrm{Map}^\sharp(X, Y)$, which restricts to the invertible 1-morphisms (2-cells).

Remark 17.4.3 In particular, the notion of simplicial homotopy for our preferred simplicial enrichment on $\mathbf{sSet}^+$ is defined using the marked 1-simplex $(\Delta^1)^\sharp$. That is, by adjunction, homotopies in $\mathbf{sSet}^+$ between a pair of maps from X to Y are marked 1-simplices in $\mathrm{Map}(X, Y)$.

Marked simplicial sets form a simplicial model category with hom-spaces $\mathrm{Map}^\sharp(X, Y)$.

Theorem 17.4.4 *There exists a left proper cofibrantly generated model structure on* $\mathbf{sSet}^+$ *whose cofibrations are monomorphisms and fibrant objects are quasi-categories with the isomorphisms marked. The fibrations between fibrant objects are those fibrations in Joyal's model structure. The weak equivalences between fibrant objects are precisely the categorical equivalences. In general, a map* $X \to Y$ *between marked simplicial sets is a weak equivalence if, for all quasi-categories* Z, *either of the equivalent conditions is satisfied:*

(i) $\mathrm{Map}^\flat(Y, Z^\natural) \to \mathrm{Map}^\flat(X, Z^\natural)$ *is an equivalence of quasi-categories.*
(ii) $\mathrm{Map}^\sharp(Y, Z^\natural) \to \mathrm{Map}^\sharp(X, Z^\natural)$ *is an equivalence of Kan complexes.*

Furthermore, these definitions give $\mathbf{sSet}^+$ *the structure of a simplicial model category with* $\mathrm{Map}^\sharp(X, Y)$ *as hom-spaces. Finally, the adjunction* $(-)^\flat : \mathbf{sSet} \rightleftarrows \mathbf{sSet}^+ : (-)^\flat$, *whose right adjoint forgets the markings, defines a Quillen equivalence between this model structure and Joyal's model structure for quasi-categories.*

Proof This theorem is a special case of a class of model structures on stratified simplicial sets [86, 113] and also a special case of results in [49, chapter 3]. Adopting the terminology of the latter, if X is a quasi-category, a p-cartesian edge with respect to $p \colon X \to *$ is precisely an isomorphism, and all maps $X \to *$ are Cartesian equivalences, immediately from the definition and the fact that degeneracies are isomorphisms. The assertions about the model structure are special cases of model structures on $\mathbf{sSet}^+/S$, where $S = *$, contained in 3.1.3.5, 3.1.3.7, 3.1.4.1, 3.1.4.4, 3.1.5.3.   □

Exercise 17.4.5 Use Proposition 17.2.8 to prove that $(i) \Rightarrow (ii)$. The other half of the equivalence is more subtle.

Now consider two cosimplicial objects $M, \tilde{M} \colon \triangle \rightrightarrows \mathbf{sSet}^+$ defined by $M^n = (\Delta^n)^\sharp$ and $\tilde{M}^n = (\tilde{\Delta}^n)^\sharp$. The canonical map $i \colon M \to \tilde{M}$ is a pointwise monomorphism.

Lemma 17.4.6 *The map* $i \colon M \to \tilde{M}$ *is a Reedy cofibration in* $(\mathbf{sSet}^+)^\triangle$.

Proof Both M and $\tilde{M}$ are unaugmentable by the same reasons as Example 14.3.9. The conclusion follows from Lemma 14.3.8.   □

We claim that $i \colon M \to \tilde{M}$ is also a Reedy, that is, pointwise, weak equivalence. Using the 2-of-3 property, we prove this by showing that $M^n = (\Delta^n)^\sharp$ and $\tilde{M}^n = (\tilde{\Delta}^n)^\sharp$ are both contractible in the marked sense; explicitly, that they admit simplicial homotopy equivalences to the point. Because $\mathbf{sSet}^+$ is a simplicial model category, at least with one choice of enrichment, simplicial homotopy equivalences are necessarily weak equivalences, and so this shows that each map $M^n \to \tilde{M}^n$ is a weak equivalence in the model structure.

Lemma 17.4.7 *The map $i \colon M \to \tilde{M}$ is also a pointwise weak equivalence.*

Proof We will do the case of $(\Delta^n)^\sharp$; the other is similar. The map $\Delta^0 \xrightarrow{0} (\Delta^n)^\sharp$ is an inverse homotopy equivalence to the unique map $(\Delta^n)^\sharp \to \Delta^0$. The desired homotopy $(\Delta^n)^\sharp \times (\Delta^1)^\sharp \to (\Delta^n)^\sharp$ is defined by $(i, 0) \mapsto 0$ and $(i, 1) \mapsto i$. □

The point of these observations is that they allow us to prove:

Proposition 17.4.8 *Let K be any simplicial set. The weighted colimit functor*

$$\mathrm{colim}^K - \colon (\mathbf{sSet}^+)^\Delta \to \mathbf{sSet}^+$$

is left Quillen when the domain is given the Reedy model structure.

Proof Because $\mathbf{sSet}^+$ is a bicomplete simplicial category, weighted colimits are computed via functor tensor products. Because it is a simplicial model category, by Theorem 14.3.1, the weighted colimit functor

$$- \otimes_\Delta - \colon (\mathbf{sSet})^{\Delta^{\mathrm{op}}} \times (\mathbf{sSet}^+)^\Delta \to \mathbf{sSet}^+$$

is a left Quillen bifunctor with respect to the Reedy model structures. Regarding K as a discrete bisimplicial set, the simplicial tensor product computes the colimit with weight K by (4.0.2). All bisimplicial sets are Reedy cofibrant by Lemma 14.3.7, so the result follows. □

How do we interpret this result? Forgetting the markings for a moment,

$$\mathrm{colim}^K \Delta^\bullet \cong \operatorname*{colim}_{\Delta^n \to K} \Delta^n \cong K,$$

using the fact that (unenriched) weighted colimits can be computed as colimits over categories of elements (cf. (7.1.8)). Because $(-)^\sharp \colon \mathbf{sSet} \to \mathbf{sSet}^+$ is a left adjoint, it preserves colimits and hence $\mathrm{colim}^K M^\bullet \cong K^\sharp$. Similarly,

$$\mathrm{colim}^K \tilde{\Delta}^\bullet \cong \operatorname*{colim}_{\Delta^n \to K} \tilde{\Delta}^n =: \tilde{K} \qquad (17.4.9)$$

is a simplicial set we will call $\tilde{K}$. By Yoneda's lemma, $\mathrm{colim}^{\Delta^n} \tilde{\Delta}^\bullet \cong \tilde{\Delta}^n$, so the new notation agrees with the old. As earlier, $\mathrm{colim}^K \tilde{M}$ marks everything in this simplicial set. The theorem together with Lemmas 17.4.6 and 17.4.7 says that $K^\sharp \to \tilde{K}^\sharp$ is a trivial cofibration in $\mathbf{sSet}^+$.

Corollary 17.4.10 *Let K be any simplicial set, and define $\tilde{K}$ to be the weighted colimit (17.4.9). The natural map $K^\sharp \to \tilde{K}^\sharp$ is a trivial cofibration of marked simplicial sets.*

Before applying this result, we use the fact that weighted colimits are cocontinuous in their weight to compute a few examples.

Example 17.4.11 Because Λ^2_1 is a pushout of two copies of Δ^1, $\tilde{\Lambda}^2_1 \cong \tilde{\Delta}^1 \sqcup_{\Delta^0} \tilde{\Delta}^1$. Although each edge in this simplicial set is an isomorphism, it is not a Kan complex because some required composites are absent.

Example 17.4.12 Let $\mathcal{C}$ be the free category on a commutative square; because $\mathcal{C} \cong 2 \times 2$, $N\mathcal{C} = \Delta^1 \times \Delta^1$. On account of the pushout diagram

$$
\begin{array}{ccc}
\Delta^1 & \xrightarrow{d^1} & \Delta^2 \\
{\scriptstyle d^1} \downarrow & & \downarrow \\
\Delta^2 & \xrightarrow{} & \Delta^1 \times \Delta^1
\end{array}
$$

we see that $\widetilde{\Delta^1 \times \Delta^1} \cong \tilde{\Delta}^2 \sqcup_{\tilde{\Delta}^1} \tilde{\Delta}^2$. This simplicial set has 10 non-degenerate 1-simplices, corresponding to the five non-degenerate 1-simplices of $\Delta^1 \times \Delta^1$.

By contrast, the free groupoid on $\mathcal{C}$ is defined by formally inverting all its arrows. This category is the product of two free-standing isomorphisms; hence its nerve is $J \times J = \tilde{\Delta}^1 \times \tilde{\Delta}^1$. This simplicial set is larger; for instance, it has 12 non-degenerate 1-simplices.

17.5 Inverting diagrams of isomorphisms

A map $K^\sharp \to X^\natural$ in $\mathbf{sSet}^+$ is precisely a map $K \to X$ of underlying simplicial sets so that every edge of K maps to an isomorphism in X. Immediately from Corollary 17.4.10:

Corollary 17.5.1 *Any map $K \to X$ from a simplicial set K to a quasi-category X that takes the edges of K to isomorphisms in X admits an extension*

Proof Maps to fibrant objects in $\mathbf{sSet}^+$ extend along any trivial cofibration

$\square$

This proves the assertion made at the end of Section 17.3. We can also apply this result to "partially invert" simplices. Suppose we have a Δ^{n+1+k} simplex in a quasi-category so that the last k edges along the spine are invertible. It follows that all edges spanned by the last $k + 1$ vertices are invertible. We show it is possible to "invert this simplex": we build a new simplex that leaves the faces connecting the first n-vertices to one of the last $k + 1$ vertices unchanged. This result is a corollary of the following lemma, whose proof is simplified by the use of markings.

Following [86], we modify the join bifunctor introduced in Section 17.1. Recall that the join bifunctor $\star\colon \mathbf{sSet}_+ \times \mathbf{sSet}_+ \to \mathbf{sSet}_+$ is the Day convolution of the ordinal sum bifunctor $\oplus\colon \Delta_+ \times \Delta_+ \to \Delta_+$. We restrict $\star$ to define a new bifunctor

$$\star' := \quad \mathbf{sSet} \times \mathbf{sSet} \xrightarrow{\ \pi_0 \times \pi_0\ } \mathbf{sSet}_+ \times \mathbf{sSet}_+ \xrightarrow{\ \star\ } \mathbf{sSet}_+ \xrightarrow{\ U\ } \mathbf{sSet},$$

cocontinuous in both variables, that augments a pair of simplicial sets with their sets of path components (see Example 1.1.10), applies the join for augmented simplicial sets, and then forgets the augmentation. If a simplicial set X is connected, its path components augmentation and trivial augmentation, the left and right adjoints to the forgetful functor $\mathbf{sSet}_+ \to \mathbf{sSet}$, coincide. When Y is also connected, $X \star Y \cong X \star' Y$. Note, however, that $X \star \emptyset = X$, whereas $X \star' \emptyset = \emptyset$.

The bifunctor $\star'$ extends to marked simplicial sets with the convention that an edge in $X \star' Y$ is marked only if it is marked in X or in Y; edges in $X \star' Y$ corresponding to the join of a vertex in X with a vertex in Y are not marked. This bifunctor is cocontinuous in both variables. Hence $(\Delta^n)^\flat \star' -\colon \mathbf{sSet}^+ \to \mathbf{sSet}^+$ admits a right adjoint $\mathrm{dec}_l((\Delta^n)^\flat, -)$, called the **décalage** functor. A k-simplex in $\mathrm{dec}_l((\Delta^n)^\flat, X^\natural)$, the object of interest, is a map $(\Delta^n)^\flat \star' (\Delta^k)^\flat \cong (\Delta^n \star \Delta^k)^\flat \to X^\natural$ of marked simplicial sets, subject to no further restrictions. Its marked edges are maps $(\Delta^n)^\flat \star' (\Delta^1)^\sharp \to X^\natural$. The domain is the $(n + 2)$-simplex with only its last non-degenerate edge marked. This encodes the "special outer horns" Λ_n^n satisfying the hypotheses of Theorem 17.1.1.

A bifunctor dec_r can be defined dually, fixing the other variable of $\star'$.

Lemma 17.5.2 *If X is a quasi-category, then $\mathrm{dec}_l((\Delta^n)^\flat, X^\natural)$ is fibrant.*

Proof Our proof is a special case of [86, 38]. To simplify notation, we drop the flats from simplicial sets with only the degeneracies marked. We must show that $\mathrm{dec}_l(\Delta^n, X^\natural)$ lifts against the following inclusions, which detect fibrant objects:

(i) $\Lambda_k^m \to \Delta^m$ for $m \geq 2, 0 < k < m$

(ii) $\Lambda_0^m \to \Delta^m$ and $\Lambda_m^m \to \Delta^m$ for all $m \geq 1$ with first and last edges marked, respectively, in both the domain and codomain

(iii) the "2-of-6" map $\Delta^3 \to (\Delta^3)^\natural$ for which the edges [02] and [13] are marked in the domain

By Lemma 11.1.5, it suffices to show that the left adjoint $\Delta^n \star' -$ preserves the weakly saturated class generated by these maps; because the domains and codomains of the generators (i)–(iii) are connected, we may replace this functor by $\Delta^n \star -$. For the case (i), the map $\Delta^n \star \Lambda_k^m \to \Delta^n \star \Delta^m$ factors through $(\emptyset \to \Delta^n)\hat{\star} j_k^m$ along the pushout of a Λ_k^m horn which attaches an m-simplex to the last $m + 1$ vertices in Δ^{n+m+1}. By Lemma 17.1.6, this composite is inner anodyne.

The join of Δ^n with the map of (iii) gives rise to a map of simplicial sets $\Delta^{n+4} \to \Delta^{n+4}$ with certain of the edges between the last four vertices marked; this map of marked simplicial sets is a pushout of the map $\Delta^3 \to (\Delta^3)^\natural$ along the obvious face map and thus remains in the same weakly saturated class.

Finally, the maps $\Delta^n \star \Lambda_0^m \to \Delta^{n+m+1}$ and $\Delta^n \star \Lambda_m^m \to \Delta^{n+m+1}$ obtained by applying $\Delta^n \star -$ to the maps of (ii) are generalized horn inclusions. For the former, we can fill the obvious Λ_0^m horn by a pushout of the generating map $\Lambda_0^m \to \Delta^m$ of marked simplicial sets. The result is the pushout-product $(\emptyset \to \Delta^n)\hat{\star} j_0^m$, which is inner anodyne by Lemma 17.1.6. For the latter, first fill the Λ_m^m horns to get each missing m-simplex containing the last edge. Then fill the Λ_{m+1}^{m+1} horns containing the last edge, and so on. $\qquad\square$

In particular, by Theorem 17.4.4, the underlying simplicial set $\mathrm{dec}_l(\Delta^k, X)$ is a quasi-category.

Corollary 17.5.3 *Given $\Delta^{n+1+k} \xrightarrow{\sigma} X$ with the last k edges along the spine invertible, there exists $\Delta^{n+1+k} \xrightarrow{\sigma'} X$, which agrees with the original simplex on the face spanned by the first $n + 1$ vertices. Furthermore, the last $k + 1$ vertices of σ' coincide with the last $k + 1$ vertices of σ but appear in reverse order, and each edge between these vertices has been replaced by an inverse isomorphism. Additionally, the faces of σ' spanned by the first $n + 1$ vertices and a single one of these last vertices coincide with the ones in σ.*

Proof Indeed, our proof implies a more robust conclusion. Apply Corollary 17.5.1 to a map $(\Delta^k)^\natural \to (\mathrm{dec}_l(\Delta^n, X))^\natural$ to obtain an extension $\tilde{\Delta}^k \to \mathrm{dec}_l(\Delta^n, X)$, giving rise to the desired map $\Delta^n \star \tilde{\Delta}^k \to X$. $\qquad\square$

17.6 A context for invertibility

The functor $(\tilde{-})\colon \mathbf{sSet} \to \mathbf{sSet}$ defined by (17.4.9) is equivalently described as the left Kan extension of the cosimplicial simplicial set $\tilde{\Delta}^{\bullet}\colon \Delta \to \mathbf{sSet}$, introduced in Theorem 17.3.1, along the Yoneda embedding. In particular, it has a right adjoint that sends a simplicial set X to the simplicial set whose n-simplices are maps $\tilde{\Delta}^n \to X$.

Theorem 17.6.1 ([42, 9.13]) *These functors define a Quillen adjunction whose left adjoint $(\tilde{-})$ is a functor from Quillen's model structure on simplicial sets to Joyal's.*

Proof By the proof of Proposition 17.4.8, the functor $- \otimes_{\Delta} \tilde{M}^{\bullet}\colon \mathbf{sSet}^{\Delta^{op}} \to \mathbf{sSet}^+$ is left Quillen with respect to the Reedy model structure on bisimplicial sets and the model structure of 17.4.4 on marked simplicial sets. In particular, a monomorphism of simplicial sets $K \to L$ becomes a monomorphism of discrete bisimplicial sets. This is also a Reedy cofibration and hence maps to a monomorphism $\tilde{K} \to \tilde{L}$, once we have forgotten the markings. This shows that $(\tilde{-})$ preserves cofibrations.

It remains to show that $(\tilde{-})$ carries weak homotopy equivalences to categorical equivalences. The natural map $K \to \tilde{K}$ induced by comparing the K-weighted colimits of $\Delta^{\bullet}$ and $\tilde{\Delta}^{\bullet}$ is a weak homotopy equivalence: To see this note that for any quasi-category Z, $\mathrm{Map}^{\sharp}(K^{\sharp}, Z^{\flat}) = (\iota Z)^K$. By the proof of Theorem 15.3.1, a map $A \to B$ is a weak homotopy equivalence if and only if the maps induced by homming into any Kan complex are weak homotopy equivalences. It follows from the characterization of the weak equivalences in Theorem 17.4.4 and from Corollary 17.4.10 that weak equivalences of marked simplicial sets are in particular weak homotopy equivalences.

Suppose given a weak homotopy equivalence $K \to L$. By the 2-of-3 property and the argument just given, $\tilde{K} \to \tilde{L}$ is a weak homotopy equivalence. We want to show that it is a categorical equivalence. By construction, every 1-simplex in $\tilde{K}$ is an isomorphism. In particular, $h\tilde{K}$ is a groupoid. By Corollary 15.3.9, a fibrant replacement K' of $\tilde{K}$ in Joyal's model structure will be a quasi-category whose homotopy category is a groupoid. By the 2-of-3 property for the weak homotopy equivalences: $K' \to L'$ is again a weak homotopy equivalence

$$
\begin{array}{ccc}
\tilde{K} & \xrightarrow{\ c.e\ } & K' \\
{\scriptstyle w.h.e.}\downarrow & & \downarrow{\scriptstyle w.h.e} \\
\tilde{L} & \xrightarrow[\ c.e.\ ]{} & L'
\end{array}
$$

By Corollary 17.2.1, this quasi-category is a Kan complex. A weak homotopy equivalence between Kan complexes is a categorical equivalence by Proposition 17.2.8. Hence the 2-of-3 property implies that $\tilde{K} \to \tilde{L}$ is a categorical equivalence, completing the proof. □

In other words, the functor that glues together "freely inverted simplices" defines a left Quillen functor from the model structure for Kan complexes to the model structure for quasi-categories.

17.7 Homotopy limits of quasi-categories

A diagram $F : \mathcal{D} \to \mathbf{qCat}$ induces a diagram $F^{\natural} : \mathcal{D} \to \mathbf{sSet}^{+}$ by marking the isomorphisms. By Theorem 17.4.4, $F^{\natural}$ is a diagram taking values in the fibrant objects of a simplicial model category. By Theorem 6.6.1, the homotopy limit is therefore the limit of $F^{\natural}$ weighted by $N(\mathcal{D}/-) : \mathcal{D} \to \mathbf{sSet}$. The proof of Theorem 15.2.5, which holds mutatis mutandis for diagrams of fibrant objects in any simplicial model category, tells us that this weighted limit is again a quasi-category with marked edges precisely the isomorphisms. In this way, we have proven:

Theorem 17.7.1   *The quasi-categories are closed under the formation of homotopy limits.*

Note that the construction of the homotopy limit of a diagram of quasi-categories, making use of the marked model structure, differs from the construction of the homotopy limit of a diagram of Kan complexes. Let us explore this via examples.

Example 17.7.2   By the argument in Example 6.5.2, the homotopy limit of a diagram $f : X \to Y$ of quasi-categories, represented by a diagram $f^{\natural} : X^{\natural} \to Y^{\natural}$, is computed via the following pullback in marked simplicial sets:

$$
\begin{array}{ccc}
\operatorname{holim} f^{\natural} & \longrightarrow & \operatorname{Map}((\Delta^{1})^{\sharp}, Y^{\natural}) \\
\downarrow & & \downarrow{\scriptstyle d^{1}} \\
\operatorname{Map}((\Delta^{0})^{\sharp}, X^{\natural}) & \longrightarrow & \operatorname{Map}((\Delta^{0})^{\sharp}, Y^{\natural})
\end{array}
$$

This $(\Delta^{0})^{\flat} \cong (\Delta^{0})^{\sharp}$ represents the identity, so the bottom horizontal map is $f^{\natural} : X^{\natural} \to Y^{\natural}$.

The underlying set of $\operatorname{Map}((\Delta^{1})^{\sharp}, Y^{\natural})$ is the full subspace ΠY of the path space $Y^{\Delta^{1}}$ on the paths $\Delta^{1} \to Y$ whose images are isomorphisms; in particular, ΠY is again a quasi-category. By the proof of Proposition 17.4.1, a marked edge in $\text{Map}((\Delta^1)^\sharp, Y^\natural)$ is a diagram $(\Delta^1)^\sharp \times (\Delta^1)^\sharp \to Y^\natural$. From this definition, we see that an edge is marked just when it is an isomorphism in the quasi-category ΠY: the details of the proof of this claim parallel those given in Lemma 18.2.3. In particular, each object in the diagram defining holim f is a naturally marked quasi-category, so we are comfortable interpreting the result in **qCat**, from which we conclude that holim f is the pullback

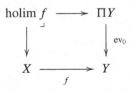

Note that if Y is a Kan complex, then $\Pi Y = Y^{\Delta^1}$, and this homotopy limit coincides with the usual mapping path space.

Example 17.7.3 Consider a diagram of quasi-categories $X \xrightarrow{f} Z \xleftarrow{g} Y$. Using the fact that $\text{Map}(\Delta^0, -)$ is the identity functor on marked simplicial sets, we see that the homotopy limit of the corresponding diagram of marked simplicial sets is the limit of the diagram

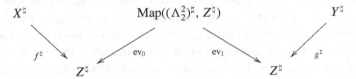

As in the previous example, the underlying simplicial set of $\text{Map}((\Lambda_2^2)^\sharp, Z^\natural)$ is the sub-quasi-category of $Z^{\Lambda_2^2}$ spanned by the Λ_2^2 horns in Z whose edges are isomorphisms. By Theorem 17.7.1, the limit is the naturally marked quasi-category whose vertices are objects $x \in X$ and $y \in Y$ together with a pair of isomorphisms $f(x) \to z \leftarrow g(y)$ in Z; whose 1-simplices are natural transformations of such; and so on.

Remark 17.7.4 By the dual of Corollary 14.3.2 interpreted in the marked model structure of Theorem 17.4.4, the homotopy pullback constructed in Example 17.7.3 is equivalent to the ordinary pullback in the case when either f or g is an **isofibration**, that is to say, a fibration in the Joyal model structure between quasi-categories.

18

A sampling of 2-categorical aspects of quasi-category theory

One of the most useful formal features of the category of simplicial sets, or here its full subcategory of quasi-categories, is that it is cartesian closed and, in particular, self-enriched. It follows by Theorem 7.5.3 that its limits and colimits all satisfy simplicially enriched universal properties. Throughout much of this text, we have exploited simplicial enrichments because of their convenience, even when our interest was in mere homotopy types.

In this final chapter, following observations of Joyal and Verity, we use the self-enrichment of the category of quasi-categories established by Corollary 15.2.3 to define a **2-category of quasi-categories** appropriate for its homotopy theory: equivalences in this 2-category are exactly equivalences of quasi-categories. This (strict) 2-category $\mathbf{qCat}_2$ is a truncation of the simplicial category of quasi-categories. The hom-spaces between quasi-categories have cells in each dimension starting from the vertices, which are ordinary maps of simplicial sets. Accordingly, in this chapter, we denote this simplicial category by $\mathbf{qCat}_\infty$. The 2-cells in $\mathbf{qCat}_2$ are homotopy classes of 1-simplices in the corresponding hom-spaces; all higher-dimensional information is discarded. Our interest in these structures is predicated on their competing enrichments; for this reason, we write $\mathbf{qCat}_2$ and $\mathbf{qCat}_\infty$ for the enriched categories, without the underline used elsewhere to signal enrichments. The category $\mathbf{qCat}$ is the common underlying category of both the 2-category $\mathbf{qCat}_2$ and the simplicial category $\mathbf{qCat}_\infty$.

We illustrate through examples how $\mathbf{qCat}_2$ can be used to determine the appropriate quasi-categorical generalizations of categorical concepts. Important to our development of the category theory of quasi-categories is that, in many cases, the proofs from classical category theory can be imported to demonstrate the corresponding quasi-categorical results. In this way, we might say our project is to develop the formal category theory of quasi-categories.

There is considerably more to be said in this vein – indeed, a large part of Joyal's unpublished treatises presents this viewpoint. Here we only have space to describe a sampling of 2-categorical aspects of quasi-category theory. A considerably more expansive continuation of this story can be found in [74, 75].

18.1 The 2-category of quasi-categories

The definition of the 2-category of quasi-categories makes use of the adjunction

$$h: \mathbf{sSet} \xrightleftharpoons{\perp} \mathbf{Cat}: N.$$

It is easy to see that the counit is an isomorphism. We can use this to prove:

Lemma 18.1.1 *The functor $h: \mathbf{sSet} \to \mathbf{Cat}$ preserves finite products.*

Proof Because h is a left adjoint and $\mathbf{sSet}$ and $\mathbf{Cat}$ are cartesian closed, the bifunctors $(h-) \times (h-)$ and $h(- \times -)$ preserve colimits in both variables. Hence it suffices to show that $h\Delta^n \times h\Delta^m \cong h(\Delta^n \times \Delta^m)$. The representable Δ^n is the nerve of the category, which we temporarily denote by $\tilde{n}$. We have isomorphisms

$$(h\Delta^n) \times (h\Delta^m) \cong (hN\tilde{n}) \times (hN\tilde{m}) \cong \tilde{n} \times \tilde{m} \cong hN(\tilde{n} \times \tilde{m}) \cong h(N\tilde{n} \times N\tilde{m})$$

$$\cong h(\Delta^n \times \Delta^m). \qquad \square$$

Remark 18.1.2 In fact, the restricted functor $h: \mathbf{qCat} \to \mathbf{Cat}$ preserves arbitrary products of quasi-categories, not just finite ones. It is clear that the canonical comparison functor $h(\prod_\alpha X^\alpha) \to \prod_\alpha hX^\alpha$ is the identity on objects, these just being products of vertices in the underlying simplicial sets. Morphisms in $h(\prod_\alpha X^\alpha)$ are products of 1-simplices up to the homotopy relation. If two such products represent the same map in $\prod_\alpha hX^\alpha$, then because each X^α is a quasi-category, it is possible to choose a 2-simplex in each coordinate that witnesses these homotopies. But then the product of these 2-simplices indicates that these maps are the same in $h(\prod_\alpha X^\alpha)$ as well. Similarly, if (f_α) and (g_α) compose to (h_α) in $\prod_\alpha hX^\alpha$, then, because each X^α is a quasi-category, there exist 2-simplices σ_α with $\sigma_\alpha d^2 = f_\alpha$, $\sigma_\alpha d^0 = g_\alpha$, and $\sigma_\alpha d^1 = h_\alpha$. Thus (σ_α) witnesses the fact that the same relation holds in $h(\prod_\alpha X^\alpha)$. The converse implication is obvious.

This lemma shows that both h and N are strong monoidal; it follows that there is an induced change of base adjunction

$$h_*: \mathbf{sCat} \xrightleftharpoons{\perp} \mathbf{2Cat}: N_*.$$

between 2-categories and simplicial categories obtained by applying the functors N and h to each hom-object [15, 85].

Definition 18.1.3 (2-category of quasi-categories)   In particular, the full simplicial subcategory $\mathbf{qCat}_\infty \hookrightarrow \underline{\mathbf{sSet}}$ spanned by the quasi-categories has an associated 2-category $\mathbf{qCat}_2 := h_*\mathbf{qCat}_\infty$ that we shall call the **2-category of quasi-categories**. Its objects are quasi-categories, its 1-cells are maps of quasi-categories, and its 2-cells are homotopy classes of homotopies.

Because any morphism in the homotopy category of a quasi-category is represented by a 1-simplex, a 2-cell $X \begin{smallmatrix} f \\ \Downarrow\alpha \\ g \end{smallmatrix} Y$ in $\mathbf{qCat}_2$ is represented by a diagram

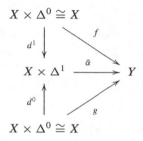

More concisely notated, a 2-cell $\alpha\colon f \Rightarrow g$ is represented by a 1-simplex $\tilde{\alpha}\colon f \to g$ in Y^X. Two such 1-simplices from f to g represent the same 2-cell if and only if they are homotopic as 1-simplices in Y^X, that is, if and only if there exists a 2-simplex with these 1-simplices as faces and either the degeneracy at f or g as the third face. The underlying category of any 2-category simply forgets its 2-cells. Hence the underlying category of $\mathbf{qCat}_2$ is the unenriched full subcategory of quasi-categories $\mathbf{qCat}$.

The 2-category $\mathbf{qCat}_2$ inherits a number of pleasing properties from the simplicial structure on $\mathbf{qCat}_\infty$.

Proposition 18.1.4 $\mathbf{qCat}_2$ *is cartesian closed.*

Proof   We show that the terminal object, products, and internal hom in $\mathbf{qCat}_\infty$ define the analogous structures in $\mathbf{qCat}_2$. In each case, we translate the $\mathbf{sSet}$-enriched universal properties to $\mathbf{Cat}$-enriched ones.

Because Δ^0 is a terminal object in the simplicially enriched sense, that is, because $(\Delta^0)^A \cong \Delta^0$, it is also terminal in the 2-categorical sense: the isomorphisms

$$\mathbf{qCat}_2(A, \Delta^0) = h((\Delta^0)^A) \cong h(\Delta^0) \cong \mathbb{1}$$

assert that the hom-category from A to Δ^0 is the terminal category. Similarly, because $X \times Y$ is a simplicially enriched product, we have isomorphisms $(X \times Y)^A \cong X^A \times Y^A$, which lead to an isomorphism of hom-categories:

$$\mathbf{qCat}_2(A, X \times Y) = h((X \times Y)^A) \cong h(X^A \times Y^A) \cong h(X^A) \times h(Y^A)$$

$$= \mathbf{qCat}_2(A, X) \times \mathbf{qCat}_2(A, Y).$$

Finally, because $Z^{X \times Y} \cong (Z^Y)^X$, we have

$$\mathbf{qCat}_2(X \times Y, Z) = h(Z^{X \times Y}) \cong h((Z^Y)^X) = \mathbf{qCat}_2(X, Z^Y),$$

which says that the exponential Z^Y defines an internal hom for the 2-category $\mathbf{qCat}_2$, just as it does for $\mathbf{qCat}_\infty$. □

Remark 18.1.5 (equivalences are equivalences) An **equivalence** in a 2-category is a pair of one cells $f : x \rightleftarrows y : g$ together with 2-cell isomorphisms $1_x \Rightarrow gf$, $fg \Rightarrow 1_y$. A 2-cell between a pair of maps from x to y is an isomorphism just when it is an isomorphism in the hom-category from x to y. In $\mathbf{qCat}_2$ a 2-cell between maps from X to Y is an isomorphism just when (any of) its representing 1-simplices in Y^X are isomorphisms. From (17.2.6), we see that an equivalence $f : X \rightleftarrows Y : g$ in $\mathbf{qCat}_2$ is precisely an equivalence of quasi-categories.

Remark 18.1.6 The functor $h : \mathbf{qCat} \to \mathbf{Cat}$ extends to a 2-functor $h : \mathbf{qCat}_2 \to \underline{\mathbf{Cat}}$. On hom-categories, this 2-functor is defined by the functor $h(Y^X) \to hY^{hX}$, which is adjunct to $h(Y^X) \times hX \cong h(Y^X \times X) \xrightarrow{h(\mathrm{ev})} hY$. Because 2-functors preserve equivalences, we see immediately that an equivalence of quasi-categories descends to an equivalence between their homotopy categories.

18.2 Weak limits in the 2-category of quasi-categories

By Proposition 18.1.4, the 2-category $\mathbf{qCat}_2$ has finite products in the 2-categorical (i.e., $\mathbf{Cat}$-enriched) sense. To develop the category theory of quasi-categories, we would like to make use of other 2-limits. Of particular importance in ordinary category theory is the fact that $\mathbf{Cat}$ admits cotensors by the walking arrow category $\mathbf{2}$. This allows us to form arrow categories and encode natural transformations as functors (see the proof of Lemma 8.5.3).

To say that $\mathbf{qCat}_2$ admits cotensors by $\mathbf{2}$ would mean that for every quasi-category X, there is some quasi-category $X^\mathbf{2}$ so that the categories

$$\mathbf{qCat}_2(A, X^\mathbf{2}) = h((X^\mathbf{2})^A) \cong (h(X^A))^\mathbf{2} = (\mathbf{qCat}_2(A, X))^\mathbf{2} \qquad (18.2.1)$$

are naturally isomorphic. By the defining universal property as a weighted limit, the cotensor would necessarily commute with the internal hom $(-)^A$. Hence, because $\mathbf{qCat}_2$ is cartesian closed, in place of the natural isomorphism (18.2.1), it would suffice to demand an isomorphism $h(X^2) \cong (hX)^2$.

By Corollary 15.2.3, $\mathbf{qCat}_\infty$ admits cotensors by any simplicial sets. With this in mind, the only reasonable guess would be to define the cotensor to be X^{Δ^1}, recalling that $N2 \cong \Delta^1$ and hence $h\Delta^1 \cong 2$. Indeed, because h is a 2-functor, there is a natural comparison

$$h(X^{\Delta^1}) \to (hX)^2. \tag{18.2.2}$$

However, this functor is *not* an isomorphism. In particular, it is not bijective on objects. However, certain properties of this functor guarantee that X^{Δ^1} is a **weak cotensor** with 2 in a sense made precise by the following lemma. In Lemma 18.3.2, we will see that these properties enable us to prove that X^{Δ^1} is really the only reasonable choice: specifically, we can show that weak cotensors by 2 are unique up to equivalence.

Lemma 18.2.3   *The canonical comparison functor $h(X^{\Delta^1}) \to (hX)^2$ is surjective on objects, full, and conservative.*

Proof   Surjectivity on objects says that every arrow in hX is represented by a 1-simplex in X, which would be false for generic simplicial sets but is true for quasi-categories.

To prove fullness, suppose given a commutative square in hX, and choose arbitrary 1-simplices representing each morphism:

$$
\begin{array}{ccc}
\cdot & \xrightarrow{\ a\ } & \cdot \\
{\scriptstyle f}\downarrow & & \downarrow{\scriptstyle g} \\
\cdot & \xrightarrow[\ b\ ]{} & \cdot
\end{array}
\tag{18.2.4}
$$

Choose a composite k of a with g and a 2-simplex witnessing this fact. The morphism k is the composite of f and b in hX; hence there is a 2-simplex witnessing this fact. These pair of 2-simplices define a map $\Delta^1 \to X^{\Delta^1}$

$$
\begin{array}{ccc}
\cdot & \xrightarrow{\ a\ } & \cdot \\
{\scriptstyle f}\downarrow & \underset{\sim}{\overset{\textstyle\diagdown}{\scriptstyle k}} & \downarrow{\scriptstyle g} \\
\cdot & \xrightarrow[\ b\ ]{} & \cdot
\end{array}
\tag{18.2.5}
$$

which represents an arrow in the category $h(X^{\Delta^1})$ whose image is the specified commutative square.

A functor is **conservative** if and only if every morphism in the domain mapping to an isomorphism in the codomain is already an isomorphism. Suppose given a map in $h(X^{\Delta^1})$ whose image (18.2.4) is an isomorphism, that is, so that a and b are isomorphisms in hX, which is the case if and only if they are isomorphisms in the quasi-category X. Choose inverse isomorphisms a^{-1} and b^{-1} and 2-simplices (17.0.1) witnessing this. We can choose a simplex ℓ so that the diagram

$$
\begin{array}{ccc}
\cdot & \xrightarrow{\ a^{-1}\ } & \cdot \\
{\scriptstyle g}\downarrow & \diagdown\ell & \downarrow{\scriptstyle f} \\
\cdot & \xrightarrow[\ b^{-1}\ ]{} & \cdot
\end{array}
\tag{18.2.6}
$$

commutes in hX, but we do not choose witnessing 2-simplices just yet. Suppose we had done so. Then this data, (18.2.5), and the 2-simplices exhibiting a^{-1} as left inverse to a and b^{-1} as left inverse to b would form a diagram

$$
\Lambda^2_1 \times \Delta^1 \coprod_{\Lambda^2_1 \times \partial \Delta^1} \Delta^2 \times \partial \Delta^1 \to X
$$

from the domain of $j^2_1 \mathbin{\hat{\times}} i_1$ to X; another such diagram would be obtained using the 2-simplices that exhibit a^{-1} and b^{-1} as right inverses.

The objective is to fill these diagrams to define maps $\Delta^2 \times \Delta^1 \rightrightarrows X$ in such a way that the missing squares $\Delta^1 \times \Delta^1$ are formed using the two degenerate 1-simplices on f and on g, respectively. This is possible if we forget the 2-simplices of (18.2.6): first apply Corollary 17.5.3 to invert one of the bs and one of the as in the s^1-degenerate images of the two 2-simplices of (18.2.5). These give one of the outer shuffles in each $\Delta^2 \times \Delta^1$. The inner shuffle is then obtained by filling an inner horn and applying Corollary 17.5.3. This process also chooses the two 2-simplices for the diagram (18.2.6). Then we apply 17.5.3 to invert one of the b^{-1}s and one of the a^{-1}s in the s^1-degenerate image of these two 2-simplices to obtain the final desired shuffles. An alternate construction produces these shuffles together with the faces of (18.2.6) using the special outer horn filling of Theorem 17.1.1. □

On account of Lemma 18.2.3, we refer to the weak 2-cotensor X^{Δ^1} as the **arrow quasi-category** of the quasi-category X.

18.3 Arrow quasi-categories in practice

Applying (18.2.2) to the quasi-category X^A, we have a functor

$$\mathbf{qCat}_2(A, X^{\Delta^1}) = h((X^{\Delta^1})^A) \cong h((X^A)^{\Delta^1}) \to h(X^A)^2 = (\mathbf{qCat}_2(A, X))^2$$

$$(18.3.1)$$

natural in A. Taking $A = X^{\Delta^1}$, the image of the identity map is a 2-cell

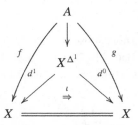

in $\mathbf{qCat}_2$. In general, the functor (18.3.1) maps a homotopy $A \to X^{\Delta^1}$, an object of the category $\mathbf{qCat}_2(A, X^{\Delta^1})$, say, from f to g, to the composite 2-cell

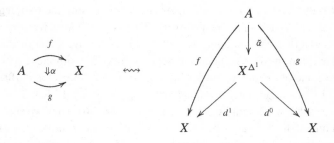

By Lemma 18.2.3, the functor (18.3.1) is surjective on objects, full, and conservative. Surjectivity says that any 2-cell in $\mathbf{qCat}_2$ is represented by a homotopy in $\mathbf{qCat}$

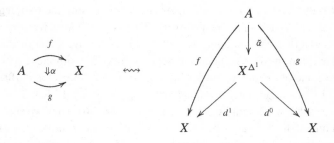

such that the whiskered composite of the 1-cell $\tilde{\alpha} \colon A \to X^{\Delta^1}$ in $\mathbf{qCat}_2$ with ι is the 2-cell α.

Note that the representing homotopy $\tilde{\alpha}$ is not unique; this is what is meant by saying that X^{Δ^1} is only a *weak* cotensor by 2. However, we can use the fact that the functors (18.3.1) are full and conservative to show that the universal properties of 18.2.3 characterize the arrow quasi-category up to equivalence.

Lemma 18.3.2 *Any quasi-category Z, for which there exists a natural transformation whose components $h(Z^A) \to h(X^A)^2$ are surjective on objects, full, and conservative, is equivalent to X^{Δ^1}.*

Proof Any such Z has a canonical 2-cell

defined by taking the image of $1_Z \in h(Z^Z)$. As for X^{Δ^1}, surjectivity on objects implies that this 2-cell has the following weak universal property: any 2-simplex $A \overset{f}{\underset{g}{\Rightarrow\!\!\!\Downarrow\alpha}} X$ factors through κ along some map $\check{\alpha} : A \to Z$. Applying the weak universal property of ι to κ, we obtain a map $\tilde{\kappa} : Z \to X^{\Delta^1}$ whose whiskered composite with ι is κ. Applying the universal property of κ to ι, we obtain a map $\check{\iota} : X^{\Delta^1} \to Z$ whose whiskered composite with κ is ι. Hence the composite $X^{\Delta^1} \overset{\check{\iota}}{\to} Z \overset{\tilde{\kappa}}{\to} X^{\Delta^1}$ gives a factorization of ι through itself. In particular, the objects $\tilde{\kappa}\check{\iota}$ and $1_{X^{\Delta^1}}$ have the same image under the functor (18.3.1). The identity defines an isomorphism between the image of these objects, so by fullness and conservativity of $h((X^{\Delta^1})^{X^{\Delta^1}}) \to h(X^{(X^{\Delta^1})})^2$, there is an isomorphism between $\tilde{\kappa}\check{\iota}$ and $1_{X^{\Delta^1}}$ in $h((X^{\Delta^1})^{X^{\Delta^1}})$. A similar argument implies the existence of an isomorphism in $h(Z^Z)$ between the $\check{\iota}\tilde{\kappa}$ and 1_Z. These isomorphisms, in turn, are represented by maps $J \to (X^{\Delta^1})^{X^{\Delta^1}}$ and $J \to Z^Z$ by Lemma 17.2.5, which shows that $\check{\iota}$ and $\tilde{\kappa}$ define an equivalence between Z and X^{Δ^1}. $\square$

Exercise 18.3.3 Generalize this argument to show that $\mathbf{qCat}_2$ admits weak cotensors by any category freely generated by a graph. It might be helpful to observe that the map from the 1-skeletal simplicial set formed by a reflexive directed graph to the nerve of the category it freely generates is inner anodyne.

18.4 Homotopy pullbacks

We will make use of other weak 2-limits with analogous universal properties. Let $E \overset{p}{\to} B \overset{q}{\leftarrow} F$ be maps of quasi-categories, with q an **isofibration**, by which we mean a fibration between fibrant objects in Joyal's model structure of Theorem 15.3.6. The 2-functor $h : \mathbf{qCat}_2 \to \underline{\mathbf{Cat}}$ induces a canonical comparison map $h(E \times_B F) \to hE \times_{hB} hF$.

Lemma 18.4.1   *For any diagram* $E \xrightarrow{p} B \xleftarrow{q} F$ *of quasi-categories with* q *an isofibration, the canonical functor* $h(E \times_B F) \to hE \times_{hB} hF$ *is bijective on objects, full, and conservative.*

Proof   Bijectivity on objects is clear. To show fullness, suppose given vertices $e, e' \in E$, $f, f' \in F$ so that $p(e) = q(f)$ and $p(e') = q(f')$. An arrow between these objects in $hE \times_{hB} hF$ is represented by 1-simplices $a \colon e \to e'$ and $b \colon f \to f'$ so that $p(a)$ and $q(b)$ are homotopic in B. The homotopy in turn is represented by a 2-simplex with second face $q(b)$, first face $p(a)$, and zeroth face degenerate. This defines a lifting problem between j_1^2 and q. The filler specifies a 1-simplex $b' \colon f \to f'$ such that $q(b') = p(a)$; the pair (a, b') defines the desired morphism in $h(E \times_B F)$.

The proof of conservativity is similar. Here we may suppose we are given $a \colon e \to e'$ and $b \colon f \to f'$ with $p(a) = q(b)$. Suppose a and b are both isomorphisms, and choose inverse isomorphisms a^{-1} and b^{-1} together with 2-simplices witnessing these inverses. The images $p(a^{-1})$ and $q(b^{-1})$ are homotopic. As earlier, we use the fact that q lifts against all inner horn inclusions to replace b^{-1} with a homotopic map b'^{-1} in the fiber over $p(a^{-1})$. This data can be used to construct lifting problems of q against 3-dimensional inner horns whose solutions give rise to 2-simplices in F that witness the isomorphism between b and b'^{-1} and lie over the image of the corresponding 2-simplices in E. This data shows that the 1-simplex $\Delta^1 \xrightarrow{(a,b)} E \times_B F$ is an isomorphism, as desired. $\square$

Remark 18.4.2   This proof only required that q is an inner fibration. This hypothesis is also necessary for the pullback to be a quasi-category. We prefer to also ask it to be an isofibration so that the pullbacks we consider are homotopy pullbacks in **qCat**. See Remark 17.7.4.

Remark 18.4.3   Because exponentials preserve pullbacks and isofibrations, we also conclude that the canonical functor $h((E \times_B F)^A) \to h(E^A) \times_{h(B^A)} h(F^A)$ is bijective on objects, full, and conservative for any simplicial set A. An argument analogous to the proof of Lemma 18.3.2 shows that the quasi-category $E \times_B F$ with the weak universal property of Lemma 18.4.1 is unique up to equivalence. We refer to any quasi-category with this universal property as a **weak homotopy pullback**.

18.5 Comma quasi-categories

We are interested in the weak universal properties of Lemma 18.2.3 and Lemma 18.4.1 because they can be combined to define **weak comma objects**. Let $E \xrightarrow{p} B \xleftarrow{q} F$ be a diagram of quasi-categories, though we no longer insist that p or q

is an isofibration. Define a quasi-category $p \downarrow q$ by forming the pullback

$$
\begin{array}{ccc}
p \downarrow q & \longrightarrow & B^{\Delta^1} \\
\downarrow & & \downarrow \\
E \times F & \xrightarrow{p \times q} & B \times B \cong B^{\partial \Delta^1}
\end{array}
\qquad (18.5.1)
$$

Because the Joyal model structure is monoidal, the right-hand vertical map is an isofibration. By the standard closure properties fibrations between fibrant objects, the left-hand map is also. In particular, $p \downarrow q$ is a quasi-category, which we refer to as a **comma quasi-category**.

Exercise 18.5.2 Use Example 7.4.5 as inspiration to define the comma quasi-category (18.5.1) as a weighted limit. Then use Theorem 8.2.2 to show that the nerve of a comma category is a weak comma object of the nerves of the constituent categories.

Exercise 18.5.3 Use Exercise 18.5.2 and Theorem 15.2.5 to give another proof that $p \downarrow q$ is a quasi-category.

Combining Lemmas 18.2.3 and 18.4.1, we conclude:

Corollary 18.5.4 *The canonical functor*

$$
h(p \downarrow q) \to h(E \times F) \times_{h(B \times B)} h(B^{\Delta^1}) \to (hE \times hF) \times_{hB \times hB} (hB)^2
$$
$$
= h(p) \downarrow h(q)
$$

is surjective on objects, full, and conservative.

Here the target category is just the usual comma category constructed in **Cat** (see Example 7.4.5). Because exponentials preserve all the constructions involved, we may conclude that for any A, the functor

$$
h((p \downarrow q)^A) \to (h(E^A) \times h(F^A)) \times_{h(B^A) \times h(B^A)} h(B^A)^2 = h(p^A) \downarrow h(q^A)
$$
$$
(18.5.5)
$$

also has these properties.

Remark 18.5.6 From the functor (18.5.5), we conclude that the quasi-category $p \downarrow q$ has the following weak universal property. The image of the identity at $p \downarrow q$ defines a 2-cell

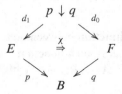

Surjectivity of (18.5.5) says that for any 2-cell

there exists a map $A \to p \downarrow q$ that defines a factorization of e_1 and e_0 through d_1 and d_0 and so that when this map is whiskered with the 2-cell χ, we get α. Combining fullness and conservativity, we see further that, for instance, if $f, g \colon A \rightrightarrows p \downarrow q$ are two maps that whisker with χ to define the same 2-cell α, then there is an isomorphism $A \overset{f}{\underset{g}{\cong\Downarrow}} p \downarrow q$, represented by a map $A \times J \to p \downarrow q$. This latter fact can be used to show that the comma quasi-category $p \downarrow q$ is unique up to equivalence by the argument used to prove Lemma 18.3.2.

18.6 Adjunctions between quasi-categories

Our goal is now to use the weak 2-limits we have constructed in $\mathbf{qCat}_2$ to begin to develop the formal category theory of quasi-categories. We might posit that quasi-categorical structures that are recognizable in the 2-category $\mathbf{qCat}_2$ are structures guaranteed to be homotopy coherent from the existence of certain low-dimensional data. One example is the notion of isomorphism: isomorphisms in quasi-categories defined in terms of 1-simplices and 2-simplices extend by Lemma 17.2.5 to diagrams on the simplicial set J, which has non-degenerate cells in each dimension. By virtue of Remark 17.2.7, these comments extend to equivalences of quasi-categories.

A non-example is the notion of a monad on a quasi-category. Just as an H-space need not be an A_∞-space, a monad in $\mathbf{qCat}_2$ need not be homotopy coherent. However, adjunctions of quasi-categories are detectable in $\mathbf{qCat}_2$; one way to explain this difference is that an adjunction is characterized by a universal property, whereas a monad is simply equationally defined. Let us now explore this definition. Another approach can be found in [26, §3.2–3].

Definition 18.6.1   An **adjunction** between quasi-categories is an adjunction in the 2-category $\mathbf{qCat}_2$. That is, an adjunction consists of quasi-categories A and B; maps $f \colon B \to A$, $u \colon A \to B$; and 2-cells $\eta \colon 1_B \Rightarrow uf$, $\epsilon \colon fu \Rightarrow 1_A$, satisfying the triangle identities.

It is clear from this definition that the 2-functor $h\colon \mathbf{qCat}_2 \to \underline{\mathbf{Cat}}$ sends an adjunction between quasi-categories to an adjunction between their homotopy categories. The nerve defines a fully faithful 2-functor $N\colon \underline{\mathbf{Cat}} \to \mathbf{qCat}_2$. So adjunctions between categories define adjunctions between their nerves.

Proposition 18.6.2 *Suppose $f\colon B \rightleftarrows A\colon u$ is an adjunction between quasi-categories. Then, for any simplicial set K and any quasi-category X, there are induced adjunctions*

$$f_*\colon B^K \rightleftarrows A^K\colon u_* \qquad\qquad u^*\colon X^A \rightleftarrows X^B\colon f^*.$$

Proof The internal hom defines a 2-functor $X^{(-)}\colon \mathbf{qCat}_2^{\mathrm{op}} \to \mathbf{qCat}_2$ because $\mathbf{qCat}_2$ is cartesian closed, carrying the adjunction $f \dashv u$ to an adjunction $u^* \dashv f^*$. The proof of Proposition 18.1.4 also demonstrates that exponentiation with any simplicial set K defines a 2-functor $(-)^K\colon \mathbf{qCat}_2 \to \mathbf{qCat}_2$, which carries the adjunction $f \dashv u$ to $f_* \dashv u_*$. $\qquad\square$

Example 18.6.3 Any simplicial Quillen adjunction between simplicial model categories descends to an adjunction between their representing quasi-categories, constructed as in Example 16.4.11. For proof, see [74].

Example 18.6.4 A **terminal** object in a quasi-category A is a vertex t with the following weak universal property: any map $\partial\Delta^n \to A$ whose nth vertex is t can be filled to an n-simplex [49, 1.2.12]. Given a terminal object t, we can choose, for any n-simplex σ in A, an $(n+1)$-simplex $\sigma^\triangleright$ whose $(n+1)$th face is σ and whose $(n+1)$th vertex is t. We do so inductively by dimension, choosing first 1-simplices for each vertex, then 2-simplices compatible with these choices for all 1-simplices in A, and so forth. When making these choices, we require that we choose the degenerate 1-simplex for $t^\triangleright$ and a compatibly degenerate $\sigma^\triangleright$ for each degenerate σ.

Using this data, we define a homotopy $A \times \Delta^1 \to A$ whose component at $\sigma \in A_n$ is the map $\Delta^n \times \Delta^1 \xrightarrow{r_R} \Delta^{n+1} \xrightarrow{\sigma^\triangleright} A$ where r_R is the "last shuffle map" defined by $(i, 0) \mapsto i$ and $(i, 1) \mapsto n + 1$. This homotopy represents the unit of an adjunction $A \xrightarrow[\substack{t}]{\overset{!}{\underset{\perp}{\rightleftarrows}}} \Delta^0$. The counit is an identity, and the only non-trivial triangle identity follows from our condition on $t^\triangleright$.

Remark 18.6.5   An **adjoint equivalence** is an adjunction in which the unit and counit 2-cells are isomorphisms. Choosing representatives

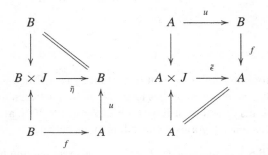

we have data witnessing that $f : B \rightleftarrows A : u$ is an equivalence. In particular, any adjunction between Kan complexes is always an adjoint equivalence by Corollary 17.2.1: all 1-simplices in the hom-spaces A^A and B^B are isomorphisms.

Given an adjunction, we form the comma quasi-categories

$$
\begin{array}{ccc}
f \downarrow A & \longrightarrow & A^{\Delta^1} \\
\downarrow & \lrcorner & \downarrow \\
B \times A & \longrightarrow & A \times A \\
& f \times 1 &
\end{array}
\qquad
\begin{array}{ccc}
B \downarrow u & \longrightarrow & B^{\Delta^1} \\
\downarrow & \lrcorner & \downarrow \\
B \times A & \longrightarrow & B \times B \\
& 1 \times u &
\end{array}
\qquad (18.6.6)
$$

By Corollary 18.5.4, these quasi-categories are equipped with 2-cells

$$
\begin{array}{c}
f \downarrow A \\
\swarrow \quad \Rightarrow\alpha \quad \searrow \\
B \xrightarrow{\quad f \quad} A
\end{array}
\qquad\qquad
\begin{array}{c}
B \downarrow u \\
\swarrow \quad \Rightarrow\beta \quad \searrow \\
B \xleftarrow{\quad u \quad} A
\end{array}
$$

satisfying the weak universal property described in Remark 18.5.6. The meaning of our phrase "the formal category theory of quasi-categories" is illustrated by the following proposition.

Proposition 18.6.7   *If $f : B \rightleftarrows A : u$ is an adjunction between quasi-categories, then the quasi-categories $f \downarrow A$ and $B \downarrow u$ are equivalent.*

Proof A pleasing feature of our proof is that it mimics the usual construction of an isomorphism of hom-sets from the unit and counit of an adjunction. By the weak universal properties of α and β, the composite 2-cells displayed in the following define maps $f \downarrow A \to B \downarrow u$ and $B \downarrow u \to f \downarrow A$ satisfying the following identities:

$$
\begin{array}{ccc}
\begin{array}{c}
B \downarrow u \\
\swarrow \quad \overset{\Rightarrow\beta}{\underset{u}{}} \quad \searrow \\
B \xleftarrow{\hspace{2em}} A \\
\searrow_{f} \quad \overset{\Rightarrow\epsilon}{} \quad \parallel \\
A
\end{array}
& = &
\begin{array}{c}
B \downarrow u \\
\downarrow \\
f \downarrow A \\
\swarrow \quad \overset{\Rightarrow\alpha}{} \quad \searrow \\
B \xrightarrow[f]{\hspace{2em}} A
\end{array}
\end{array}
$$

$$
\begin{array}{ccc}
\begin{array}{c}
f \downarrow A \\
\swarrow \quad \overset{\Rightarrow\alpha}{\underset{f}{}} \quad \searrow \\
B \xrightarrow{\hspace{2em}} A \\
\parallel \quad \overset{\Rightarrow\eta}{} \quad \swarrow_{u} \\
B
\end{array}
& = &
\begin{array}{c}
f \downarrow A \\
\downarrow \\
B \downarrow u \\
\swarrow \quad \overset{\Rightarrow\beta}{} \quad \searrow \\
B \xleftarrow[u]{\hspace{2em}} A
\end{array}
\end{array}
$$

Composing the maps, we have identities

$$
\begin{array}{ccc}
\begin{array}{c}
f \downarrow A \\
\downarrow \\
B \downarrow u \\
\downarrow \\
f \downarrow A \\
\swarrow \quad \overset{\Rightarrow\alpha}{} \quad \searrow \\
B \xrightarrow[f]{\hspace{2em}} A
\end{array}
& = &
\begin{array}{c}
f \downarrow A \\
\downarrow \\
B \downarrow u \\
\swarrow \quad \overset{\Rightarrow\beta}{\underset{u}{}} \quad \searrow \\
B \xleftarrow{\hspace{2em}} A \\
\searrow_{f} \quad \overset{\Rightarrow\epsilon}{} \quad \parallel \\
A
\end{array}
& =
\end{array}
$$

$$
\begin{array}{ccc}
\begin{array}{c}
f \downarrow A \\
\swarrow \quad \overset{\Rightarrow\alpha}{\underset{f}{}} \quad \searrow \\
B \xrightarrow{\hspace{2em}} A \\
\parallel \quad \overset{\Rightarrow\eta}{} \quad \swarrow_{u} \quad \parallel \\
B \quad \overset{\Rightarrow\epsilon}{} \\
\searrow_{f} \qquad \parallel \\
A
\end{array}
& = &
\begin{array}{c}
f \downarrow A \\
\swarrow \quad \overset{\Rightarrow\alpha}{} \quad \searrow \\
B \xrightarrow[f]{\hspace{2em}} A
\end{array}
\end{array}
$$

(18.6.8)

and similarly for $B \downarrow u \to f \downarrow A \to B \downarrow u$. The diagram (18.6.8) shows that $f \downarrow A \to B \downarrow u \to f \downarrow A$ and the identity at $f \downarrow A$ have the same image under (18.5.5). By fullness and conservativity, the identity map between their images lifts to a 2-cell isomorphism. A similar isomorphism exists for the other composite and the identity at $B \downarrow u$. The lifted isomorphisms show that the maps $f \downarrow A \rightleftarrows B \downarrow u$ define an equivalence. $\square$

Remark 18.6.9   Note the equivalence constructed in the proof of Proposition 18.6.7 commutes with the projections to $B \times A$. Conversely, if $f : B \to A$ and $u : A \to B$ are any pair of maps between quasi-categories so that the comma quasi-categories $f \downarrow A$ and $B \downarrow u$ are equivalent over $B \times A$, then f and u form an adjunction. The proof of this is more subtle, so we save the details for [74].

Let $a \in A$ and $b \in B$ be any vertices. Pulling back, we form comma objects

$$
\begin{array}{ccccc}
fb \downarrow a & \longrightarrow & f \downarrow A & \longrightarrow & A^{\Delta^1} \\
\downarrow & & \downarrow & & \downarrow \\
* & \xrightarrow[b \times a]{} & B \times A & \xrightarrow[f \times 1]{} & A \times A
\end{array}
\qquad
\begin{array}{ccccc}
b \downarrow ua & \longrightarrow & B \downarrow u & \longrightarrow & B^{\Delta^1} \\
\downarrow & & \downarrow & & \downarrow \\
* & \xrightarrow[b \times a]{} & B \times A & \xrightarrow[1 \times u]{} & B \times B
\end{array}
$$

The pullbacks $fb \downarrow a$ and $b \downarrow ua$ are the mapping spaces $\mathrm{Hom}_A(fb, a)$ and $\mathrm{Hom}_B(b, ua)$ introduced in Section 15.4.

Corollary 18.6.10   *The Kan complexes* $\mathrm{Hom}_A(fb, a)$ *and* $\mathrm{Hom}_B(b, ua)$ *are equivalent.*

Proof   The rightmost maps in each pullback are isofibrations. Hence, by Remark 17.7.4, each pullback is a homotopy pullback, and the equivalence $f \downarrow A \simeq B \downarrow u$ pulls back to an equivalence $fb \downarrow a \simeq b \downarrow ua$. $\square$

Example 18.6.11   Suppose given an adjunction $A \underset{t}{\overset{!}{\rightleftarrows}} \Delta^0$ of quasi-categories. Noting that $! \downarrow \Delta^0 = A$, Proposition 18.6.7 gives an equivalence

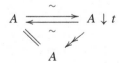

over A. Pulling back along $a \in A$, we see from Corollary 18.6.10 that each hom-space $\mathrm{Hom}_A(a, t)$ is equivalent to a point.

Combining Example 18.6.11 and the 2-of-3 property, we conclude that the isofibration $A \downarrow t \to A$ is a trivial fibration. The quasi-category $A \downarrow t$ is reminiscent of the slice quasi-category $A_{/t}$ defined in (17.1.3): an n-simplex in the former is a cylinder $\Delta^n \times \Delta^1 \to A$ with the n-simplex at one end degenerate at t while an n-simplex in the latter is an $(n + 1)$-simplex in A with final vertex t.

The forgetful functor $A_{/t} \to A$ defined in (17.1.4) is an isofibration: the proof, left to the reader, is essentially contained in Lemma 17.1.6 and Exercise 17.1.7. We would like to conclude that the isofibration $A_{/t} \to A$ is a trivial fibration, for this says exactly that $t \in A$ is a terminal object, as defined in Example 18.6.4. This claim will follow from a geometric argument that proves that there is an equivalence $A \downarrow t \simeq A_{/t}$ over A.

This sort of argument is essential to the translation between the 2-categorically derived definitions – which ultimately derive from the simplicial category $\mathbf{qCat}_\infty$ and hence have a cylindrical shape – and the décalage-type definitions found in [40] and [49]. An extended version, which proves that the comma quasi-category of "cones" over a diagram of arbitrary shape is equivalent to the analogous slice category, is given in [74].

18.7 Essential geometry of terminal objects

Our geometric argument, which is similar to but, sadly, not derivable from the proof of Proposition 15.4.7, again makes use of Reedy category theory.

Lemma 18.7.1 *Let $f : X^\bullet \to Y^\bullet$ be a map of Reedy cofibrant cosimplicial objects in a left proper model category. If the relative latching maps $X^n \coprod_{L^n X} L^n Y \to Y^n$ are weak equivalences, then f is a pointwise weak equivalence.*

Proof The proof is by induction: $f^0 : X^0 \to Y^0$ is the zeroth relative latching map. By Theorem 14.3.1, Lemma 14.3.7, and Ken Brown's Lemma 11.3.14, the functor $L^n = \mathrm{colim}^{\partial \Delta^n}$ preserves pointwise weak equivalences between Reedy cofibrant cosimplicial objects. Its use here seems circular, except that, by the argument used to prove Lemma 4.4.3, the nth latching object can be defined to be the weighted colimit of the $(n - 1)$-truncation. Hence, the maps $L^n X \to L^n Y$ are weak equivalences. We have

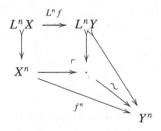

which by left properness (see Digression 14.3.5) and the 2-of-3 property implies the desired result. ☐

Theorem 18.7.2   *For any quasi-category B with vertex $b \in B$, the natural quotient map $B \downarrow b \to B/b$ is an equivalence of quasi-categories.*

Proof   We regard B as an object of **sSet**$_*$ with basepoint b and consider a pair of cosimplicial objects $D^\bullet, C^\bullet \colon \Delta \rightrightarrows$ **sSet**$_*$. Define $D^n = \Delta^{n+1}$, with the final vertex serving as the basepoint, and define C^n to be the pushout

$$
\begin{array}{ccc}
\Delta^n & \longrightarrow & \Delta^0 \\
{\scriptstyle 1 \times d^0} \downarrow & & \downarrow \\
\Delta^n \times \Delta^1 & \overset{\ulcorner}{\longrightarrow} & C^n
\end{array}
$$

with the last vertex as basepoint again. The quotient C^n is a cylinder with its last end n-simplex degenerated to a point.

Both cosimplicial objects are easily seen to be Reedy cofibrant using Lemma 14.3.8. There is a canonical quotient map $C^n \to D^n$ descended from the map $r_R \colon \Delta^n \times \Delta^1 \to \Delta^{n+1}$ defined in 18.6.4. We call the image of the section to r_R the "last shuffle"; it is the $(n + 1)$-simplex spanned by the vertices $(i, 0)$ and $(n, 1)$. By Lemma 15.4.9 and the argument presented immediately prior, the conclusion follows once we show that the map $C^\bullet \to D^\bullet$ is a pointwise weak equivalence. By Lemma 18.7.1, it suffices to show that the relative latching maps

$$
P^n := C^n \coprod_{L^n C^\bullet} L^n D^\bullet \to D^n = \Delta^{n+1}
$$

are weak equivalences.

By inspection, $L^n D^\bullet = \Lambda^{n+1}_{n+1}$ and $L^n C^\bullet$ is the quotient of $\partial \Delta^n \times \Delta^1 \cup_{\{1\}} \Delta^n$ that collapses the end Δ^n to a point. There is an inner anodyne extension $\partial(\Delta^n \times \Delta^1) \to H^n$, where H^n is the subset of $\Delta^n \times \Delta^1$ that contains everything but the last shuffle. At the first stage, we attach a Λ^{n+1}_1 horn, then a Λ^{n+1}_2, then a $\Lambda^{n+1}_3, \ldots, \Lambda^{n+1}_n$, with each horn filler attaching one of the other n shuffles in accordance with the total order of 15.0.3. By Remark 14.3.4, the relative latching map, the dotted arrow displayed in the following, is a categorical equivalence because it is the comparison between two pointwise categorically equivalent homotopy pushouts.

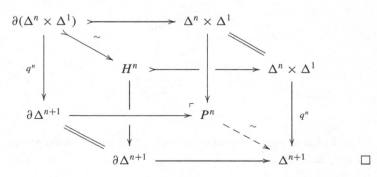

This geometry aside, the following results are absurdly easy to prove.

Corollary 18.7.3 *If* $A \overset{!}{\underset{t}{\rightleftarrows}} \Delta^0$ *is an adjunction between quasi-categories, then t is a terminal vertex of A.*

Proof By Proposition 18.6.7, interpreted as in Example 18.6.11, $A \downarrow t \to A$ is a trivial fibration. From Theorem 18.7.2, there is an equivalence $A \downarrow t \simeq A_{/t}$ over A, which means that $A_{/t} \to A$ is also a trivial foundation. The lifting property

$$
\begin{array}{ccc}
\partial\Delta^n & \longrightarrow & A_{/t} \\
\downarrow & \nearrow & \downarrow \\
\Delta^n & \longrightarrow & A
\end{array}
$$

says exactly that an $(n+1)$-sphere in A whose final vertex is t can be filled, which was the definition of terminal object given in Example 18.6.4. $\square$

Example 18.7.4 For instance, we can show that the degenerate 1-simplex $bs^0 \colon b \to b$ is a terminal object in $B \downarrow b$ by proving that there is an adjunction $! \colon B \downarrow b \rightleftarrows \Delta^0 \colon bs^0$. The counit is trivial. The unit is a map $(B \downarrow b) \times \Delta^1 \to (B \downarrow b)$ defined on $\Delta^n \times \Delta^1 \to B$ by precomposing with $\Delta^n \times \Delta^1 \times \Delta^1 \xrightarrow{1 \times \max} \Delta^n \times \Delta^1$. Here the map $\max \colon \Delta^1 \times \Delta^1 \to \Delta^1$ is defined in the evident way on vertices. The only non-trivial triangle identity requires that when we restrict the unit along $bs^0 \colon \Delta^0 \to B \downarrow b$, we get the degenerate 1-simplex $\Delta^1 \to (B \downarrow b)$ on bs^0, and indeed this is the case.

Proposition 18.7.5 *Right adjoints preserve terminal objects.*

Proof By 18.6.4, a terminal object $t \in A$ induces an adjunction $! \dashv t$. By Corollary 18.7.3, the composite adjunction

$$
B \overset{f}{\underset{u}{\rightleftarrows}} A \overset{!}{\underset{t}{\rightleftarrows}} \Delta^0
$$

implies that $ut \in B$ is terminal. $\square$

The following result also admits a direct proof, but we prefer this one.

Corollary 18.7.6 *Equivalences preserve terminal objects.*

Proof By a standard 2-categorical argument, any equivalence can be promoted to an adjoint equivalence at the cost of changing either the unit or counit. $\square$

In another consequence of Theorem 18.7.2, we obtain a "special outer horn filler universal property" for the unit and counit of an adjunction. As in the proof of Corollary 18.6.10, the equivalence arising from an adjunction pulls back along $B \xrightarrow{1 \times a} B \times A$ to an equivalence $f \downarrow a \rightleftarrows B \downarrow ua$ over B. By Example 18.7.4, $ua \to ua$ is terminal in $B \downarrow ua$. The image in $f \downarrow a$ is a 1-simplex $fua \to a$ in A that we call ϵ_a. By Corollary 18.7.6, $\epsilon_a \colon fua \to a$ is a terminal object of $f \downarrow a$. It follows that the fibration $(f_{/a})_{/\epsilon_a} \to f_{/a}$ is a trivial fibration. Unpacking, this statement encodes the following universal property: given a map $\Lambda_n^n \to A$ whose last edge is ϵ_a and so that the boundary of the missing face is the image of a specified $\partial \Delta^{n-1} \to B$, there exists a filler for this sphere together with a filler for the resulting sphere $\partial \Delta^n \to A$.

This universal property is an essential ingredient in the proof that any adjunction between quasi-categories is automatically homotopy coherent. This means that the data of f and u and either the unit or counit (but not both) can be extended to a simplicial functor whose codomain is $\mathbf{qCat}_\infty$ and whose domain is a cofibrant simplicial category encoding the shape of the free homotopy coherent adjunction. See [75] for a precise statement and proof.

Bibliography

[1] T. Athorne. The coalgebraic structure of cell complexes. *Theory Appl. Categ.*, 26(11):304–330, 2012.

[2] T. Barthel, J. P. May, and E. Riehl. Six model structures for dg-modules over dgas: Model category theory in homological action. Preprint, arXiv:1310.1159 [math.CT], 2013.

[3] T. Barthel and E. Riehl. On the construction of functorial factorizations for model categories. *Algebr. Geom. Topol.*, 13(2):1089–1124, 2013.

[4] T. Beke. Sheafifiable homotopy model categories. *Math. Proc. Cambridge Philos. Soc.*, 129(3):447–475, 2000.

[5] C. Berger and I. Moerdijk. On an extension of the notion of Reedy category. *Math. Z.*, 269(3–4):977–1004, 2011.

[6] J. E. Bergner. A model category structure on the category of simplicial categories. *Trans. Amer. Math. Soc.*, 359(5):2043–2058, 2007.

[7] A. J. Blumberg and E. Riehl. Homotopical resolutions associated to deformable adjunctions. Preprint, arXiv:1208.2844 [math.AT], 2012.

[8] J. M. Boardman and R. M. Vogt. *Homotopy invariant algebraic structures on topological spaces. Lecture Notes in Mathematics*, Vol. 347. Springer, Berlin, 1973.

[9] F. Borceux. *Handbook of categorical algebra. 2. Encyclopedia of Mathematics and Its Applications*, vol. 51. Cambridge University Press, Cambridge, 1994.

[10] A. K. Bousfield and D. M. Kan. *Homotopy limits, completions and localizations. Lecture Notes in Mathematics*, Vol. 304. Springer, Berlin, 1972.

[11] M. C. Bunge. Relative functor categories and categories of algebras. *J. Algebra*, 11:64–101, 1969.

[12] J. D. Christensen and M. Hovey. Quillen model structures for relative homological algebra. *Math. Proc. Cambridge Philos. Soc.*, 133(2):261–293, 2002.

[13] M. Cole. Mixing model structures. *Topology Appl.*, 153(7):1016–1032, 2006.

[14] J.-M. Cordier and T. Porter. Maps between homotopy coherent diagrams. *Topology Appl.*, 28(3):255–275, 1988.

[15] G. S. H. Cruttwell. *Normed spaces and the change of base for enriched categories.* PhD thesis, Dalhousie University, 2008.

[16] E. J. Dubuc. *Kan extensions in enriched category theory. Lecture Notes in Mathematics*, Vol. 145. Springer, Berlin, 1970.

[17] D. Dugger. Replacing model categories with simplicial ones. *Trans. Amer. Math. Soc.*, 353(12):5003–5027, 2001.

[18] D. Dugger. A primer on homotopy colimits. Preprint in progress, available from http://pages.uoregon.edu/ddugger/, 2008.

[19] D. Dugger and D. C. Isaksen. Topological hypercovers and $\mathbb{A}^1$-realizations. *Math. Z.*, 246(4):667–689, 2004.

[20] D. Dugger and D. I. Spivak. Mapping spaces in quasi-categories. *Algebr. Geom. Topol.*, 11(1):263–325, 2011.

[21] D. Dugger and D. I. Spivak. Rigidification of quasi-categories. *Algebr. Geom. Topol.*, 11(1):225–261, 2011.

[22] W. G. Dwyer, P. S. Hirschhorn, D. M. Kan, and J. H. Smith. *Homotopy limit functors on model categories and homotopical categories. Mathematical Surveys and Monographs*, vol. 113. American Mathematical Society, Providence, RI, 2004.

[23] W. G. Dwyer and D. M. Kan. A classification theorem for diagrams of simplicial sets. *Topology*, 23(2):139–155, 1984.

[24] W. G. Dwyer and J. Spaliński. Homotopy theories and model categories. In *Handbook of algebraic topology*, pages 73–126. North-Holland, Amsterdam, 1995.

[25] A. D. Elmendorf, I. Kříž, M. A. Mandell, and J. P. May. Modern foundations for stable homotopy theory. In *Handbook of algebraic topology*, pages 213–253. North-Holland, Amsterdam, 1995.

[26] T. M. Fiore and W. Lück. Waldhausen additivity: Classical and quasicategorical. Preprint, arXiv:1207.6613 [math.AT], 2012.

[27] R. Fritsch and R. A. Piccinini. *Cellular structures in topology. Cambridge Studies in Advanced Mathematics*, vol. 19. Cambridge University Press, Cambridge, 1990.

[28] P. Gabriel and M. Zisman. *Calculus of fractions and homotopy theory. Ergebnisse der Mathematik und ihrer Grenzgebiete*, vol. 35. Springer, New York, 1967.

[29] N. Gambino. Weighted limits in simplicial homotopy theory. *J. Pure Appl. Algebra*, 214(7):1193–1199, 2010.

[30] R. Garner. Cofibrantly generated natural weak factorisation systems. Preprint, arXiv:math/0702290 [math.CT], 2007.

[31] R. Garner. Understanding the small object argument. *Appl. Categ. Structures*, 17(3):247–285, 2009.

[32] P. G. Goerss and J. F. Jardine. *Simplicial homotopy theory. Progress in Mathematics*, vol. 174. Birkhäuser, Basel, 1999.

[33] M. Grandis and W. Tholen. Natural weak factorization systems. *Arch. Math. (Brno)*, 42(4):397–408, 2006.

[34] B. Gray. *Homotopy theory*. Pure and Applied Mathematics, Vol. 64. Academic Press, New York, 1975.

[35] A. Hatcher. *Algebraic topology*. Cambridge University Press, Cambridge, 2002.

[36] P. S. Hirschhorn. *Model categories and their localizations. Mathematical Surveys and Monographs*, vol. 99. American Mathematical Society, Providence, RI, 2003.

[37] J. Hollender and R. M. Vogt. Modules of topological spaces, applications to homotopy limits and E_∞ structures. *Arch. Math. (Basel)*, 59(2):115–129, 1992.

[38] M. Hovey. *Model categories. Mathematical Surveys and Monographs*, vol. 63. American Mathematical Society, Providence, RI, 1999.

[39] M. Hovey, B. Shipley, and J. Smith. Symmetric spectra. *J. Amer. Math. Soc.*, 13(1):149–208, 2000.

[40] A. Joyal. Quasi-categories and Kan complexes. *J. Pure Appl. Algebra*, 175(1–3):207–222, 2002. Special volume celebrating the 70th birthday of Professor Max Kelly.

[41] A. Joyal. The theory of quasi-categories and its applications. Preprint in progress, 2008.

[42] A. Joyal. The theory of quasi-categories I. Preprint in progress, 2008.

[43] A. Joyal and M. Tierney. Quasi-categories vs Segal spaces. In *Categories in algebra, geometry and mathematical physics. Contemp. Math.*, vol. 431, pages 277–326. American Mathematical Society, Providence, RI, 2007.

[44] G. M. Kelly. Doctrinal adjunction. In *Category Seminar (Proc. Sem., Sydney, 1972/1973)*. Lecture Notes in Mathematics, Vol. 420, pages 257–280. Springer, Berlin, 1974.

[45] G. M. Kelly. A unified treatment of transfinite constructions for free algebras, free monoids, colimits, associated sheaves, and so on. *Bull. Austral. Math. Soc.*, 22(1):1–83, 1980.

[46] G. M. Kelly. *Basic concepts of enriched category theory*. Reprints in Theory and Applications of Categories. Reprint of the 1982 original by Cambridge University Press, Cambridge.

[47] G. M. Kelly and R. Street. Review of the elements of 2-categories. In *Category Seminar (Proc. Sem., Sydney, 1972/1973)*. Lecture Notes in Mathematics, Vol. 420, pages 75–103. Springer, Berlin, 1974.

[48] S. Lack. Homotopy-theoretic aspects of 2-monads. *J. Homotopy Relat. Struct.*, 2(2):229–260, 2007.

[49] J. Lurie. *Higher topos theory. Annals of Mathematics Studies*, vol. 170. Princeton University Press, Princeton, NJ, 2009.

[50] S. Mac Lane. *Categories for the working mathematician*, 2nd ed. *Graduate Texts in Mathematics*, vol. 5. Springer, New York, 1998.

[51] P. J. Malraison Jr. Fibrations as triple algebras. *J. Pure Appl. Algebra*, 3:287–293, 1973.

[52] G. Maltsiniotis. Le théorème de Quillen, d'adjonction des foncteurs dérivés, revisité. *C. R. Math. Acad. Sci. Paris*, 344(9):549–552, 2007.

[53] M. A. Mandell and J. P. May. Equivariant orthogonal spectra and S-modules. *Mem. Amer. Math. Soc.*, 159(755):x+108, 2002.

[54] M. A. Mandell, J. P. May, S. Schwede, and B. Shipley. Model categories of diagram spectra. *Proc. London Math. Soc.*, 82(2):441–512, 2001.

[55] J. P. May. *Simplicial objects in algebraic topology*. Chicago Lectures in Mathematics. Reprint of the 1967 original by University of Chicago Press, Chicago, IL.

[56] J. P. May. Classifying spaces and fibrations. *Mem. Amer. Math. Soc.*, 1(1, 155):xiii+98, 1975.

[57] J. P. May. *A concise course in algebraic topology*. Chicago Lectures in Mathematics. University of Chicago Press, Chicago, IL, 1999.

[58] J. P. May and K. Ponto. *More concise algebraic topology: Localization, completion, and model categories*. Chicago Lectures in Mathematics. University of Chicago Press, Chicago, IL, 2012.

[59] J. P. May and J. Sigurdsson. *Parametrized homotopy theory. Mathematical Surveys and Monographs*, vol. 132. American Mathematical Society, Providence, RI, 2006.

[60] M. C. McCord. Classifying spaces and infinite symmetric products. *Trans. Amer. Math. Soc.*, 146:273–298, 1969.

[61] D. McDuff. On the classifying spaces of discrete monoids. *Topology*, 18(4):313–320, 1979.

[62] J.-P. Meyer. Bar and cobar constructions. I. *J. Pure Appl. Algebra*, 33(2):163–207, 1984.

[63] J. R. Munkres. *Topology: A first course*. Prentice Hall, Englewood Cliffs, NJ, 1975.

[64] M. C. Pedicchio and S. Solomini. On a "good" dense class of topological spaces. *J. Pure Appl. Algebra*, 24:287–295, 1986.

[65] D. G. Quillen. *Homotopical algebra*. Lecture Notes in Mathematics, vol. 43. Springer, Berlin, 1967.

[66] A. Radulescu-Banu. *Cofibrance and completion*. PhD thesis, Massachusetts Institute of Technology. arXiv:0612.203 [math.AT], 1999.

[67] G. Raptis. Homotopy theory of posets. *Homology, Homotopy Appl.*, 12(2):211–230, 2010.

[68] C. L. Reedy. Homotopy theory of model categories. Unpublished manuscript, available at the Hopf Topology Archive, ftp://hopf.math.purdue.edu/pub/Reedy/reedy.dvi, 1974.

[69] C. Rezk, S. Schwede, and B. Shipley. Simplicial structures on model categories and functors. *Amer. J. Math.*, 123(3):551–575, 2001.

[70] E. Riehl. A model structure for quasi-categories. Unpublished expository manuscript, available at http://www.math.harvard.edu/~eriehl, 2008.

[71] E. Riehl. Algebraic model structures. *New York J. Math.*, 17:173–231, 2011.

[72] E. Riehl. On the structure of simplicial categories associated to quasi-categories. *Math. Proc. Cambridge Philos. Soc.*, 150(3):489–504, 2011.

[73] E. Riehl. Monoidal algebraic model structures. *J. Pure Appl. Algebra*, 217(6):1069–1104, 2013.

[74] E. Riehl and D. R. B. Verity. The 2-category theory of quasi-categories. Preprint, arXiv:1306.5144 [math.CT], 2013.

[75] E. Riehl and D. R. B. Verity. Homotopy coherent adjunctions and the formal theory of monads. Preprint, arXiv:1310.8279 [math.CT], 2013.

[76] E. Riehl and D. R. B. Verity. The theory and practice of Reedy categories. Preprint, arXiv:1304.6871 [math.CT], 2013.

[77] G. Segal. Configuration-spaces and iterated loop-spaces. *Invent. Math.*, 21:213–221, 1973.

[78] G. Segal. Categories and cohomology theories. *Topology*, 13:293–312, 1974.

[79] M. Shulman. Homotopy limits and colimits and enriched homotopy theory. Preprint, arXiv:math/0610194 [math.AT], 2009.

[80] M. Shulman. Comparing composites of left and right derived functors. *New York J. Math.*, 17:75–125, 2011.

[81] N. E. Steenrod. A convenient category of topological spaces. *Michigan Math. J.*, 14:133–152, 1967.

[82] A. Strøm. The homotopy category is a homotopy category. *Arch. Math. (Basel)*, 23:435–441, 1972.

[83] D. Sullivan. *Geometric topology. Part I*. Massachusetts Institute of Technology, Cambridge, MA, 1971. Localization, periodicity, and Galois symmetry, revised version.

[84] W. P. Thurston. On proof and progress in mathematics. *Bull. Amer. Math. Soc. (N.S.)*, 30(2):161–177, 1994.

[85] D. R. B. Verity. Enriched categories, internal categories and change of base. *Repr. Theory Appl. Categ.*, (20):1–266, 1992.

[86] D. R. B. Verity. Weak complicial sets. I. Basic homotopy theory. *Adv. Math.*, 219(4):1081–1149, 2008.

[87] R. M. Vogt. Convenient categories of topological spaces for homotopy theory. *Arch. Math. (Basel)*, 22:545–555, 1971.

Glossary of Notation

General notational conventions are described in the preface.

Index